LONDON MATHEMATICAL SOCIETY LECTURE NOTE SERIES

The titles below are available from booksellers, or, in case of difficulty, from Cambridge University Press.

34 Representation theory of Lie groups, M.F. ATIYAH *et al*
46 p-adic analysis: a short course on recent work, N. KOBLITZ
50 Commutator calculus and groups of homotopy classes, H.J. BAUES
59 Applicable differential geometry, M. CRAMPIN & F.A.E. PIRANI
66 Several complex variables and complex manifolds II, M.J. FIELD
69 Representation theory, I.M. GELFAND *et al*
76 Spectral theory of linear differential operators and comparison algebras, H.O. CORDES
77 Isolated singular points on complete intersections, E.J.N. LOOIJENGA
83 Homogeneous structures on Riemannian manifolds, F. TRICERRI & L. VANHECKE
86 Topological topics, I.M. JAMES (ed)
87 Surveys in set theory, A.R.D. MATHIAS (ed)
88 FPF ring theory, C. FAITH & S. PAGE
89 An F-space sampler, N.J. KALTON, N.T. PECK & J.W. ROBERTS
90 Polytopes and symmetry, S.A. ROBERTSON
92 Representation of rings over skew fields, A.H. SCHOFIELD
93 Aspects of topology, I.M. JAMES & E.H. KRONHEIMER (eds)
94 Representations of general linear groups, G.D. JAMES
95 Low-dimensional topology 1982, R.A. FENN (ed)
96 Diophantine equations over function fields, R.C. MASON
97 Varieties of constructive mathematics, D.S. BRIDGES & F. RICHMAN
98 Localization in Noetherian rings, A.V. JATEGAONKAR
99 Methods of differential geometry in algebraic topology, M. KAROUBI & C. LERUSTE
100 Stopping time techniques for analysts and probabilists, L. EGGHE
104 Elliptic structures on 3-manifolds, C.B. THOMAS
105 A local spectral theory for closed operators, I. ERDELYI & WANG SHENGWANG
107 Compactification of Siegel moduli schemes, C-L. CHAI
108 Some topics in graph theory, H.P. YAP
109 Diophantine analysis, J. LOXTON & A. VAN DER POORTEN (eds)
110 An introduction to surreal numbers, H. GONSHOR
113 Lectures on the asymptotic theory of ideals, D. REES
114 Lectures on Bochner-Riesz means, K.M. DAVIS & Y-C. CHANG
115 An introduction to independence for analysts, H.G. DALES & W.H. WOODIN
116 Representations of algebras, P.J. WEBB (ed)
118 Skew linear groups, M. SHIRVANI & B. WEHRFRITZ
119 Triangulated categories in the representation theory of finite-dimensional algebras, D. HAPPEL
121 Proceedings of *Groups - St Andrews 1985*, E. ROBERTSON & C. CAMPBELL (eds)
122 Non-classical continuum mechanics, R.J. KNOPS & A.A. LACEY (eds)
125 Commutator theory for congruence modular varieties, R. FREESE & R. MCKENZIE
126 Van der Corput's method of exponential sums, S.W. GRAHAM & G. KOLESNIK
128 Descriptive set theory and the structure of sets of uniqueness, A.S. KECHRIS & A. LOUVEAU
129 The subgroup structure of the finite classical groups, P.B. KLEIDMAN & M.W. LIEBECK
130 Model theory and modules, M. PREST
131 Algebraic, extremal & metric combinatorics, M-M. DEZA, P. FRANKL & I.G. ROSENBERG (eds)
132 Whitehead groups of finite groups, ROBERT OLIVER
133 Linear algebraic monoids, MOHAN S. PUTCHA
134 Number theory and dynamical systems, M. DODSON & J. VICKERS (eds)
135 Operator algebras and applications, 1, D. EVANS & M. TAKESAKI (eds)
136 Operator algebras and applications, 2, D. EVANS & M. TAKESAKI (eds)
137 Analysis at Urbana, I, E. BERKSON, T. PECK, & J. UHL (eds)
138 Analysis at Urbana, II, E. BERKSON, T. PECK, & J. UHL (eds)
139 Advances in homotopy theory, S. SALAMON, B. STEER & W. SUTHERLAND (eds)
140 Geometric aspects of Banach spaces, E.M. PEINADOR & A. RODES (eds)
141 Surveys in combinatorics 1989, J. SIEMONS (ed)
144 Introduction to uniform spaces, I.M. JAMES
145 Homological questions in local algebra, JAN R. STROOKER
146 Cohen-Macaulay modules over Cohen-Macaulay rings, Y. YOSHINO
147 Continuous and discrete modules, S.H. MOHAMED & B.J. MÜLLER
148 Helices and vector bundles, A.N. RUDAKOV *et al*
149 Solitons, nonlinear evolution equations and inverse scattering, M. ABLOWITZ & P. CLARKSON

150 Geometry of low-dimensional manifolds 1, S. DONALDSON & C.B. THOMAS (eds)
151 Geometry of low-dimensional manifolds 2, S. DONALDSON & C.B. THOMAS (eds)
152 Oligomorphic permutation groups, P. CAMERON
153 L-functions and arithmetic, J. COATES & M.J. TAYLOR (eds)
154 Number theory and cryptography, J. LOXTON (ed)
155 Classification theories of polarized varieties, TAKAO FUJITA
156 Twistors in mathematics and physics, T.N. BAILEY & R.J. BASTON (eds)
157 Analytic pro-*p* groups, J.D. DIXON, M.P.F. DU SAUTOY, A. MANN & D. SEGAL
158 Geometry of Banach spaces, P.F.X. MÜLLER & W. SCHACHERMAYER (eds)
159 Groups St Andrews 1989 volume 1, C.M. CAMPBELL & E.F. ROBERTSON (eds)
160 Groups St Andrews 1989 volume 2, C.M. CAMPBELL & E.F. ROBERTSON (eds)
161 Lectures on block theory, BURKHARD KÜLSHAMMER
162 Harmonic analysis and representation theory, A. FIGA-TALAMANCA & C. NEBBIA
163 Topics in varieties of group representations, S.M. VOVSI
164 Quasi-symmetric designs, M.S. SHRIKANDE & S.S. SANE
165 Groups, combinatorics & geometry, M.W. LIEBECK & J. SAXL (eds)
166 Surveys in combinatorics, 1991, A.D. KEEDWELL (ed)
167 Stochastic analysis, M.T. BARLOW & N.H. BINGHAM (eds)
168 Representations of algebras, H. TACHIKAWA & S. BRENNER (eds)
169 Boolean function complexity, M.S. PATERSON (ed)
170 Manifolds with singularities and the Adams-Novikov spectral sequence, B. BOTVINNIK
171 Squares, A.R. RAJWADE
172 Algebraic varieties, GEORGE R. KEMPF
173 Discrete groups and geometry, W.J. HARVEY & C. MACLACHLAN (eds)
174 Lectures on mechanics, J.E. MARSDEN
175 Adams memorial symposium on algebraic topology 1, N. RAY & G. WALKER (eds)
176 Adams memorial symposium on algebraic topology 2, N. RAY & G. WALKER (eds)
177 Applications of categories in computer science, M. FOURMAN, P. JOHNSTONE, & A. PITTS (eds)
178 Lower K- and L-theory, A. RANICKI
179 Complex projective geometry, G. ELLINGSRUD *et al*
180 Lectures on ergodic theory and Pesin theory on compact manifolds, M. POLLICOTT
181 Geometric group theory I, G.A. NIBLO & M.A. ROLLER (eds)
182 Geometric group theory II, G.A. NIBLO & M.A. ROLLER (eds)
183 Shintani zeta functions, A. YUKIE
184 Arithmetical functions, W. SCHWARZ & J. SPILKER
185 Representations of solvable groups, O. MANZ & T.R. WOLF
186 Complexity: knots, colourings and counting, D.J.A. WELSH
187 Surveys in combinatorics, 1993, K. WALKER (ed)
188 Local analysis for the odd order theorem, H. BENDER & G. GLAUBERMAN
189 Locally presentable and accessible categories, J. ADAMEK & J. ROSICKY
190 Polynomial invariants of finite groups, D.J. BENSON
191 Finite geometry and combinatorics, F. DE CLERCK *et al*
192 Symplectic geometry, D. SALAMON (ed)
193 Computer algebra and differential equations, E. TOURNIER (ed)
194 Independent random variables and rearrangement invariant spaces, M. BRAVERMAN
195 Arithmetic of blowup algebras, WOLMER VASCONCELOS
196 Microlocal analysis for differential operators, A. GRIGIS & J. SJÖSTRAND
197 Two-dimensional homotopy and combinatorial group theory, C. HOG-ANGELONI, W. METZLER & A.J. SIERADSKI (eds)
198 The algebraic characterization of geometric 4-manifolds, J.A. HILLMAN
199 Invariant potential theory in the unit ball of $\mathbf{C}^n$, MANFRED STOLL
200 The Grothendieck theory of dessins d'enfant, L. SCHNEPS (ed)
201 Singularities, JEAN-PAUL BRASSELET (ed)
202 The technique of pseudodifferential operators, H.O. CORDES
203 Hochschild cohomology of von Neumann algebras, A. SINCLAIR & R. SMITH
204 Combinatorial and geometric group theory, A.J. DUNCAN, N.D. GILBERT & J. HOWIE (eds)
205 Ergodic theory and its connections with harmonic analysis, K. PETERSEN & I. SALAMA (eds)
206 Lectures on noncommutative geometry, J. MADORE
207 Groups of Lie type and their geometries, W.M. KANTOR & L. DI MARTINO (eds)
208 Vector bundles in algebraic geometry, N.J. HITCHIN, P. NEWSTEAD & W.M. OXBURY (eds)
209 Arithmetic of diagonal hypersurfaces over finite fields, F.Q. GOUVÊA & N. YUI
210 Hilbert C*-modules, E.C. LANCE
211 Groups 93 Galway / St Andrews I, C.R. CAMPBELL *et al*
212 Groups 93 Galway / St Andrews II, C.R. CAMPBELL *et al*
215 Number theory, S. DAVID (ed)
216 Stochastic partial differential equations, A. ETHERIDGE (ed)
221 Harmonic approximation, S.J. GARDINER
222 Advances in linear logic, J.-Y. GIRARD, Y. LAFONT & L. REGNIER (eds)

London Mathematical Society Lecture Note Series. 222

Advances in Linear Logic

Edited by

Jean-Yves Girard
Laboratoire de Mathématique Discrètes, Marseille

Yves Lafont
Laboratoire de Mathématique Discrètes, Marseille

Laurent Regnier
Laboratoire de Mathématique Discrètes, Marseille

Published by the Press Syndicate of the University of Cambridge
The Pitt Building, Trumpington Street, Cambridge CB2 1RP
40 West 20th Street, New York, NY 10011-4211, USA
10 Stamford Road, Oakleigh, Melbourne 3166, Australia

First published 1995

Library of Congress cataloging in publication data available

British Library cataloguing in publication data available

ISBN 0 521 55961 8 paperback

Transferred to digital printing 2003

Contents

Preface

This volume is based to a large extent on the *Linear Logic Workshop* held June 14-18, 1993 at the MSI[1] and partially supported by the US Army Research Office and the US Office of Naval Research. The workshop was attended by about 70 participants from the USA, Canada, Europe, and Japan. The workshop program committee was chaired by A. Scedrov (University of Pennsylvania) and included S. Abramsky (Imperial College, London), J.-Y. Girard (CNRS, Marseille), D. Miller (University of Pennsylvania), and J. Mitchell (Stanford). The principal speakers at the workshop were J.-M. Andreoli, A. Blass, V. Danos, J.-Y. Girard, A. Joyal, Y. Lafont, J. Lambek, P. Lincoln, M. Moortgat, R. Pareschi, and V. Pratt. There were also a number of invited 30 minute talks and several software demonstration sessions.

Our intention was not only to publish a volume of proceedings. We also wanted to give an overview of a topic that started almost 10 years ago and that is of interest for mathematicians as well as for computer scientists. For these reasons, the book is divided into 5 parts:

1. Categories and Semantics
2. Complexity and Expressivity
3. Proof Theory
4. Proof Nets
5. Geometry of Interaction

The five parts are preceded by a general introduction to Linear Logic by J.-Y. Girard. Furthermore, parts 2 and 4 start with survey papers by P. Lincoln and Y. Lafont. We hope this book can be useful for those who work in this area as well as for those who want to learn about it. All papers have been refereed and the editors are grateful to A. Scedrov who took care of the refereeing process for the papers written by the the editors themselves.

January 1995

Jean-Yves Girard
Yves Lafont
Laurent Regnier

[1] Mathematical Sciences Institute, Cornell University, Ithaca, New York, USA. MSI is a US Army Center of Excellence.

Linear logic : its syntax and semantics

Jean-Yves Girard
Laboratoire de Mathématiques Discrètes
UPR 9016 – **CNRS**
163, Avenue de Luminy, Case 930
F-13288 Marseille Cedex 09

girard@lmd.univ-mrs.fr

1 THE SYNTAX OF LINEAR LOGIC

1.1 The connectives of linear logic

Linear logic is not an alternative logic ; it should rather be seen as an extension of usual logic. Since there is no hope to modify the extant classical or intuitionistic connectives [1], linear logic introduces new connectives.

1.1.1 Exponentials : actions vs situations

Classical and intuitionistic logics deal with stable truths :

if A and $A \Rightarrow B$, then B, *but* A *still holds.*

This is perfect in mathematics, but wrong in real life, since real implication is *causal.* A causal implication cannot be iterated since the conditions are modified after its use ; this process of modification of the premises (conditions) is known in physics as *reaction.* For instance, if A is to spend \$1 on a pack of cigarettes and B is to get them, you lose \$1 in this process, and you cannot do it a second time. The reaction here was that \$1 went out of your pocket. The first objection to that view is that there are in mathematics, in real life, cases where reaction does not exist or can be neglected : think of a lemma which is forever true, or of a Mr. Soros, who has almost an infinite amount of dollars. Such cases are *situations* in the sense of stable truths. Our logical refinements should not prevent us to cope with situations, and there will be a specific kind

1. Witness the fate of non-monotonic "logics" who tried to tamper with logical rules without changing the basic operations ...

of connectives (*exponentials*, "!" and "?") which shall express the iterability of an action, i.e. the absence of any reaction ; typically $!A$ means to spend as many dollars as one needs. If we use the symbol $\multimap$ (*linear implication*) for causal implication, a usual intuitionistic implication $A \Rightarrow B$ therefore appears as

$$A \Rightarrow B \quad = \quad (!A) \multimap B$$

i.e. A implies B exactly when B is caused by some iteration of A. This formula is the essential ingredient of a faithful translation of intuitionistic logic into linear logic ; of course classical logic is also faithfully translatable into linear logic [2], so nothing will be lost ... It remains to see what is gained.

1.1.2 The two conjunctions

In linear logic, two conjunctions $\otimes$ (*times*) and & (*with*) coexist. They correspond to two radically different uses of the word "and". Both conjunctions express the availability of two actions ; but in the case of $\otimes$, both will be done, whereas in the case of &, only one of them will be performed (but we shall decide which one). To understand the distinction consider A,B,C :

A : to spend \$1,
B : to get a pack of Camels,
C : to get a pack of Marlboro.

An action of type A will be a way of taking \$1 out of one's pocket (there may be several actions of this type since we own several notes). Similarly, there are several packs of Camels at the dealer's, hence there are several actions of type B. An action type $A \multimap B$ is a way of replacing any specific dollar by a specific pack of Camels.

Now, given an action of type $A \multimap B$ and an action of type $A \multimap C$, there will be no way of forming an action of type $A \multimap B \otimes C$, since for \$1 you will never get what costs \$2 (there will be an action of type $A \otimes A \multimap B \otimes C$, namely getting two packs for \$2). However, there will be an action of type $A \multimap B \& C$, namely the superimposition of both actions. In order to perform this action, we have first to choose which among the two possible actions we want to perform, and then to do the one selected. This is an exact analogue of the computer instruction **if** ... **then** ... **else** ... : in this familiar case, the parts **then** ... and **else** ... are available, but only one of them will be done. Although "&" has obvious disjunctive features, it would be technically wrong to view it as a disjunction : the formulas $A \& B \multimap A$ and $A \& B \multimap B$ are both provable (in the same way "⅋", to be introduced below, is technically a disjunction, but has prominent conjunctive features). There is a very important property, namely

2. With some problems, see 2.2.7

the equivalence [3] between $!(A\&B)$ and $!A\otimes!B$.

By the way, there are two disjunctions in linear logic :

- "$\oplus$" (*plus*) which is the dual of "&", expresses the choice of one action between two possible types ; typically an action of type $A \multimap B \oplus C$ will be to get one pack of Marlboro for the dollar, another one is to get the pack of Camels. In that case, we can no longer decide which brand of cigarettes we shall get. In terms of computer science, the distinction $\&/\oplus$ corresponds to the distinction outer/inner non determinism.

- "$\parr$" (*par*) which is the dual of "$\otimes$", expresses a dependency between two types of actions ; the meaning of $\parr$ is not that easy, let us just say — anticipating on the introduction of linear negation — that $A \parr B$ can either be read as $A^\perp \multimap B$ or as $B^\perp \multimap A$, i.e. "$\parr$" is a symmetric form of "$\multimap$" ; in some sense, "$\parr$" is the constructive contents of classical disjunction.

1.1.3 Linear negation

The most important linear connective is linear negation $(\cdot)^\perp$ (*nil*). Since linear implication will eventually be rewritten as $A^\perp \parr B$, "nil" is the only negative operation of logic. Linear negation behaves like transposition in linear algebra ($A \multimap B$ will be the same as $B^\perp \multimap A^\perp$), i.e. it expresses a *duality*, that is a change of standpoint :

$$\textit{action of type } A = \textit{reaction of type } A^\perp$$

(other aspects of this duality are output/input, or answer/question).

The main property of $(\cdot)^\perp$ is that $A^{\perp\perp}$ can, without any problem, be identified with A like in classical logic. But (as we shall see in Section 2) linear logic has a very simple *constructive* meaning, whereas the constructive contents of classical logic (which exists, see 2.2.7) is by no means ... obvious. The involutive character of "nil" ensures De Morgan-like laws for all connectives and quantifiers, e.g.

$$\exists x A \quad = \quad (\forall x A^\perp)^\perp$$

which may look surprising at first sight, especially if we keep in mind that the existential quantifier of linear logic is *effective* : typically, if one proves $\exists x A$, then one proves $A[t/x]$ for a certain term t. This exceptional behaviour of "nil" comes from the fact that $A^\perp$ negates (i.e. reacts to) a single action of type A, whereas usual negation only negates some (unspecified) iteration of A, what usually leads to a Herbrand disjunction of unspecified length, whereas the idea of linear negation is not connected to anything like a Herbrand disjunction.

3. This is much more than an equivalence, this is a denotational isomorphism, see 2.2.5

Linear negation is therefore more primitive, but also stronger (i.e. more difficult to prove) than usual negation.

1.1.4 States and transitions

A typical consequence of the excessive focusing of logicians on mathematics is that the notion of *state* of a system has been overlooked.

We shall consider below the example of states in (summary !) chemistry, consisting of lists of molecules involved in a reaction (but a similar argumentation can be applied to Petri nets, as first observed by Asperti [4], — a state being a distribution of tokens — or the game of chess — a state being the current position during a play — etc.)

Observe that summary chemistry is modelled according to precise protocols, hence can be formalized : it can eventually be written in mathematics. But in all cases, one will have to introduce an extraneous temporal parameter, and the formalization will explain, in classical logic, how to pass from the state S (modelled as (S,t)) to a new one (modelled as $(S',t+1)$). This is very awkward, and it would be preferable to ignore this *ad hoc* temporal parameter.

In fact, one would like to represent states by formulas, and transitions by means of implications of states, in such a way that S' is accessible from S exactly when $S \multimap S'$ is provable from the transitions, taken as axioms. But here we meet the problem that, with usual logic, the phenomenon of *updating* cannot be represented. For instance take the chemical equation

$$2\mathrm{H_2} + \mathrm{O_2} \rightarrow 2\mathrm{H_2O}.$$

A paraphrase of it in current language could be

$$\mathrm{H_2} \text{ and } \mathrm{H_2} \text{ and } \mathrm{O_2} \text{ imply } \mathrm{H_2O} \text{ and } \mathrm{H_2O}.$$

Common sense knows how to manipulate this as a logical inference ; but this common sense knows that the sense of "and" here is not idempotent (because the proportions are crucial) and that once the starting state has been used to produce the final one, it cannot be reused. The features which are needed here are those of "$\otimes$" to represent "and" and "$\multimap$" to represent "imply" ; a correct representation will therefore be

$$\mathrm{H_2} \otimes \mathrm{H_2} \otimes \mathrm{O_2} \quad \multimap \quad \mathrm{H_2O} \otimes \mathrm{H_2O}$$

and it turns out that if we take chemical equations written in this way as axioms, then the notion of linear consequence will correspond to the notion of accessible state from an initial one. In this example we see that it is crucial that the two following principles of classical logic

$$A \wedge B \Rightarrow A \quad (\textit{weakening})$$
$$A \Rightarrow A \wedge A \quad (\textit{contraction})$$

become wrong when $\Rightarrow$ is replaced by $\multimap$ and $\wedge$ is replaced by $\otimes$ (contraction would say that the proportions do not matter, whereas weakening would enable us to add an atom of carbon to the left, that would not be present on the right).

To sum up our discussion about states and transitions : the familiar notion of theory — classical logic + axioms — should therefore be replaced by :

$$\textit{theory} = \textit{linear logic} + \textit{axioms} + \textit{current state.}$$

The axioms are there forever ; but the current state is available for a single use : hence once it has been used to prove another state, then the theory is updated, i.e. this other state becomes the next current state. The axioms can of course be replaced by formulas $!A$.

This remark is the basis for potential applications to AI, see [10], this volume : in linear logic certain informations can be logically erased, i.e. the process of revision can be performed by means of logical consequence. What makes it possible is the distinction between formulas $!A$ that speak of stable facts (like the rule of a game) and ordinary ones (that speak about the current state). The impossibility of doing the same thing in classical logic comes from the fact that this distinction makes no sense classically, so any solution to the updating of states would *ipso facto* also be a solution to the updating of the rule of the game [4].

These dynamical features have been fully exploited in Linear Logic Programming, as first observed in [3]. The basic idea is that the resolution method for linear logic (i.e. proof-search in linear sequent calculus) updates the context, in sharp contrast to intuitionistic proof-search, for which the contexts are monotonic. Updating, inheritance, parallelism are the main features of linear logic programming.

1.1.5 The expressive power of linear logic

Due to the presence of exponentials, linear logic is as expressive as classical or intuitionistic logic. In fact it is more expressive. Here we must be cautious :

4. In particular it would update classical mathematics : can anybody with a mathematical background imagine a minute that commutative algebra can be updated into non-commutative algebra ?

in some sense everything that can be expressed can be expressed in classical logic ... so what ? In fact we have the same problem with intuitionistic logic, which is also "more expressive" than classical logic.

The basic point is that linear logic connectives can express features that classical logic could only handle through complex and *ad hoc* translations. Typically the update of the position m of a pawn inside a chess game with current board M into m' (yielding a new current board M') can be classically handled by means of an implication involving M and M' (and additional features, like temporal markers), whereas the linear implication $m \multimap m'$ will do exactly the same job. The introduction of new connectives is therefore the key to a more manageable way of formalizing ; also the restriction to various fragments opens the area of languages with specific expressive power, e.g. with a given computational complexity.

It is in fact surprising how easily various kinds of abstract machines (besides the pioneering case of Petri nets) can be faithfully translated in linear logic. This is perhaps the most remarkable feature in the study of the complexity of various fragments of linear logic initiated in [24]. See [23], this volume. It is to be remarked that these theorems strongly rely on cut-elimination.

1.1.6 A Far West : non-commutative linear logic

In summary chemistry, all the molecules which contribute to a given state are simultaneously available ; however one finds other kinds of problems in which this is not the case. Typically think of a *stack* $a_0 \ldots a_n$ in which a_{n-1} is "hidden" by a_n : if we represent such a state by a conjunction then another classical principle, namely

$$A \wedge B \Rightarrow B \wedge A \quad (\textit{exchange})$$

fails, which suggests yet a more drastic modification, i.e. *non-commutative linear logic*. By the way there is an interesting prefiguration of linear logic in the literature, namely Lambek's *syntactic calculus*, introduced in 1958 to cope with certain questions of linguistic, see [22], this volume. This system is based on a non-commutative $\otimes$, which in turn induces two linear implications. There would be no problems to enrich the system with additives & and $\oplus$, but the expressive power remains extremely limited. The missing items are exponentials and negation :

- Exponentials stumble on the question of the equivalence between $!(A\&B)$ and $!A\otimes!B$, which is one of the main highway of linear logic : since & is commutative, the "Times" should be commutative in this case ... or should one have simultaneously a commutative "Times", in which case the relation between both types of conjunctions should be understood.

- Linear negation is delicate, since there are several possibilities, e.g. a single negation, like in *cyclic linear logic* as expounded in [25] or two negations, like the two linear implications, in which case the situation may become extremely intricate. Abrusci, see [2], this volume, proposed an interesting solution with two negations.

The problem of finding "the" non-commutative system is delicate, since although many people will agree that non-commutativity makes sense, non-trivial semantics of non-commutativity are not manyfold. In particular a convincing denotational semantics should be set up. By the way, it has been observed from the beginning that non-commutative proof-nets should be planar, which suggests either a planarity restriction or the introduction of braids. Besides the introduction of a natural semantics, the methodology for acknowledging a non-commutative system would also include the gain of expressive power w.r.t. the commutative case.

1.2 Linear sequent calculus

1.2.1 Structural rules

In 1934 Gentzen introduced *sequent calculus*, which is a basic synthetic tool for studying the laws of logic. This calculus is not always convenient to build proofs, but it is essential to study their properties. (In the same way, Hamilton's equations in mechanics are not very useful to solve practical problems of motion, but they play an essential role when we want to discuss the very principles of mechanics.) Technically speaking, Gentzen introduced *sequents*, i.e. expressions $\Gamma \vdash \Delta$ where Γ $(= A_1, \ldots, A_n)$ and Δ $(= B_1, \ldots, B_m)$ are finite sequences of formulas. The intended meaning of $\Gamma \vdash \Delta$ is that

$$A_1 \text{ and } \ldots \text{ and } A_n \quad \text{imply} \quad B_1 \text{ or } \ldots \text{ or } B_m$$

but the sense of "and", "imply", "or" has to be clarified. The calculus is divided into three groups of rules (identity, structural, logical), among which the structural block has been systematically overlooked. In fact, a close inspection shows that the actual meaning of the words "and", "imply", "or", is wholly in the structural group and it is not too excessive to say that a logic is essentially a set of structural rules ! The structural rules considered by Gentzen (respectively *weakening, contraction, exchange*)

$$\frac{\Gamma \vdash \Delta}{\Gamma \vdash A, \Delta} \qquad \frac{\Gamma \vdash \Delta}{\Gamma, A \vdash \Delta}$$

$$\frac{\Gamma \vdash A, A, \Delta}{\Gamma \vdash A, \Delta} \qquad \frac{\Gamma, A, A \vdash \Delta}{\Gamma, A \vdash \Delta}$$

$$\frac{\Gamma \vdash \Delta}{\sigma(\Gamma) \vdash \tau(\Delta)}$$

are the sequent calculus formulation of the three classical principles already met and criticized. Let us detail them.

Weakening. — Weakening opens the door for fake dependencies : from a sequent $\Gamma \vdash \Delta$ we can get another one $\Gamma' \vdash \Delta'$ by extending the sequences Γ, Δ. Typically, it speaks of causes without effect, e.g. spending $1 to get nothing —not even smoke—; but it is an essential tool in mathematics (from B deduce $A \Rightarrow B$) since it allows us not to use all the hypotheses in a deduction. It will rightly be rejected from linear logic.

Anticipating on linear sequent calculus, we see that the rule says that $\otimes$ is stronger than & :

$$\cfrac{\cfrac{\cfrac{A \vdash A}{A, B \vdash A} \quad \cfrac{B \vdash B}{A, B \vdash B}}{A, B \vdash A \& B}}{A \otimes B \vdash A \& B}$$

Affine linear logic is the system of linear logic enriched (?) with weakening. There is no much use for this system since the affine implication between A and B can be faithfully mimicked by $1 \& A \multimap B$. Although the system enjoys cut-elimination, it has no obvious denotational semantics, like classical logic.

Contraction. — Contraction is the fingernail of infinity in propositional calculus : it says that what you have, will always keep, no matter how you use it. The rule corresponds to the replacement of $\Gamma \vdash \Delta$ by $\Gamma' \vdash \Delta'$ where Γ' and Δ' come from Γ and Δ by identifying several occurrences of the same formula (on the same side of "$\vdash$"). To convince oneself that the rule is about infinity (and in fact that without it there is no infinite at all in logic), take the formula $I : \forall x \exists y\ x < y$ (together with others saying that $<$ is a strict order). This axiom has only infinite models, and we show this by exhibiting 1, 2, 3, 4, ... distinct elements ; but, if we want to exhibit 27 distinct elements, we are actually using I 26 times, and without a principle saying that 26 I can be contracted into one, we would never make it ! In other terms infinity does not mean *many*, but *always*. Another infinitary feature of the rule is that it is the only responsible for undecidability [5] : Gentzen's subformula property yields a decision method for predicate calculus, provided we can bound the length of the sequents involved in a cut-free proof, and this is obviously the case in the absence of contraction.

In linear logic, both contraction and weakening will be forbidden *as structural rules*. But linear logic is not logic without weakening and contraction, since it would be nonsense not to recover them in some way : we have introduced a new interpretation for the basic notions of logic (actions), but we do not want to abolish the old one (situations), and this is why special connectives (exponentials "!" and "?") will be introduced, with the two missing structurals as their main rules. The main difference is that we now *control* in many cases the use of contraction, which —one should not forget it— means controlling the length of Herbrand disjunctions, of proof-search, normalization procedures, etc.

Whereas the meaning of weakening is the fact that "⊗" is stronger than "&", contraction means the reverse implication : using contraction we get :

$$\dfrac{\dfrac{\dfrac{A \vdash A}{A\&B \vdash A} \quad \dfrac{B \vdash B}{A\&B \vdash B}}{A\&B, A\&B \vdash A \otimes B}}{A\&B \vdash A \otimes B}$$

It is difficult to find any evidence of such an implication outside classical logic. The problem is that if we accept contraction without accepting weakening too, we arrive at a very confusing system, which would correspond to an imperfect analysis of causality : consider a petrol engine, in which petrol causes the motion ($P \vdash M$) ; weakening would enable us to call any engine a petrol engine (from $\vdash M$ deduce $P \vdash M$), which is only dishonest, but contraction would be miraculous : from $P \vdash M$ we could deduce $P \vdash P \otimes M$, i.e. that the petrol is not consumed in the causality. This is why the attempts of philosophers to build various *relevance logics* out of the only rejection of weakening were never very convincing [6]

5. If we stay first-order : second-order linear logic is undecidable in the absence of exponentials, as recently shown by Lafont (unpublished), see also [23].
6. These systems are now called substructural logics, which is an abuse, since most of the calculi associated have no cut-elimination ....

Intuitionistic logic accepts contraction (and weakening as well), but only on the left of sequents : this is done in (what can now be seen as) a very hypocritical way, by restricting the sequents to the case where Δ consists of one formula, so that we are never actually in position to apply a right structural rule. So, when we have a cut-free proof of $\vdash A$, the last rule must be logical, and this has immediate consequences, e.g. if A is $\exists y\, B$ then $B[t]$ has been proved for some t, etc. These features, that just come from the absence of right contraction, will therefore be present in linear logic, in spite of the presence of an involutive negation.

Exchange. — Exchange expresses the commutativity of multiplicatives : we can replace $\Gamma \vdash \Delta$ with $\Gamma' \vdash \Delta'$ where Γ' and Δ' are obtained from Γ and Δ by permutations of their formulas.

1.2.2 Linear sequent calculus

In order to present the calculus, we shall adopt the following notational simplification : formulas are written from literals p, q, r, $p^\perp$, $q^\perp$, $r^\perp$, etc., and constants $\mathbf{1}$, $\perp$, $\top$, $\mathbf{0}$ by means of the connectives $\otimes$, $⅋$, $\&$, $\oplus$ (binary), $!$, $?$ (unary), and the quantifiers $\forall x$, $\exists x$. Negation is *defined* by De Morgan equations, and linear implication is also a defined connective :

$$
\begin{array}{rclcrcl}
\mathbf{1}^\perp & := & \perp & \qquad & \perp^\perp & := & \mathbf{1} \\
\top^\perp & := & \mathbf{0} & & \mathbf{0}^\perp & := & \top \\
(p)^\perp & := & p^\perp & & (p^\perp)^\perp & := & p \\
(A \otimes B)^\perp & := & A^\perp ⅋ B^\perp & & (A ⅋ B)^\perp & := & A^\perp \otimes B^\perp \\
(A \& B)^\perp & := & A^\perp \oplus B^\perp & & (A \oplus B)^\perp & := & A^\perp \& B^\perp \\
(!A)^\perp & := & ?A^\perp & & (?A)^\perp & := & !A^\perp \\
(\forall x\, A)^\perp & := & \exists x\, A^\perp & & (\exists x\, A)^\perp & := & \forall x\, A^\perp
\end{array}
$$

$$A \multimap B := A^\perp ⅋ B$$

The connectives $\otimes$, $⅋$, $\multimap$, together with the neutral elements $\mathbf{1}$ (w.r.t. $\otimes$) and $\perp$ (w.r.t. $⅋$) are called *multiplicatives* ; the connectives $\&$ and $\oplus$, together with the neutral elements $\top$ (w.r.t. $\&$) and $\mathbf{0}$ (w.r.t $\oplus$) are called *additives* ; the connectives ! and ? are called *exponentials.* The notation has been chosen for its mnemonic virtues : we can remember from the notation that $\otimes$ is multiplicative and conjunctive, with neutral $\mathbf{1}$, $\oplus$ is additive and disjunctive, with neutral $\mathbf{0}$, that $⅋$ is disjunctive with neutral $\perp$, and that $\&$ is conjunctive with neutral $\top$; the distributivity of $\otimes$ over $\oplus$ is also suggested by our notation.

Sequents are right-sided, i.e. of the form $\vdash \Delta$; general sequents $\Gamma \vdash \Delta$ can be mimicked as $\vdash \Gamma^\perp, \Delta$.

Identity / Negation

$$\frac{}{\vdash A, A^{\perp}}\ (identity) \qquad \frac{\vdash \Gamma, A \quad \vdash A^{\perp}, \Delta}{\vdash \Gamma, \Delta}\ (cut)$$

Structure

$$\frac{\vdash \Gamma}{\vdash \Gamma'}\ (exchange : \Gamma' \text{ is a permutation of } \Gamma)$$

Logic

$$\frac{}{\vdash \mathbf{1}}\ (one) \qquad \frac{\vdash \Gamma}{\vdash \Gamma, \bot}\ (false)$$

$$\frac{\vdash \Gamma, A \quad \vdash B, \Delta}{\vdash \Gamma, A \otimes B, \Delta}\ (times) \qquad \frac{\vdash \Gamma, A, B}{\vdash \Gamma, A \parr B}\ (par)$$

$$\frac{}{\vdash \Gamma, \top}\ (true) \qquad (no\ rule\ for\ zero)$$

$$\frac{\vdash \Gamma, A \quad \vdash \Gamma, B}{\vdash \Gamma, A \& B}\ (with) \qquad \frac{\vdash \Gamma, A}{\vdash \Gamma, A \oplus B}\ (left\ plus) \qquad \frac{\vdash \Gamma, B}{\vdash \Gamma, A \oplus B}\ (right\ plus)$$

$$\frac{\vdash ?\Gamma, A}{\vdash ?\Gamma, !A}\ (of\ course) \qquad \frac{\vdash \Gamma}{\vdash \Gamma, ?A}\ (weakening)$$

$$\frac{\vdash \Gamma, A}{\vdash \Gamma, ?A}\ (dereliction) \qquad \frac{\vdash \Gamma, ?A, ?A}{\vdash \Gamma, ?A}\ (contraction$$

$$\frac{\vdash \Gamma, A}{\vdash \Gamma, \forall x\, A}\ (for\ all : x \text{ is not free in } \Gamma) \qquad \frac{\vdash \Gamma, A[t/x]}{\vdash \Gamma, \exists x\, A}\ (there\ is)$$

1.2.3 Comments

The rule for "$\parr$" shows that the comma behaves like a hypocritical "$\parr$" (on the left it would behave like "$\otimes$") ; "and", "or", "imply" are therefore read as "$\otimes$", "$\parr$", "$\multimap$".

In a two-sided version the identity rules would be

$$\frac{}{A \vdash A} \qquad \frac{\Gamma \vdash \Delta, A \quad A, \Lambda \vdash \Pi}{\Gamma, \Lambda \vdash \Delta, \Pi}$$

and we therefore see that the ultimate meaning of the identity group (and the only principle of logic beyond criticism) is that "A is A" ; in fact the two rules say that A on the left (represented by $A^{\perp}$ in the right-sided formulation) is stronger (resp. weaker) than A on the right. The meaning of the identity group is to some extent blurred by our right-sided formulation, since the group may also be seen as the negation group.

The logical group must be carefully examined :

- *multiplicatives and additives* : notice the difference between the rule for $\otimes$ and the rule for & : $\otimes$ requires disjoint contexts (which will never be identified unless ? is heavily used) whereas & works with twice the same context. If we see the contexts of A as the price to pay to get A, we recover our informal distinction between the two conjunctions. In a similar way, the two disjunctions are very different, since $\oplus$ requires one among the premises, whereas ⅋ requires both).

- *exponentials* : ! and ? are modalities : this means that $!A$ is simultaneously defined on all formulas : the *of course* rule mentions a context with $?\Gamma$, which means that $?\Gamma$ (or $!\Gamma^{\perp}$) is known. $!A$ indicates the possibility of using A *ad libitum*; it only indicates a *potentiality,* in the same way that a piece of paper on the slot of a copying machine *can* be copied ... but nobody would identify a copying machine with all the copies it produces ! The rules for the dual (*why not*) are precisely the three basic ways of *actualizing* this potentiality : erasing (*weakening*), making a single copy (*dereliction*), duplicate ... the machine (*contraction*). It is no wonder that the first relation of linear logic to computer science was the relation to memory pointed out by Yves Lafont in [20].

- *quantifiers* : they are not very different from what they are in usual logic, if we except the disturbing fact that $\exists x$ is now the exact dual of $\forall x$. It is important to remark that $\forall x$ is very close to & (and that $\exists x$ is very close to $\oplus$).

1.3 Proof-nets

1.3.1 The determinism of computation

For classical and intuitionistic logics, we have an essential property, which dates back to Gentzen (1934), known as the *Hauptsatz*, or cut-elimination theorem ; the *Hauptsatz* presumably traces the borderline between *logic* and the wider notion of *formal system*. It goes without saying that linear logic enjoys cut-elimination [7].

Theorem 1

There is an algorithm transforming any proof of a sequent $\vdash \Gamma$ *in linear logic into a cut-free proof of the same sequent.*

PROOF. — The proof basically follows the usual argument of Gentzen ; but due to our very cautious treatment of structural rules, the proof is in fact much simpler. There is no wonder, since linear logic comes from a proof-theoretical analysis of usual logic ! □

We have now to keep in mind that the *Hauptsatz* — under various disguises, e.g. normalization in λ-calculus — is used as possible theoretical foundation for computation. For instance consider a text editor : it can be seen as a set of general lemmas (the various subroutines about bracketing, the size of pages etc.), that we can apply to a concrete input, let us say a given page that I write from the keyboard ; observe that the number of such inputs is practically infinite and that therefore our lemmas are about the infinite. Now when I feed the program with a concrete input, there is no longer any reference to infinity ... In mathematics, we could content ourselves with something *implicit* like "your input is correct", whereas we would be mad at a machine which answers "I can do it" to a request. Therefore, the machine does not only check the correctness of the input, it also demonstrates it by exhibiting the final result, which no longer mentions abstractions about the quasi-infinite *potentiality* of all possible pages. Concretely this elimination of infinity is done by systematically making all concrete replacements — in other terms by running the program. But this is exactly what the algorithm of cut-elimination does.

This is why the *structure* of the cut-elimination procedure is essential. And this structure is quite problematic, since we get problems of *permutation of rules*.

Let us give an example : when we meet a configuration

$$\dfrac{\dfrac{\vdash \Gamma, A}{\vdash \Gamma', A}\,(r) \qquad \dfrac{\vdash A^{\perp}, \Delta}{\vdash A^{\perp}, \Delta'}\,(s)}{\vdash \Gamma', \Delta'}\,(cut)$$

there is no natural way to eliminate this cut, since the unspecified rules (r) and (s) do not act on A or $A^{\perp}$; then the idea is to forward the cut upwards :

7. A sequent calculus without cut-elimination is like a car without engine

$$\dfrac{\dfrac{\dfrac{\vdash \Gamma, A \quad \vdash A^\perp, \Delta}{\vdash \Gamma, \Delta}\,(cut)}{\vdash \Gamma', \Delta}\,(r)}{\vdash \Gamma', \Delta'}\,(s)$$

But, in doing so, we have decided that rule (r) should now be rewritten *before* rule (s), whereas the other choice

$$\dfrac{\dfrac{\dfrac{\vdash \Gamma, A \quad \vdash A^\perp, \Delta}{\vdash \Gamma, \Delta}\,(cut)}{\vdash \Gamma, \Delta'}\,(s)}{\vdash \Gamma', \Delta'}\,(r)$$

would have been legitimate too. The bifurcation starting at this point is usually irreversible : unless (r) or (s) is later erased, there is no way to interchange them. Moreover the problem stated was completely symmetrical w.r.t. left and right, and we can of course arbitrate between the two possibilities by many bureaucratical tricks ; we can decide that *left* is more important than right, but this choice will at some moment conflict with negation (or implication) whose behaviour is precisely to mimic left by right ... Let's be clear : the *taxonomical* devices that force us to write (r) before (s) or (s) before (r) are not more respectable than the alphabetical order in a dictionary. One should try to get rid of them, or at least, ensure that their effect is limited. In fact *denotational semantics*, see chapter 2 is very important in this respect, since the two solutions proposed have the same denotation. In some sense the two answers — although irreversibly different — are *consistent.* This means that if we eliminate cuts in a proof of an intuitionistic disjunction $\vdash A \vee B$ (or a linear disjunction $\vdash A \oplus B$) and eventually get "a proof of A or a proof of B", the side (A or B) is not affected by this bifurcation. However, we would like to get better, namely to have a syntax in which such bifurcations do not occur. In intuitionistic logic (at least for the fragment $\Rightarrow$, $\wedge$, $\forall$) this can be obtained by replacing sequent calculus by natural deduction. Typically the two proofs just written will get the same associated deduction ... In other terms natural deduction enjoys a *confluence* (or *Church-Rosser*) property : if $\pi \mapsto \pi', \pi''$ then there is π''' such that $\pi', \pi'' \mapsto \pi'''$, i.e. bifurcations are not irreversible.

1.3.2 Limitations of natural deduction

Let us assume that we want to use natural deduction to deal with proofs in linear logic ; then we run into problems.

(1) Natural deduction is not equipped to deal with classical symmetry : several hypotheses and one (distinguished) conclusion. To cope with symmetrical systems one should be able to accept several conclusions at once ... But then one immediately loses the tree-like structure of natural deductions, with its obvious advantage : a well-determined last rule. Hence natural deduction cannot answer the question. However it is still a serious candidate for an intuitionistic version of linear logic ; we shall below only discuss the fragment ($\otimes$, $\multimap$), for which there is an obvious natural deduction system :

$$\frac{\begin{array}{c}[A]\\ \vdots \\ B\end{array}}{A \multimap B}\ (\multimap\textit{-intro}) \qquad\qquad \frac{A \quad A \multimap B}{B}\ (\multimap\textit{-elim})$$

$$\frac{A \quad B}{A \otimes B}\ (\otimes\textit{-intro}) \qquad\qquad \frac{A \otimes B \quad \begin{array}{c}[A][B]\\ \vdots \\ C\end{array}}{C}\ (\otimes\textit{-elim})$$

As usual a formula between brackets indicates a *discharge* of hypothesis ; but here the discharge should be *linear*, i.e. exactly one occurrence is discharged (discharging zero occurrence is weakening, discharging two occurrences is contraction). Although this system succeeds in identifying a terrific number of interversion-related proofs, it is not free from serious defects, more precisely :

(2) In the elimination rules the formula which bears the symbol ($\otimes$ or $\multimap$) is written as a hypothesis ; this is user-friendly, but goes against the actual mathematical structure. Technically this "premise" is in fact the actual conclusion of the rule (think of *main hypotheses* or *headvariables*), which is therefore written upside down. However this criticism is very inessential.

(3) Due to discharge, the introduction rule for $\multimap$ (and the elimination rule for $\otimes$) does not apply to a formula, but to a whole proof. This *global* character of the rule is quite a serious defect.

(4) Last but not least, the elimination rule for $\otimes$ mentions an extraneous formula C which has nothing to do with $A \otimes B$. In intuitionistic natural deduction, we have the same problem with the rules for disjunction and existence which mention an extraneous formula C ; the theory of normalization ("commutative conversions") then becomes extremely complex and awkward.

1.3.3 The identity links

We shall find a way of fixing defects (1)–(4) in the context of the multiplicative fragment of linear logic, i.e. the only connectives ⊗ and ⅋ (and also implicitly ⊸). The idea is to put everything in conclusion ; however, when we pass from a hypothesis to a conclusion we must indicate the change by means of a negation symbol. There will be two basic *links* enabling one to replace a hypothesis with a conclusion and *vice versa*, namely

By far the best explanation of these two links can be taken from *electronics*. Think of a sequent Γ as the interface of some electronic equipment, this interface being made of plugs of various forms $A_1, \ldots, A_n$; the negation corresponds to the complementarity between male and female plugs. Now a proof of Γ can be seen as any equipment with interface Γ. For instance the axiom link is such a unit and it exists in everyday life as the *extension cord* :

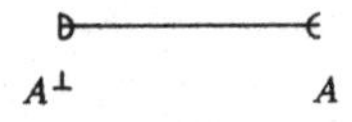

Now, the cut link is well explained as a plugging :

$$\Gamma \cdots \qquad \cdots \Delta$$
$$A \quad A^{\perp}$$

The main property of the extension cord is that

$$\Gamma \cdots$$

behaves like

$$\Gamma \cdots$$

It seems that the ultimate, deep meaning of cut-elimination is located there. Moreover observe that common sense would forbid self-plugging of an extension cord :

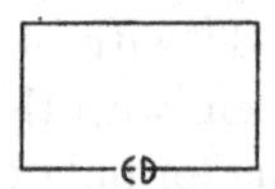

which would correspond, in terms of proof-nets to the incestuous configuration :

$$\boxed{A \qquad A^{\perp}}$$

which is not acknowledged as a proof-net ; in fact in some sense the ultimate meaning of the *correctness* criterion that will be stated below is to forbid such a configuration (and also disconnected ones).

1.3.4 Proof-structures

If we accept the additional links :

$$\frac{A \quad B}{A \otimes B} \ (\textit{times link}) \qquad\qquad \frac{A \quad B}{A ⅋ B} \ (\textit{par link})$$

then we can associate to any proof of $\vdash \Gamma$ in linear sequent calculus a graph-like *proof-structure* with as conclusions the formulas of Γ. More precisely :

1. To the identity axiom associate an axiom link.

2. Do not interpret the exchange rule (this rule does not affect conclusions ; however, if we insist on writing a proof-structure on a plane, the effect of the rule can be seen as introducing *crossings* between axiom links ; planar proof-structures will therefore correspond to proofs in some non-commutative variants of linear logic).

3. If a proof-structure β ending with Γ, A and B has been associated to a proof π of $\vdash \Gamma, A, B$ and if one now applies a "par" rule to this proof to get a proof π' of $\vdash \Gamma, A ⅋ B$, then the structure β' associated to π' will be obtained from β by linking A and B via a *par* link : therefore A and B are no longer conclusions, and a new conclusion $A ⅋ B$ is created.

4. If π_1 is a proof of $\vdash \Gamma, A$ and π_2 is a proof of $\vdash B, \Delta$ to which proof-structures β_1 and β_2 have been associated, then the proof π' obtained from π_1 and π_2 by means of a times rule is interpreted by means of the proof structure β obtained from β_1 and β_2 by linking A and B together via a *times* link. Therefore A and B are no longer conclusions and a new conclusion $A \otimes B$ is created.

5. If π_1 is a proof of $\vdash \Gamma, A$ and π_2 is a proof of $\vdash A^{\perp}, \Delta$ to which proof-structures β_1 and β_2 have been associated, then the proof π' obtained from π_1 and π_2 by means of a cut rule is interpreted by means of the proof structure β obtained from β_1 and β_2 by linking A and $A^{\perp}$ together via a *cut* link. Therefore A and $A^{\perp}$ are no longer conclusions.

An interesting exercise is to look back at the natural deduction of linear logic and to see how the four rules can be mimicked by proof-structures :

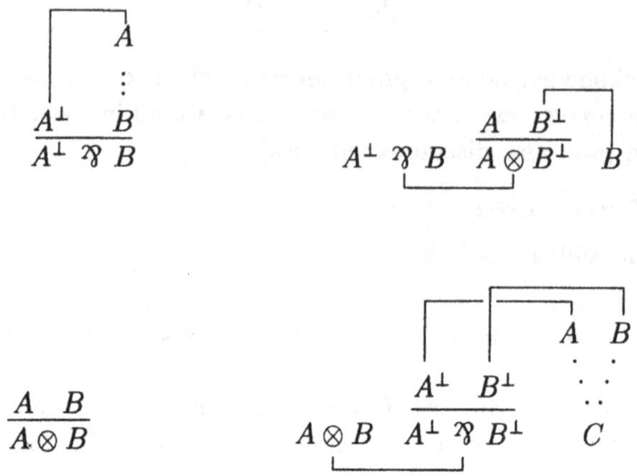

$$\dfrac{A^{\perp} \quad B}{A^{\perp} \mathbin{⅋} B} \qquad\qquad A^{\perp} \mathbin{⅋} B \quad \dfrac{A \quad B^{\perp}}{A \otimes B^{\perp}} \quad B$$

$$\dfrac{A \quad B}{A \otimes B} \qquad\qquad A \otimes B \quad \dfrac{A^{\perp} \quad B^{\perp}}{A^{\perp} \mathbin{⅋} B^{\perp}} \qquad C$$

This shows that —once everything has been put in conclusion—

$\multimap$-intro = $\otimes$-elim = par link ;

$\multimap$-elim = $\otimes$-intro = times link.

1.3.5 Proof-nets

A proof-structure is nothing but a graph whose vertices are (occurrences of) formulas and whose edges are links ; moreover each formula is the conclusion of exactly one link and the premise of at most one link. The formulas which are not premises are the *conclusions* of the structure. Inside proof-structures, let us call *proof-nets* those which can be obtained as the interpretation of sequent calculus proofs. Of course most structures are not nets : typically the definition of a proof-structure does not distinguish between $\otimes$-links and ⅋-links whereas conjunction is surely different from disjunction.

The question which now arises is to find an independent characterization of proof-nets. Let us explain why this is essential :

1. If we define proof-nets from sequent calculus, this means that we work with a proof-structure together with a *sequentialization*, in other terms a step by step construction of this net. But this sequentialization is far from being unique, typically there might be several candidates for the "last rule" of a given proof-net. In practice, we may have a proof-net with a given sequentialization but we may need to use another one : this means that we will spend all of our energy on problems of commutation of rules, as with old

sequent calculus, and we will not benefit too much from the new approach. Typically, if a proof-net ends with a *splitting* $\otimes$-link, (i.e. a link whose removal induces two disconnected structures), we would like to conclude that the last rule can be chosen as $\otimes$-rule ; working with a sequentialization this can be proved, but the proof is long and boring, whereas, with a criterion, the result is immediate, since the two components inherit the criterion.

2. The distinction between "and" and "or" has always been explained in semantical terms which ultimately use "and" and "or" ; a purely geometrical characterization would therefore establish the distinction on more intrinsic grounds.

The survey of Yves Lafont [22] (this volume) contains the correctness criterion (first proved in [12] and simplified by Danos and Regnier in [9]) and the sequentialization theorem. From the proof of the theorem, one can extract a quadratic algorithm checking whether or not a given multiplicative proof-structure is a proof-net. Among the uses of multiplicative proof-nets, let us mention the questions of coherence in monoidal closed categories [6].

1.3.6 Cut-elimination for proof-nets

The crucial test for the new syntax is the possibility to handle syntactical manipulations directly at the level of proof-nets (therefore completely ignoring sequent calculus). When we meet a cut link

$$A \qquad A^{\perp}$$

we look at links whose conclusions are A and $A^{\perp}$:

(1) One of these links is an axiom link, typically :

$$A^{\perp} \qquad A \qquad A^{\perp}$$

such a configuration can be replaced by

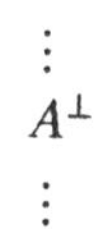

however the graphism is misleading, since it cannot be excluded that the two occurrences of $A^{\perp}$ in the original net are the same ! But this would correspond to a configuration

$$A \qquad A^{\perp}$$

in β, and such configurations are excluded by the correctness criterion.

(2) If both formulas are conclusions of logical links for ⊗ and ⅋, typically

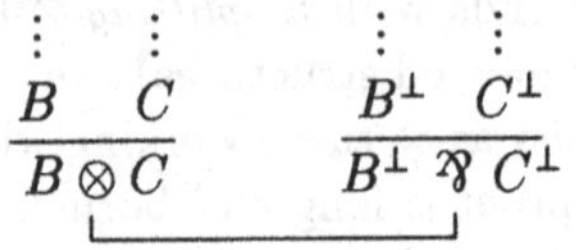

$$\frac{\begin{matrix}\vdots & \vdots\\ B & C\end{matrix}}{B \otimes C} \qquad \frac{\begin{matrix}\vdots & \vdots\\ B^\perp & C^\perp\end{matrix}}{B^\perp ⅋ C^\perp}$$

then we can replace it by

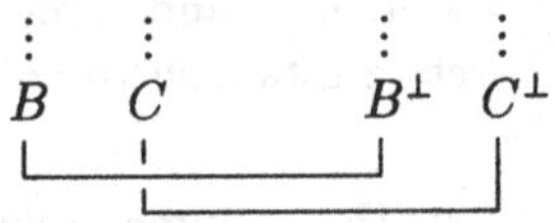

$$\begin{matrix}\vdots & \vdots & & \vdots & \vdots\\ B & C & & B^\perp & C^\perp\end{matrix}$$

and it is easily checked that the new structure still enjoys the correctness criterion. This cut-elimination procedure has very nice features :

1. It enjoys a Church-Rosser property (immediate).
2. It is linear in time : simply observe that the proof-net shrinks with any application of steps (1) and (2) ; this linearity is the start of a line of applications to computational complexity.
3. The treatment of the multiplicative fragment is purely local ; in fact all cut-links can be simultaneously eliminated. This must have something to do with parallelism and recently Yves Lafont developed his *interaction nets* as a kind of parallel machine working like proof-nets [21], this volume.

1.3.7 Extension to full linear logic

Proof-nets can be extended to full linear logic. In the case of quantifiers one uses unary links :

$$\frac{A[y/x]}{\forall x A} \qquad\qquad \frac{A[t/x]}{\exists x A}$$

in the $\forall x$-link an *eigenvariable* y must be chosen ; each $\forall x$-link must use a distinct eigenvariable (as the name suggests). The sequentialization theorem can be extended to quantifiers, with an appropriate extension of the correctness criterion.

Additives and neutral elements also get their own notion of proof-nets [16], as well as the part of the exponential rules which does not deal with "!". Although this extension induces a tremendous simplification of the familiar sequent calculus, it is not as satisfactory as the multiplicative/quantifier case.

Eventually, the only rule which resists to the proof-net technology is the !-rule. For such a rule, one must use a box, see [21]. The box structure has a deep meaning, since the nesting of boxes is ultimately responsible for cut-elimination.

1.4 Is there a unique logic ?

1.4.1 LU

By the turn of the century the situation concerning logic was quite simple : there was basically one logic (classical logic) which could be used (by changing the set of proper axioms) in various situations. Logic was about *pure* reasoning. Brouwer's criticism destroyed this dream of unity : classical logic was not adapted to constructive features and therefore lost its universality. By the end of the century we are now faced with an incredible number of logics — some of them only named "logics" by antiphrasis, some of them introduced on serious grounds —. Is still logic about pure reasoning? In other terms, could there be a way to reunify logical systems — let us say those systems with a good sequent calculus — into a single sequent calculus. Could we handle the (legitimate) distinction classical/intuitionistic not through a change of system, but through a change of formulas? Is it possible to obtain classical effects by restricting one to classical formulas? etc.

Of course there are surely ways to achieve this by cheating, typically by considering a disjoint union of systems ... All these jokes will be made impossible if we insist on the fact that the various systems represented should freely communicate (and for instance a classical theorem could have an intuitionistic corollary and *vice versa*).

In the unified calculus **LU** see [13], classical, linear, and intuitionistic logics appear as *fragments*. This means that one can define notions of *classical*, *intuitionistic*, or *linear* sequents and prove that a cut-free proof of a sequent in one of these fragments is wholly inside the fragment ; of course a proof with cuts has the right to use arbitrary sequents, i.e. the fragments can freely communicate.

Perhaps the most interesting feature of this new system is that the classical, intuitionistic and linear fragments of **LU** are better behaved than the original sequent calculi. In **LU** the distinction between several styles of maintenance (e.g. "rule of the game" vs "current state") is particularly satisfactory. But after all, **LU** is little more than a clever reformulation of linear sequent calculus.

1.4.2 LLL and ELL

This dream of unity stumbles on a new fact : the recent discovery [15] of two systems which definitely diverge from classical or intuitionistic logic, **LLL** (*light linear logic*) and **ELL** (elementary linear logic). They come from the basic remark that, in the absence of exponentials, cut-elimination can be performed in linear time. This result (which is conspicuous from a proof-net argument [8]), holds for *lazy* cut-elimination, which does not normalize above &-rules, and which is enough for algorithmic purposes ; notice that the result holds without regards for the kind of quantifiers —first or second order— used. However the absence of exponentials renders such systems desperately inexpressive. The first attempt to expand this inexpressive system while keeping interesting complexity bounds was not satisfactory : **BLL** (bounded linear logic) [18] had to keep track of polynomial I/O bounds that induced polytime complexity effects, but the price paid was obviously too much.

These new systems have been found by means of the following methodology : naive set-theory is just the extreme kind of higher order quantification, and its familiar inconsistency should be seen as the fact that the cut-elimination complexity becomes so high that termination fails. The problem to solve was therefore the following : which restriction(s) on the exponentials do ensure

- cut-elimination, (hence consistency) for naive set-theory, i.e. full comprehension
- the familiar equivalence between $!(A\&B)$ and $!A\otimes!B$

Two systems have been found, both based on the idea that normalization should respect the depth of formulas (with respect to the nesting of !-boxes). Normalization is performed in linear time at depth 0, and induces some duplication of bigger depths, then it is performed at depth 1, etc. and eventually stops, since the total depth does not change. The global complexity therefore depends on the (fixed) global depth and on the number of duplications operated by the "cleansing" of a given level.

- In **LLL** the sizes of inner boxes are multiplied by a factor corresponding to the outer size. The global procedure is therefore done in a time which is a polynomial in the size (with a degree depending of the total depth). Conversely every polytime algorithm can be represented in **LLL**.
- In **ELL** the sizes of inner boxes are exponentiated by a power corresponding to the outer size. The global procedure is therefore done in a time which is a elementary in the size (the runtime is a tower of exponentials whose height

8. Remember that the size of a proof-net shrinks during cut-elimination

depends on the total depth). Conversely every elementary algorithm can be represented in **ELL**. **ELL** differs from **LLL** only in the extra principle $!A \otimes !B \multimap !(A \otimes B)$.

LLL may have interesting applications in complexity theory ; **ELL** is expressive enough to accommodate a lot of current mathematics.

2 THE SEMANTICS OF LINEAR LOGIC

2.1 The phase semantics of linear logic

The most traditional, and also the less interesting semantics of linear logic associates values to formulas, in the spirit of classical model theory. Therefore it only modelizes provability, and not proofs.

2.1.1 Phase spaces

A *phase space* is a pair $(M, \perp)$, where M is a commutative monoid (usually written multiplicatively) and $\perp$ is a subset of M. Given two subsets X and Y of M, one can define $X \multimap Y := \{m \in M \,;\, \forall n \in X \quad mn \in Y\}$. In particular, we can define for each subset X of M its *orthogonal* $X^\perp := X \multimap \perp$. A *fact* is any subset of M equal to its biorthogonal, or equivalently any subset of the form $Y^\perp$. It is immediate that $X \multimap Y$ is a fact as soon as Y is a fact.

2.1.2 Interpretation of the connectives

The basic idea is to interpret all the operations of linear logic by operations on facts : once this is done the interpretation of the language is more or less immediate. We shall use the same notation for the interpretation, hence for instance $X \otimes Y$ will be the fact interpreting the tensorization of two formulas respectively interpreted by X and Y. This suggests that we already know how to interpret $\perp$, linear implication and linear negation.

1. **times** : $X \otimes Y := \{mn \,;\, m \in X \wedge n \in Y\}^{\perp\perp}$

2. **par** : $X \parr Y := (X^\perp \otimes Y^\perp)^\perp$

3. **1** : $\mathbf{1} := \{1\}^{\perp\perp}$, where 1 is the neutral element of M

4. **plus** : $X \oplus Y := (X \cup Y)^{\perp\perp}$

5. **with** : $X \& Y := X \cap Y$

6. **zero** : $\mathbf{0} := \emptyset^{\perp\perp}$

7. **true** : $\top := M$

8. **of course** : $!X := (X \cap I)^{\perp\perp}$, where I is the set of idempotents of M which belong to **1**

9. **why not** : $?X := (X^{\perp} \cap I)^{\perp}$

(The interpretation of exponentials is an improvement of the original definition of [11] which was awfully *ad hoc*). This is enough to define what is a model of propositional linear logic. This can easily be extended to yield an interpretation of quantifiers (intersection and biorthogonal of the union). Observe that the definitions satisfy the obvious De Morgan laws relating $\otimes$ and $⅋$ etc. A non-trivial exercise is to prove the associativity of $\otimes$.

2.1.3 Soundness and completeness

It is easily seen that the semantics is sound and complete :

Theorem 2
A formula A of linear logic is provable iff for any interpretation (involving a phase space $(M, \perp)$), the interpretation A^* of A contains the neutral element 1.

PROOF. — Soundness is proved by a straightforward induction. Completeness involves the building of a specific phase space. In fact we can take as M the monoid of contexts (i.e. multisets of formulas), whose neutral element is the empty context, and we define $\perp := \{\Gamma \ ; \ \vdash \ \Gamma \quad provable\}$. If we consider the sets $A^* := \{\Gamma \ ; \ \vdash \ \Gamma, A \quad provable\}$, then these sets are easily shown to be facts. More precisely, one can prove (using the identity group) that $A^{\perp *} = A^{*\perp}$. It is then quite easy to prove that in fact A^* is the value of A in a given model : this amounts to prove commutations of the style $(A \otimes B)^* = A^* \otimes B^*$ (these proofs are simplified by the fact that in any De Morgan pair one commutation implies the other, hence we can choose the friendlier commutation). Therefore, if $1 \in A^*$, it follows that $\vdash \ A$ is provable. □
As far as I know there is no applications for completeness, due to the fact that there is no known concrete phase spaces to which one could restrict and still have completeness. Soundness is slightly more fruitful : for instance Yves Lafont (unpublished, 1994) proved the undecidability of second order propositional linear logic without exponentials by means of a soundness argument. This exploits the fact that a phase semantics is not defined as any algebraic structure enjoying the laws of linear logic, but that it is fully determined from the choice of a commutative monoid and the interpretation $\perp$, as soon as the atoms are interpreted by facts.

2.2 The denotational semantics of linear logic

2.2.1 Implicit versus explicit

First observe that the cut rule is a way to formulate *modus ponens.* It is the essential ingredient of any proof. If I want to prove B, I usually try to prove a useful lemma A and, assuming A, I then prove B. All proofs in nature, including the most simple ones, are done in this way. Therefore, there is an absolute evidence that the cut rule is the only rule of logic that cannot be removed : without cut it is no longer possible to *reason.*

Now against common sense Gentzen proved his *Hauptsatz* ; for classical and intuitionistic logics (and remember that can be extended to linear logic without problems). This result implies that we can make proofs without cut, i.e. without lemmas (i.e. without modularity, without ideas, etc.). For instance if we take an intuitionistic disjunction $A \vee B$ (or a linear plus $A \oplus B$) then a cut-free proof of it must contain a proof of A or a proof of B. We see at once that this is artificial : who in real life would state $A \vee B$ when he has proved A? If we want to give a decent status to proof-theory, we have to explain this contradiction.

Formal reasoning (any reasoning) is about *implicit* data. This is because it is more convenient to *forget.* So, when we prove $A \vee B$, we never know which side holds. However, there is — inside the sequent calculus formulation — a completely artificial use of the rules, i.e. to prove without the help of cut ; this artificial subsystem is completely *explicit.* The result of Gentzen is a way to replace a proof by another without cut, which makes explicit the contents of the original proof. Variants of the Gentzen procedure (normalization in natural deduction or in λ-calculus) should also be analysed in that way.

2.2.2 Generalities about denotational semantics

The purpose of *denotational* semantics is precisely to analyse this implicit contents of proofs. The name comes from the old Fregean opposition *sense/denotation* : the denotation is what is implicit in the sense.

The kind of semantics we are interested in is concrete, i.e. to each proof π we associate a set π^*. This map can be seen as a way to define an equivalence $\approx$ between proofs ($\pi \approx \pi'$ iff $\pi^* = \pi'^*$) of the same formulas (or sequents), which should enjoy the following :

1. if π normalizes to π', then $\pi \approx \pi'$;
2. $\approx$ is non-degenerated, i.e. one can find a formula with at least two non-equivalent proofs ;

3. $\approx$ is a congruence : this means that if π and π' have been obtained from λ and λ' by applying the same logical rule, and if $\lambda \approx \lambda'$, then $\pi \approx \pi'$;

4. certain canonical isomorphisms are satisfied ; among those which are crucial let us mention :

   - involutivity of negation (hence De Morgan),
   - associativity of "par" (hence of "times").

Let us comment these points :

- (1) says that $\approx$ is about cut-elimination.

- (2) : of course if all proofs of the same formula are declared to be equivalent, the contents of $\approx$ is empty.

- (3) is the analogue of a Church-Rosser property, and is the key to a modular approach to normalization.

- (4) : another key to modularity is commutation, which means that certain sequences of operations on proofs are equivalent w.r.t. $\approx$. It is clear that the more commutation we get the better, and that we cannot ask too much a priori. However, the two commutations mentioned are a strict minimum without which we would get a mess :

  - involutivity of negation means that we have not to bother about double negations ; in fact this is the semantical justification of our choice of a *defined* negation.
  - associativity of "par" means that the bracketing of a ternary "par" is inessential ; furthermore, associativity renders possible the identification of $A \multimap (B \multimap C)$ with $(A \otimes B) \multimap C$.

The denotational semantics we shall present is a simplification of Scott domains which has been obtained by exploiting the notion of stability due to Berry (see [17] for a discussion). These drastically simplified Scott domains are called *coherent spaces* ; these spaces were first intended as denotational semantics for intuitionistic logic, but it turned out that there were a lot of other operations hanging around. Linear logic first appeared as a kind of linear algebra built on coherent spaces ; then linear sequent calculus was extracted out of the semantics. Recently Ehrhard, see [9], this volume, refined coherent semantics into *hypercoherences*, with applications to the question of sequentiality.

2.2.3 Coherent spaces

Definition 1

A coherent space is a reflexive undirected graph. In other terms it consists of a set $|X|$ *of* atoms *together with a* compatibility *or* coherence relation *between atoms, noted* $x \frown\!\!\!\smile y$ *or* $x \frown\!\!\!\smile y$ [mod X] *if there is any ambiguity as to* X.

A clique a *in* X *(notation* $a \sqsubset X$*) is a subset* a *of* X *made of pairwise coherent atoms :* $a \sqsubset X$ *iff* $\forall x \forall y\ (x \in a \wedge y \in a \Rightarrow x \frown\!\!\!\smile y)$*. In fact a coherent space can be also presented as a set of cliques ; when we want to emphasize the underlying graph* $(|X|, \frown\!\!\!\smile)$ *we call it the* web *of* X.

Besides coherence we can also introduce

- *strict coherence* : $x \frown y$ iff $x \frown\!\!\!\smile y$ and $x \neq y$,
- *incoherence* : $x \asymp y$ iff $\neg(x \frown y)$,
- *strict incoherence* : $x \smile y$ iff $\neg(x \frown\!\!\!\smile y)$.

Any of these four relations can serve as a definition of coherent space. Observe fact that $\asymp$ is the negation of $\frown$ and not of $\frown\!\!\!\smile$; this due to the reflexivity of the web.

Definition 2

Given a coherent space X*, its* linear negation $X^\perp$ *is defined by*

- $|X^\perp| = |X|$,
- $x \frown\!\!\!\smile y$ [mod $X^\perp$] *iff* $x \asymp y$ [mod X].

In other terms, linear negation is nothing but the exchange of coherence and incoherence. It is obvious that linear negation is involutive :
$X^{\perp\perp} = X$.

Definition 3

Given two coherent spaces X *and* Y*, the* multiplicative connectives $\otimes$, $\parr$, $\multimap$ *define a new coherent space* Z *with* $|Z| = |X| \times |Y|$ *; coherence is defined by*

- $(x, y) \frown\!\!\!\smile (x', y')$ [mod $X \otimes Y$] *iff*
  $x \frown\!\!\!\smile x'$ [mod X] *and* $y \frown\!\!\!\smile y'$ [mod Y],
- $(x, y) \frown (x', y')$ [mod $X \parr Y$] *iff*
  $x \frown x'$ [mod X] *or* $y \frown y'$ [mod Y],
- $(x, y) \frown (x', y')$ [mod $X \multimap Y$] *iff*
  $x \frown\!\!\!\smile x'$ [mod X] *implies* $y \frown y'$ [mod Y].

Observe that $\otimes$ is defined in terms of $\frown$ but $⅋$ and $\multimap$ in terms of $\frown$. A lot of useful isomorphisms can be obtained

1. De Morgan equalities : $(X \otimes Y)^\perp = X^\perp ⅋ Y^\perp$; $(X ⅋ Y)^\perp = X^\perp \otimes Y^\perp$; $X \multimap Y = X^\perp ⅋ Y$;
2. commutativity isomorphisms : $X \otimes Y \simeq Y \otimes X$; $X ⅋ Y \simeq Y ⅋ X$; $X \multimap Y \simeq Y^\perp \multimap X^\perp$;
3. associativity isomorphisms : $X \otimes (Y \otimes Z) \simeq (X \otimes Y) \otimes Z$; $X ⅋ (Y ⅋ Z) \simeq (X ⅋ Y) ⅋ Z$; $X \multimap (Y \multimap Z) \simeq (X \otimes Y) \multimap Z$; $X \multimap (Y ⅋ Z) \simeq (X \multimap Y) ⅋ Z$.

Definition 4

Up to isomorphism there is a unique coherent space whose web consists of one atom 0, this space is self dual, i.e. equal to its linear negation. However the algebraic isomorphism between this space and its dual is logically meaningless, and we shall, depending on the context, use the notation $\mathbf{1}$ *or the notation* $\perp$ *for this space, with the convention that* $\mathbf{1}^\perp = \perp$, $\perp^\perp = \mathbf{1}$.

This space is neutral w.r.t. multiplicatives, namely $X \otimes \mathbf{1} \simeq X$, $X ⅋ \perp \simeq X$, $\mathbf{1} \multimap X \simeq X$, $X \multimap \perp \simeq X^\perp$.
This notational distinction is mere preciosity ; one of the main drawbacks of denotational semantics is that it interprets logically irrelevant properties ... but nobody is perfect.

Definition 5

Given two coherent spaces X and Y the additive connectives *& and* $\oplus$, *define a new coherent space Z with* $|Z| = |X| + |Y|$ $(= |X| \times \{0\} \cup |Y| \times \{1\})$ *; coherence is defined by*

- $(x, 0) \frown (x', 0)$ [mod Z] *iff* $x \frown x'$ [mod X],
- $(y, 1) \frown (y', 1)$ [mod Z] *iff* $y \frown y'$ [mod Y],
- $(x, 0) \frown (y, 1)$ [mod $X \& Y$],
- $(x, 0) \smile (y, 1)$ [mod $X \oplus Y$].

A lot of useful isomorphisms are immediately obtained :

- De Morgan equalities : $(X \& Y)^\perp = X^\perp \oplus Y^\perp$; $(X \oplus Y)^\perp = X^\perp \& Y^\perp$;
- commutativity isomorphisms : $X \& Y \simeq Y \& X$; $X \oplus Y \simeq Y \oplus X$;
- associativity isomorphisms : $X \& (Y \& Z) \simeq (X \& Y) \& Z$; $X \oplus (Y \oplus Z) \simeq (X \oplus Y) \oplus Z$;

▸ distributivity isomorphisms : $X \otimes (Y \oplus Z) \simeq (X \otimes Y) \oplus (X \otimes Z)$; $X ⅋ (Y \& Z) \simeq (X ⅋ Y)\&(X ⅋ Z)$; $X \multimap (Y\&Z) \simeq (X \multimap Y)\&(X \multimap Z)$; $(X \oplus Y) \multimap Z \simeq (X \multimap Z)\&(Y \multimap Z)$.

The other distributivities fail ; for instance $X \otimes (Y\&Z)$ is not isomorphic to $(X \otimes Y)\&(X \otimes Z)$.

Definition 6

There is a unique coherent space with an empty web. Although this space is also self dual, we shall use distinct notations for it and its negation : $\top$ *and* $\mathbf{0}$.

These spaces are neutral w.r.t. additives : $X \oplus \mathbf{0} \simeq X$, $X\&\top \simeq X$, and absorbing w.r.t. multiplicatives $X \otimes \mathbf{0} \simeq \mathbf{0}$, $X ⅋ \top \simeq \top$, $\mathbf{0} \multimap X \simeq \top$, $X \multimap \top \simeq \top$.

2.2.4 Interpretation of MALL

MALL is the fragment of linear logic without the exponentials "!" and "?". In fact we shall content ourselves with the propositional part and omit the quantifiers. If we wanted to treat the quantifiers, the idea would be to essentially interpret $\forall x$ and $\exists x$ respectively "big" & and $\oplus$ indexed by the domain of interpretation of variables ; the precise definition involves considerable bureaucracy for something completely straightforward. The treatment of second-order quantifiers is of course much more challenging and will not be explained here. See for instance [11].

Once we decided to ignore exponentials and quantifiers, everything is ready to interpret formulas of **MALL** : more precisely, if we assume that the atomic propositions $p, q, r, \ldots$ of the language have been interpreted by coherent spaces $p^*, q^*, r^*, \ldots$, then any formula A of the language is interpreted by a well-defined coherent space A^* ; moreover this interpretation is consistent with the definitions of linear negation and implication (i.e. $A^{\perp *} = A^{*\perp}$, $(A \multimap B)^* = A^* \multimap B^*$). It remains to interpret sequents ; the idea is to interpret $\vdash \Gamma$ $(= \vdash A_1, \ldots, A_n)$ as $A_1^* ⅋ \cdots ⅋ A_n^*$. More precisely

Definition 7

If $\Vdash \Xi$ $(= \Vdash X_1, \ldots, X_n)$ *is a formal sequent made of coherent spaces, then the coherent space* $\Vdash \Xi$ *is defined by*

1. $|\Vdash \Xi| = |X_1| \times \cdots \times |X_n|$; *we use the notation* $x_1 \ldots x_n$ *for the atoms of* $\Vdash \Xi$.

2. $x_1 \ldots x_n \frown y_1 \ldots y_n \Leftrightarrow \exists i \; x_i \frown y_i$.

If $\vdash \Gamma$ ($= \vdash A_1, \ldots, A_n$) is a sequent of linear logic, then $\Vdash \Gamma^*$ will be the coherent space $\Vdash A_1^*, \ldots, A_n^*$.

The next step is to interpret proofs ; the idea is that a proof π of $\vdash \Gamma$ will be interpreted by a *clique* $\pi^* \sqsubset \Vdash \Gamma^*$. In particular (since sequent calculus is eventually about proofs of singletons $\vdash A$) a proof π of $\vdash A$ is interpreted by a clique of $\Vdash A^*$ i.e. a clique in A^*.

Definition 8

1. *The identity axiom $\vdash A, A^\perp$ of linear logic is interpreted by the set $\{xx \,;\, x \in |A^*|\}$.*
2. *Assume that the proofs π of $\vdash \Gamma, A$ and λ of $\vdash A^\perp, \Delta$ have been interpreted by cliques π^* and λ^* in the associated coherent spaces ; then the proof ρ of $\vdash \Gamma, \Delta$ obtained by means of a cut rule between π and λ is interpreted by the set $\rho^* = \{\underline{x}\underline{x}' \,;\, \exists z\, (\underline{x}z \in \pi^* \wedge z\underline{x}' \in \lambda^*)\}$.*
3. *Assume that the proof π of $\vdash \Gamma$ has been interpreted by a clique $\pi^* \sqsubset \Vdash \Gamma^*$, and that ρ is obtained from π by an exchange rule (permutation σ of Γ); then ρ^* is obtained from π^* by applying the same permutation $\rho^* = \{\sigma(\underline{x}) \,;\, \underline{x} \in \pi^*\}$.*

All the sets constructed by our definition are cliques ; let us remark that in the case of cut, the atom z of the formula is uniquely determined by $\underline{x}$ and $\underline{x}'$.

Definition 9

1. *The axiom $\vdash \mathbf{1}$ of linear logic is interpreted by the clique $\{0\}$ of $\mathbf{1}$ (if we call 0 the only atom of $\mathbf{1}$).*
2. *The axioms $\vdash \Gamma, \top$ of linear logic are interpreted by void cliques (since $\top$ has an empty web, the spaces $(\vdash \Gamma, \top)^*$ have empty webs as well).*
3. *If the proof ρ of $\vdash \Gamma, \perp$ comes from a proof π of $\vdash \Gamma$ by a falsum rule, then we define $\rho^* = \{\underline{x}0 \,;\, \underline{x} \in \pi^*\}$.*
4. *If the proof ρ of $\vdash \Gamma, A \parr B$ comes from a proof π of $\vdash \Gamma, A, B$ by a par rule, we define $\rho^* = \{\underline{x}(y, z) \,;\, \underline{x}yz \in \pi^*\}$.*
5. *If the proof ρ of $\vdash \Gamma, A \otimes B, \Delta$ comes from proofs π of $\vdash \Gamma, A$ and λ of $\vdash B, \Delta$ by a times rule, then we define $\rho^* = \{\underline{x}(y, z)\underline{x}' \,;\, \underline{x}y \in \pi^* \wedge z\underline{x}' \in \lambda^*\}$.*
6. *If the proof ρ of $\vdash \Gamma, A \oplus B$ comes from a proof π of $\vdash \Gamma, A$ by a left plus rule, then we define $\rho^* = \{\underline{x}(y, 0) \,;\, \underline{x}y \in \pi^*\}$; if the proof ρ of $\vdash \Gamma, A \oplus B$ comes from a proof π of $\vdash \Gamma, B$ by a right plus rule, then we define $\rho^* = \{\underline{x}(y, 1) \,;\, \underline{x}y \in \pi^*\}$.*

7. *If the proof* ρ *of* $\vdash \Gamma, A\&B$ *comes from proofs* π *of* $\vdash \Gamma, A$ *and* λ *of* $\vdash \Gamma, B$ *by a* with rule, *then we define*
$\rho^* = \{\underline{x}(y,0)\ ;\ \underline{x}y \in \pi^*\} \cup \{\underline{x}(y,1)\ ;\ \underline{x}y \in \lambda^*\}$.

Observe that (4) is mainly a change of bracketing, i.e. does strictly nothing ; if $|A| \cap |B| = \varnothing$ then one can define $A\&B$, $A \oplus B$ as unions, in which case (6) is read as $\rho^* = \pi^*$ in both cases, and (7) is read $\rho^* = \pi^* \cup \lambda^*$.
It is of interest (since this is deeply hidden in Definition 9) to stress the relation between *linear implication* and *linear maps* :

Definition 10

Let X *and* Y *be coherent spaces ; a* linear map *from* X *to* Y *consists in a function* F *such that*

1. *if* $a \sqsubset X$ *then* $F(a) \sqsubset Y$,
2. *if* $\bigcup b_i = a \sqsubset X$ *then* $F(a) = \bigcup F(b_i)$,
3. *if* $a \cup b \sqsubset X$, *then* $F(a \cap b) = F(a) \cap F(b)$.

The last two conditions can be rephrased as the preservation of disjoint unions.

Proposition 1

There is a 1-1 correspondence between linear maps from X *to* Y *and cliques in* $X \multimap Y$ *; more precisely*

- *to any linear* F *from* X *to* Y, *associate* $\mathrm{Tr}(F) \sqsubset X \multimap Y$ *(the* trace *of* F*)*

$$\mathrm{Tr}(F) = \{(x,y)\ ;\ y \in F(\{x\})\},$$

- *to any* $A \sqsubset X \multimap Y$ *associate a linear function* $A(\cdot)$ *from* X *to* Y

$$\text{if } a \sqsubset X, \text{ then } A(a) = \{y\ ;\ \exists x \in a\, (x,y) \in A\}.$$

PROOF. — The proofs that $\mathrm{Tr}(A(\cdot)) = A$ and $\mathrm{Tr}(F)(\cdot) = F$ are left to the reader. In fact the structure of the space $X \multimap Y$ has been obtained so as to get this property and not the other way around. □

2.2.5 Exponentials

Definition 11

Let X *be a coherent space ; we define* $\mathcal{M}(X)$ *to be the free commutative monoid generated by* $|X|$. *The elements of* $\mathcal{M}(X)$ *are all the formal expressions* $[x_1, \ldots, x_n]$ *which are finite multisets of elements of* $|X|$. *This*

means that $[x_1, \ldots, x_n]$ *is a sequence in* $|X|$ *defined up to the order. The difference with a subset of* $|X|$ *is that repetitions of elements matter. One easily defines the sum of two elements of* $\mathcal{M}(X)$ *:*
$[x_1, \ldots, x_n] + [y_1, \ldots, y_n] = [x_1, \ldots, x_n, y_1, \ldots, y_n]$, *and the sum is generalized to any finite set. The neutral element of* $\mathcal{M}(X)$ *is written* $[\,]$.

If X *is a coherent space, then* $!X$ *is defined as follows :*

- $|!X| = \{[x_1, \ldots, x_n] \in \mathcal{M}(X) \; ; \; x_i \bigcirc x_j$ *for all* i *and* $j\}$,
- $\sum[x_i] \bigcirc \sum[y_j]$ [mod $!X$] *iff* $x_i \bigcirc y_j$ *for all indices* i *and* j.

If X *is a coherent space, then* $?X$ *is defined as follows :*

- $|?X| = \{[x_1, \ldots, x_n] \in \mathcal{M}(X) \; ; \; x_i \asymp x_j$ *for all* i *and* $j\}$,
- $\sum[x_i] \frown \sum[y_j]$ [mod $?X$] *iff* $x_i \frown y_j$ *for some pair of indices* i *and* j.

Among remarkable isomorphisms let us mention

- De Morgan equalities : $(!X)^\perp = ?(X^\perp)$; $(?X)^\perp = !(X^\perp)$;
- the exponentiation isomorphisms : $!(X \& Y) \simeq (!X) \otimes (!Y)$; $?(X \oplus Y) \simeq (?X) \parr (?Y)$, together with the "particular cases" $!\top \simeq \mathbf{1}$; $?\mathbf{0} \simeq \perp$.

Definition 12

1. *Assume that the proof* π *of* $\vdash ?\Gamma, A$ *has been interpreted by a clique* π^* *; then the proof* ρ *of* $\vdash ?\Gamma, !A$ *obtained from* π *by an* of course *rule is interpreted by the set*

$$\rho^* = \{(\underline{x}_1 + \cdots + \underline{x}_k)[a_1, \ldots, a_k] \; ; \; \underline{x}_1 a_1, \ldots, \underline{x}_k a_k \in \pi^*\}.$$

   About the notation : if $?\Gamma$ *is* $?B^1, \ldots, ?B^n$ *then each* $\underline{x}_i$ *is* $x_i^1, \ldots, x_i^n$ so $\underline{x}_1 + \cdots + \underline{x}_k$ *is the sequence* $x_1^1 + \cdots + x_k^1, \ldots, x_1^n + \cdots + x_k^n$; $[a_1, \ldots, a_k]$ *refers to a multiset. What is implicit in the definition (but not obvious) is that we take only those expressions* $(\underline{x}_1 + \cdots + \underline{x}_k)[a_1, \ldots, a_k]$ *such that* $\underline{x}_1 + \cdots + \underline{x}_k \in \Vdash ?\Gamma$ *(this forces* $[a_1, \ldots, a_k] \in |!A|$*).*
2. *Assume that the proof* π *of* $\vdash \Gamma$ *has been interpreted by a clique* π^* *; then the proof* ρ *of* $\vdash \Gamma, ?A$ *obtained from* π *by a* weakening *rule is interpreted by the set* $\rho^* = \{\underline{x}[\,] \; ; \; \underline{x} \in \pi^*\}$.
3. *Assume that the proof* π *of* $\vdash \Gamma, ?A, ?A$ *has been interpreted by a clique* π^* *; then the proof* ρ *of* $\vdash \Gamma, ?A$ *obtained from* π *by a* contraction rule *is interpreted by the set* $\rho^* = \{\underline{x}(a+b) \; ; \; \underline{x}ab \in \pi^* \wedge a \asymp b\}$.
4. *Assume that the proof* π *of* $\vdash \Gamma, A$ *has been interpreted by a clique* π^* *; then the proof* ρ *of* $\vdash \Gamma, ?A$ *obtained from* π *by a* dereliction rule *is interpreted by the set* $\rho^* = \{\underline{x}[a] \; ; \; \underline{x}a \in \pi^*\}$.

2.2.6 The bridge with intuitionism

First the version just given for the exponentials is not the original one, which was using sets instead of multisets. The move to multisets is a consequence of recent progress on classical logic [12] for which this replacement has deep consequences. But as far as linear and intuitionistic logic are concerned, we can work with sets, and this is what will be assumed here. In particular $\mathcal{M}(X)$ will be replaced by the monoid of finite subsets of X, and sum will be replaced by union. The web of $!X$ will be the set X_{fin} of all finite cliques of X.

Definition 13

Let X and Y be coherent spaces ; a stable map *from X to Y is a function F such that*

1. *if $a \sqsubset X$ then $F(a) \sqsubset Y$,*
2. *assume that $a = \bigcup b_i$, where b_i is directed with respect to inclusion, then*
   $F(a) = \bigcup F(b_i)$,
3. *if $a \cup b \sqsubset X$, then $F(a \cap b) = F(a) \cap F(b)$.*

Definition 14

Let X and Y be coherent spaces ; then we define the coherent space $X \Rightarrow Y$ as follows :

- $|X \Rightarrow Y| = X_{fin} \times |Y|$,
- $(a, y) \supset\!\!\!\subset (a', y')$ *iff (1) and (2) hold :*
  1. $a \cup a' \sqsubset X \;\Rightarrow\; y \supset\!\!\!\subset y'$,
  2. $a \cup a' \sqsubset X \;\wedge\; a \neq a' \;\Rightarrow\; y \frown y'$.

Proposition 2

There is a 1-1 correspondence between stable maps from X to Y and cliques in $X \Rightarrow Y$; more precisely

1. *to any stable F from X to Y, associate $\mathrm{Tr}(F) \sqsubset X \Rightarrow Y$ (the* trace *of F)*

$$\mathrm{Tr}(F) = \{(a, y) \; ; \; a \sqsubset X \;\wedge\; y \in F(a) \;\wedge \forall a' \subset a\,(y \in F(a') \Rightarrow a' = a)\}$$

2. *to any $A \sqsubset X \Rightarrow Y$, associate a stable function $A(\cdot)$ from X to Y*

$$\text{if } a \sqsubset X, \text{ then } A(a) = \{y \; ; \; \exists b \subset a\,((b, y) \in A)\}.$$

PROOF. — The essential ingredient is the normal form theorem below. □

Theorem 3

Let F be a stable function from X to Y, let $a \sqsubset X$, let $y \in F(a)$; then

1. *there exists $a_0 \subset a$, a_0 finite such that $y \in F(a_0)$,*
2. *if a_0 is chosen minimal w.r.t. inclusion, then it is unique.*

PROOF. — (1) follows from $a = \bigcup a_i$, the directed union of its finite subsets ; $z \in F(\bigcup a_i) = \bigcup F(a_i)$ hence $z \in F(a_i)$ for some i.
(2) : given two solutions a_0, a_1 included in a, we get $z \in F(a_0) \cap F(a_1) = F(a_0 \cap a_1)$; if a_0 is minimal w.r.t. inclusion, this forces $a_0 \cap a_1 = a_0$, hence $a_0 \subset a_1$. □

This establishes the basic bridge with linear logic, since $X \Rightarrow Y$ is strictly the same thing as $!X \multimap Y$ (if we use sets instead of multisets). In fact one can translate intuitionistic logic into linear logic as follows :

$$p^* := p \quad (p \text{ atomic}),$$
$$(A \Rightarrow B)^* := !A^* \multimap B^*,$$
$$(A \wedge B)^* := A^* \& B^*,$$
$$(\forall x\, A)^* := \forall x\, A^*,$$
$$(A \vee B)^* := !A^* \oplus !B^*,$$
$$(\exists x\, A)^* := \exists x\, !A^*,$$
$$(\neg A)^* := !A^* \multimap \mathbf{0}.$$

and prove the following result: $\Gamma \vdash A$ is intuitionistically provable iff $!\Gamma^* \vdash A^*$ (i.e. $\vdash ?\Gamma^{*\perp}, A^*$) is linearily provable. The possibility of such a faithful translation is of course a major evidence for linear logic, since it links it with intuitionistic logic in a strong sense. In particular linear logic can *at least* be accepted as a way of analysing intuitionistic logic.

2.2.7 The bridge with classical logic

Let us come back to exponentials ; the space $!X$ is equipped with two maps :

$$c \in !X \multimap (!X \otimes !X) \qquad\qquad w \in !X \multimap 1$$

corresponding to contraction and weakening. We can see these two maps as defining a structure of *comonoid* : intuitively this means the contraction map behaves like a commutative/associative law and that the weakening map behaves like its neutral element. The only difference with a usual monoid is that the arrows are in the wrong direction. A comonoid is therefore a triple (X, c, w)

A	B	$A \wedge B$	$A \vee B$	$A \Rightarrow B$	$\neg A$	$\forall x\, A$	$\exists x\, A$
+1	+1	+1	+1	−1	−1	−1	+1
−1	+1	+1	−1	+1	+1	−1	+1
+1	−1	+1	−1	−1			
−1	−1	−1	−1	−1			

Table 1: Polarities for classical connectives.

satisfying conditions of (co)-associativity, commutativity and neutrality. There are many examples of monoids among coherent spaces, since monoids are closed under $\otimes$, $\oplus$ and existential quantification (this means that given monoids, the above constructions can canonically be endowed with monoidal structures). Let us call them *positive correlation spaces.*
Dually, spaces $?X$ are equipped with maps :

$$c \in (?X \parr ?X) \multimap ?X \qquad\qquad w \in \bot \multimap ?X$$

enjoying dual conditions, and that should be called "cocomonoids", but we prefer to call them *negative correlation spaces*[9]. Negative correlation spaces are closed under $\parr$,& and universal quantification.
The basic idea to interpret classical logic will be to assign *polarities* to formulas, positive or negative, so that a given formula will be interpreted by a correlation space of the same polarity. The basic idea behind this assignment is that a positive formula has the right to structural rules on the left and a negative formula has the right to structural rules on the right of sequents. In other terms, putting everything to the right, either A or $A^\perp$ has structural rules for free. A classical sequent $\vdash \Gamma, \Delta$ with the formulas in Γ positive and the formulas in Δ negative is interpreted in linear logic as $\vdash ?\Gamma, \Delta$: the symbol ? in front of Γ is here to compensate the want of structural rule for positive formulas.
This induces a denotational semantics for classical logic. However, we easily see that there are many choices (using the two conjunctions and the two exponentials) when we want to interpret classical conjunction, similarly for disjunction, see [8], this volume. However, we can restrict our attention to choices enjoying an optimal amount of denotational isomorphisms. This is the reason behind the tables shown on next page.

9. The dual of a comonoid is not a monoid

A	B	$A \wedge B$	$A \vee B$	$A \Rightarrow B$	$\neg A$	$\forall x\, A$	$\exists x\, A$
$+1$	$+1$	$A \otimes B$	$A \oplus B$	$A \multimap ?B$	$A^{\perp}$	$\forall x\, ?A$	$\exists x\, A$
-1	$+1$	$!A \otimes B$	$A \parr ?B$	$A^{\perp} \oplus B$	$A^{\perp}$	$\forall x\, A$	$\exists x\, !A$
$+1$	-1	$A \otimes !B$	$?A \parr B$	$A \multimap B$			
-1	-1	$A \,\&\, B$	$A \parr B$	$!A \multimap B$			

Table 2: Classical connectives : definition in terms of linear logic.

It is easily seen that in terms of isomorphisms, negation is involutive, conjunction is commutative and associative, with a neutral element **V** of polarity +1, symmetrically for disjunction. Certain denotational distributivities $\wedge / \vee$ or $\vee / \wedge$ are satisfied, depending on the respective polarities.

Polarities are obviously a way to cope with the basic undeterminism of classical logic, since they operate a choice between the basic protocols of cut-elimination. However, this is still not enough to provide a deterministic version of Gentzen's classical calculus **LK**. The reason lies in the fact that the rule of introduction of conjunction is problematic : from cliques in respectively $?X$ and $?Y$, when both X and Y are positive, there are two ways to get a clique in $?(X \otimes Y)$. This is why one must replace **LK** with another calculus **LC**, see [12] for more details, in which a specific positive formula may be distinguished. **LC** has a denotational semantics, but the translation from **LK** to **LC** is far from being deterministic. This is why we consider that our approach is still not absolutely convincing ... typically one cannot exclude the existence of a non-deterministic denotational semantics for classical logic, but God knows how to get it !

LC is indeed fully compatible with linear logic : it is enough to add a new polarity 0 (neutral) for those formulas which are not canonically equipped with a structure of correlation space. The miracle is that this combination of classical with intuitionistic features accommodates intuitionistic logic for free, and this eventually leads to the system **LU** of *unified logic*, see [13].

2.3 Geometry of interaction

At some moment we indicated an electronic analogy ; in fact the analogy was good enough to explain step (1) of cut-elimination (see subsection 1.3.6 by the fact that an extension cord has no action (except perhaps a short delay, which corresponds to the cut-elimination step). But what about the other links ?

Let us first precise the nature of our (imaginary) plugs ; there are usually several pins in a plug. We shall restrict ourselves to one-pin plugs ; this does not contradict the fact that there may be a huge variety of plugs, and that the only allowed plugging is between complementary ones, labelled A and $A^\perp$.

The interpretation of the rules for $\otimes$ and $⅋$ both use the following well-known fact : two pins can be reduced to one (typical example : stereophonic broadcast).

- $\otimes$-*rule* : from units π, λ with respective interfaces $\vdash \Gamma, A$ and $\vdash \Delta, B$, we can built a new one by merging plugs A and B into another one (labelled $A \otimes B$) by means of an encoder.

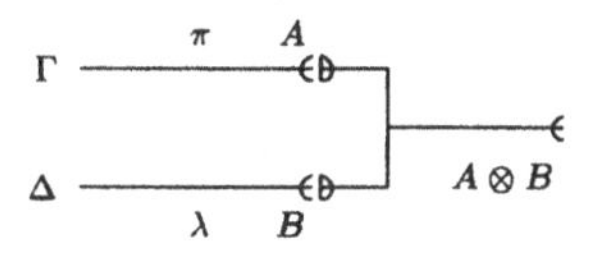

- $⅋$-*rule* : from a unit μ with an interface $\vdash C, D, \Lambda$, we can built a new one by merging plugs C and D into a new one (labelled $C ⅋ D$) by means of an encoder :

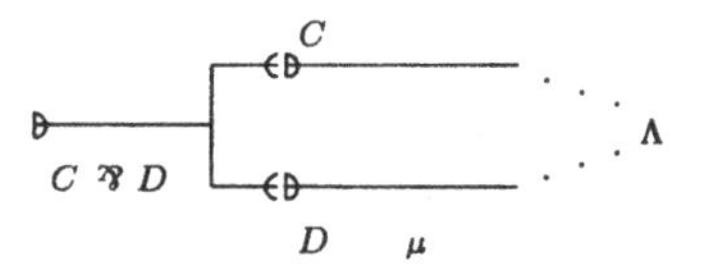

To understand what happens, let us assume that $C = A^\perp$, $D = B^\perp$; then $A^\perp ⅋ B^\perp = (A \otimes B)^\perp$, so there is the possibility of plugging. We therefore obtain

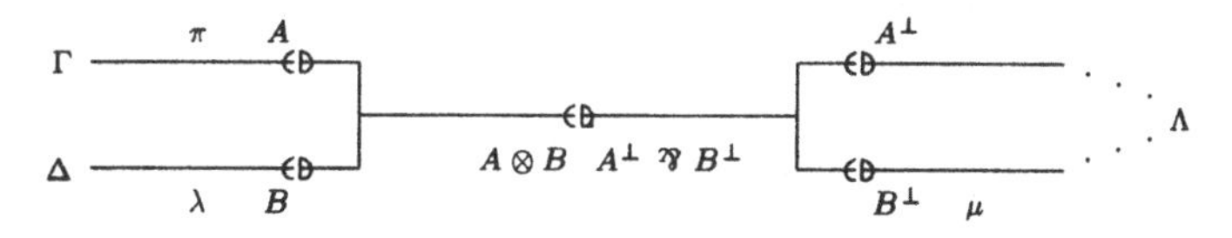

But the configuration

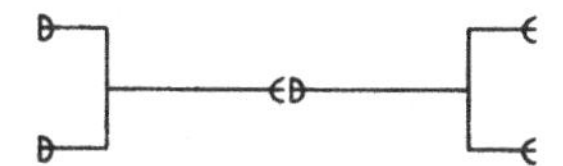

is equivalent to (if the coders are the same)

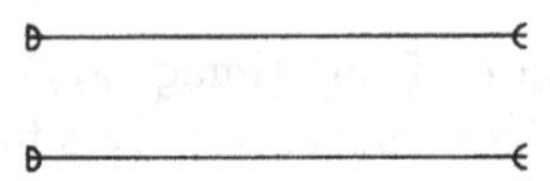

and therefore our plugging can be mimicked by two pluggings

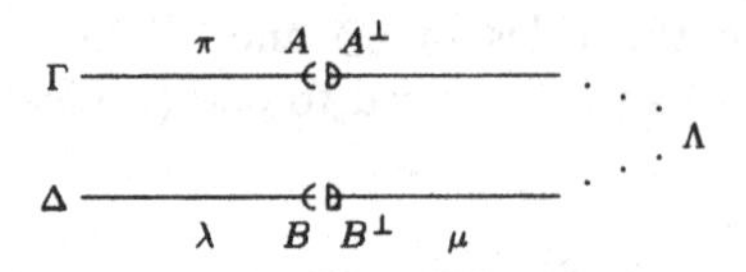

If we interpret the encoder as ⊗- or ⅋-link, according to the case, we get a very precise modelization of cut-elimination in proof-nets. Moreover, if we remember that coding is based on the development by means of Fourier series (which involves the Hilbert space) everything that was done can be formulated in terms of operator algebras. In fact the operator algebra semantics enables us to go beyond multiplicatives and quantifiers, since the interpretation also works for exponentials. We shall not go into this, which requires at least some elementary background in functional analysis ; however, we can hardly resist mentioning the formula for cut-elimination

$$\mathrm{EX}(u, \sigma) := (1 - \sigma^2)u(1 - \sigma u)^{-1}(1 - \sigma^2)$$

which gives the interpretation of the elimination of cuts (represented by σ) in a proof represented by u. Termination of the process is interpreted as the nilpotency of σu, and the part $u(1 - \sigma u)^{-1}$ is a candidate for the execution. See [15], this volume, for more details. One of the main novelties of this paper is the use of *dialects*, i.e. data which are defined up to isomorphism. The distinction between the two conjunctions can be explained by the possible ways of merging dialects : this is a new insight in the theory of parallel computation.

Geometry of interaction also works for various λ-calculi, for instance for pure λ-calculus, see [7, 26]. It has also been applied to the problem of optimal reduction in λ-calculus, see [20].

Let us end this chapter by yet another refutation of weakening and contraction :

1. If we have a unit with interface $\vdash \Gamma$, it would be wrong to add another plug A ; such a plug (since we know nothing about the inside of the unit) must be a mock plug, with no actual connection with the unit ... Imagine a plug on which it is written "danger, 220V", you expect to get some result if you plug something with it : here nothing will happen !

2. If we have a unit with a repetitive interface $\vdash \Gamma, A, A$, it would be wrong to merge the two similar plugs into a single one : in real life, we have such a situation with the stereophonic output plugs of an amplifier, which have exactly the same specification. There is no way to merge these two plugs into one and still respect the specification. More precisely, one can try to plug a single loudspeaker to the two outputs plugs simultaneously ; maybe it works, maybe it explodes, but anyway the behaviour of such an experimental plugging is not covered by the guarantee ...

2.4 Game semantics

Recently Blass introduced a semantics of linear logic, see [5], this volume. The semantics is far from being complete (i.e. it accepts additional principles), but this direction is promising.
Let us forget the state of the art and let us focus on what could be the general pattern of a convincing game semantics.

2.4.1 Plays, strategies etc.

Imagine a game between two players **I** and **II** ; the rule determines which is playing first, and it may happen that the same player plays several consecutive moves. The game eventually terminates and produces a numerical output for both players, e.g. a real number. There are some distinguished outputs for **I** for which he is declared to be the winner, similarly for **II**, but they cannot win the same play. Let us use the letter σ for a strategy for player **I**, and the letter τ for a strategy for **II**. We can therefore denote by $\sigma * \tau$ the resulting play and by $< \sigma, \tau >$ the output. The idea is to interpret formulas by games (i.e. by the rule), and a proof by a winning strategy. Typically linear negation is nothing but the interchange of players etc.

2.4.2 The three layers

We can consider three kinds of invariants :

1. Given the game A, consider all inputs for **I** of all possible plays : this vaguely looks like a phase semantics (but the analogy is still rather vague) ;

2. Given the game A and a strategy σ for **I** consider the set $\mid \sigma \mid$ of all plays $\sigma * \tau$, when τ varies among all possible strategies for **II**. This is an analogue of denotational semantics : we could similarly define the interpretation $\mid \tau \mid$ of a strategy for **II** and observe that $\mid \sigma \mid \cap \mid \tau \mid = \{\sigma * \tau\}$ (this is analogue to the fact that a clique in X and a clique in $X^\perp$ intersect in at most one point) ;

3. We could concentrate on strategies and see how they dynamically combine : this is
analogous to geometry of interaction.

Up to the moment this is pure science-fiction. By the way we are convinced that although games are a very natural approach to semantics, they are not primitive, i.e. that the game is rather a phenomenon, and that the actual semantics is a more standard mathematical object (but less friendly). Anyway, whatever is the ultimate status of games w.r.t. logic, this is an essential intuition : typically game semantics of linear logic is the main ingredient in the recent solution of the problem of full abstraction for the language **PCF**, see [1].

2.4.3 The completeness problem

The main theoretical problem at stake is to find a complete semantics for (first order) linear logic. Up to now, the only completeness is achieved at the level of provability (by phase spaces) which is rather marginal. Typically a complete game semantics would yield winning strategies only for those formulas which are provable. The difficulty is to find some semantics which is not contrived (in the same way that the phase semantics is not contrived : it does uses, under disguise, the principles of linear logic).
A non-contrived semantics for linear logic would definitely settle certain general questions, in particular which are the possible rules. It is not to be excluded that the semantics suggests tiny modifications of linear rules (e.g. many semantics accept the extra principle
$A \otimes B \multimap A \parr B$, known as *mix*), (and which can be written as a structural rule), or accepts a wider spectrum of logics (typically it could naturally be non-commutative, and then set up the delicate question of non-commutativity in logic). Surely it would give a stable foundation for constructivity.

Acknowledgements
The author is deeply indebted to Daniel Dzierzgowski and Philippe de Groote for producing LaTeX versions of a substantial part of this text, especially figures. Many thanks also to Yves Lafont for checking the final version.

REFERENCES

[1] S. Abramsky, R. Jagadeesan, and P. Malacaria. Full abstraction for pcf (extended abstract). In Masami Hagiya and John C. Mitchell, ed-

itors, *Theoretical Aspects of Computer Software. International Symposium, TACS 94, Sendai, Japan.* Springer Verlag, April 1994. Lecture Note in Computer Science 789.

[2] V. M. Abrusci. Non-commutative proof-nets. In *this volume*, 1995.

[3] J.-M. Andreoli and R. Pareschi. Linear objects: logical processes with built-in inheritance. *New Generation Computing*, 9(3–4):445–473, 1991.

[4] A. Asperti. A logic for concurrency. Technical report, Dipartimento di Informatica, Pisa, 1987.

[5] A. Blass. A game semantics for linear logic. *Annals of Pure and Applied Logic*, 56:183–220, 1992.

[6] R. Blute. Linear logic, coherence and dinaturality. *Theoretical Computer Science*, 93:3–41, 1993.

[7] V. Danos. *La logique linéaire appliquée à l'étude de divers processus de normalisation et principalement du λ-calcul.* PhD thesis, Université Paris VII, 1990.

[8] V. Danos, J.-B. Joinet, and H. Schellinx. LKQ and LKT : sequent calculi for second order logic based upon dual linear decompositions of classical implication. In *this volume*, 1995.

[9] V. Danos and L. Regnier. The structure of multiplicatives. *Archive for Mathematical Logic*, 28:181–203, 1989.

[10] T. Ehrhard. Hypercoherences : a strongly stable model of linear logic. In *this volume*, 1995.

[11] C. Fouqueré and J. Vauzeilles. Inheritance with exceptions : an attempt at formalization with linear connectives in unified logic. In *this volume*, 1995.

[12] J.-Y. Girard. Linear logic. *Theoretical Computer Science*, 50:1–102, 1987.

[13] J.-Y. Girard. A new constructive logic : classical logic. *Mathematical structures in Computer Science*, 1:255–296, 1991.

[14] J.-Y. Girard. On the unity of logic. *Annals of Pure and Applied Logic*, 59:201–217, 1993.

[15] J.-Y. Girard. Geometry of interaction III : accommodating the additives. In *this volume*, 1995.

[16] J.-Y. Girard. Light linear logic. *in preparation*, 1995.

[17] J.-Y. Girard. Proof-nets : the parallel syntax for proof-theory. In *Logic and Algebra*, New York, 1995. Marcel Dekker.

[18] J.-Y. Girard, Y. Lafont, and P. Taylor. *Proofs and types*, volume 7 of *Cambridge tracts in theoretical computer science*. Cambridge University Press, 1990.

[19] J.-Y. Girard, P.J. Scott, and A. Scedrov. Bounded linear logic: a modular approach to polynomial time computability. In S.R. Buss and P.J. Scott, editors, *Feasible mathematics*, pages 195–209, Boston, 1990. Birkhäuser.

[20] G. Gonthier, M. Abadi, and J.-J. Levy. The geometry of optimal lambda-reduction. In ACM Press, editor, *POPL'92*, Boston, 1992. Birkhäuser.

[21] Y. Lafont. The linear abstract machine. *Theoretical Computer Science*, 59:95–108, 1990.

[22] Y. Lafont. From proof-nets to interaction nets. In *this volume*, 1995.

[23] J. Lambek. Bilinear logic in algebra and linguistics. In *this volume*, 1995.

[24] P. Lincoln. Deciding provability of linear logic fragments. In *this volume*, 1995.

[25] P. Lincoln, J.C. Mitchell, J.C. Shankar, and A. Scedrov. Decision problems for propositional linear logic. In *Proceedings of 31st IEEE symposium on foundations of computer science, volume 2*, pages 662–671. IEEE Computer Society Press, 1990.

[26] Laurent Regnier. *Lambda-Calcul et Réseaux*. Thèse de doctorat, Université Paris 7, 1992.

[27] D.N. Yetter. Quantales and non-commutative linear logic. *Journal of Symbolic Logic*, 55:41–64, 1990.

Bilinear logic in algebra and linguistics [0)]

J. Lambek, McGill University, Montreal

Abstract

The *syntactic calculus*, a fragment of noncommutative linear logic, was introduced in 1958 because of its hoped for linguistic application. Working with a Gentzen style presentation, one was led to the problem of finding all derivations $f : A_1 \ldots A_n \to B$ in the free syntactic calculus generated by a contextfree grammar $\mathcal{G}$ (with arrows reversed) and to the problem of determining all equations $f = g$ between two such derivations. The first problem was solved by showing that f is equal to a derivation in *normal form*, whose construction involves no identity arrows and no cuts (except those in $\mathcal{G}$) and the second problem is solved by reducing both f and g to normal form.

The original motivation for the syntactic calculus came from multilinear algebra and a categorical semantics was given by the calculus of bimodules. Bimodules ${}_RF_S$ may be viewed as additive functors $R \to$ Mod S, where R and S are rings (of several objects). It is now clear that Lawvere's generalized bimodules will also provide a semantics for what may be called *labeled bilinear logic*.

1 Introduction.

I was asked to talk about one precursor of linear logic that I happened to be involved in, even though it anticipated only a small fraction of what goes on in the linear logic enterprise. I would now call this system "bilinear logic", meaning "non-commutative linear logic" or "logic without Gentzen's three structural rules". My original name had been "syntactic calculus", because of its intended application to linguistics; but actually it had been developed in collaboration with George Findlay [1955] in an attempt to understand some basic homological algebra. Alas, our paper was rejected on the grounds that most of our results were to appear in a book by Cartan and Eilenberg [1956], the publication of which had been delayed because of a paper shortage. All I was able to salvage from this wrecked endeavour was the notation in my book "Lectures on rings and modules" [1966], to which I shall return later. Unfortunately, this notation did not catch on with ring theorists.

The *syntactic calculus* [1958] dealt with the binary connectives $\otimes$ (*tensor*), $/$ (*over*) and $\backslash$ (*under*), to which I later added the constant I (*identity*). It was presented as a deductive system expressed with the help of an *entailment* relation $\rightarrow$ satisfying the following axioms and rules of inference:

(1) $A \rightarrow A$,

(2) $\dfrac{A \rightarrow B \quad B \rightarrow C}{A \rightarrow C}$,

(3) $A \otimes I \leftrightarrow A \leftrightarrow I \otimes A$,

(4) $(A \otimes B) \otimes C \leftrightarrow A \otimes (B \otimes C)$,

(5) $A \otimes B \rightarrow C$ iff $A \rightarrow C/B$ iff $B \rightarrow A\backslash C$.

Here $\leftrightarrow$ indicates that the entailment goes both ways. Optionally, one may also incorporate the usual lattice operations $\top, \wedge$ (*infimum*), $\bot$ and $\vee$ (*supremum*) into the syntactic calculus [1989b].

How does this system differ from ordinary logic? If we were to add the following additional axioms[1)], equivalent to Gentzen's three structural rules (see Section 4),

$$\text{(a)}\quad A \otimes B \rightarrow B \otimes A, \qquad \text{(b)}\quad A \rightarrow A \otimes A, \qquad \text{(c)}\quad A \rightarrow I,$$

we would be able to prove

$$I \leftrightarrow \top, \quad A \otimes B \leftrightarrow A \wedge B, \quad C/B \leftrightarrow B\backslash C.$$

Moreover, in view of (5), it would be reasonable to write the last expression as $B \Rightarrow C$ and we would recapture the intuitionistic propositional calculus.

For application to syntax, the additional axioms were rejected. Even without them, however, one could prove a number of theorems, for example the following, accompanied by names currently used by linguists.

(6) $A \otimes (A\backslash C) \rightarrow C$ (Ajdukiewicz),

(6') $(C/B) \otimes B \rightarrow C$,

(7) $(A\backslash B)/C \leftrightarrow A\backslash(B/C)$ (associativity),

(8) $A \rightarrow C/(A\backslash C)$ (type raising),

(8') $B \rightarrow (C/B)\backslash C$,

(9) $(C\backslash B) \otimes (B\backslash A) \rightarrow C\backslash A$ (composition),

(9') $(A/B) \otimes (B/C) \rightarrow A/C$,

(10) $B\backslash A \rightarrow (C\backslash B)\backslash(C\backslash A)$ (Geach),

(10') $A/B \rightarrow (A/C)/(B/C)$.

Following Grishin [1983], one can also introduce additional operations O, $\oplus$, $\dot{-}$ and $\dot{-}$ dual to $I, \otimes, /$ and $\backslash$ as in my [1993b], but I shall refrain from doing so here.

2 Linguistic application.

The intended linguistic application of the syntactic calculus [1958, 1959, 1961] went back to an arithmetical notation introduced by Ajdukiewicz [1937] and Bar-Hillel [1953]. The idea was to assign to each English word a *type*, constructed from a finite set of *basic types*, $N, S, \ldots$ by means of the connectives $\otimes, /$ and $\backslash$. Here N is the type of names and S the type of declarative sentences. Thus *John* has type N and *John snores* has type S.

We assign to *snores* the type $N\backslash S$, because we can prove that $N \otimes (N\backslash S) \rightarrow S$, an instant of theorem (6), the rule originally used by Ajdukiewicz. For the same reason, in *(John snores) loudly*, we are led to assign to *loudly* the type $S\backslash S$. On the other hand, if we bracket this sentence *John (snores loudly)*, we are led to assign to *snores* the type $(N\backslash S)\backslash(N\backslash S)$. Must our dictionary list both types for *snores*? Not really, because we can prove $S\backslash S \rightarrow (N\backslash S)\backslash(N\backslash S)$, an instant of theorem (10), later to be called "Geach's rule".

Now let us look at *John (likes Jane)*. From the way this is bracketed, clearly *likes Jane* should have type $N\backslash S$, the same as *snores*. But then *likes* should have type $(N\backslash S)/N$, because we can prove that $((N\backslash S)/N) \otimes N \rightarrow N\backslash S$, an instant of theorem (6′). On the other hand, had we bracketed this sentence as *(John likes) Jane*, we would have arrived at the type $N\backslash(S/N)$ for *likes*. Fortunately, in view of theorem (7), this does not matter.

What about *she* in *she snores*? We cannot assign type N to this pronoun, because English grammar forbids **John likes she*. However, we can assign $S/(N\backslash S)$ to *she*, since we can prove $(S/(N\backslash S)) \otimes (N\backslash S) \rightarrow S$, an instant of (6′). But, for the same reason, *John* should also have type $S/(N\backslash S)$, in addition to N. Fortunately, we are saved from having to assign both these types to *John* in the dictionary, because we can prove $N \rightarrow S/(N\backslash S)$, an instant of theorem (8), called "type raising" by linguists.

The original program had been to shove all of English grammar into the dictionary, which was to list a finite set of "syntactic types" for each word. These types were terms of the syntactic calculus freely generated by the set $\{S, N, \ldots\}$. Then, looking at a string of English words, one would try to prove that the string is a well-formed declarative sentence, by checking that $A_1 \otimes \ldots \otimes A_n \rightarrow S$ is a theorem of the syntactic calculus, where A_i is some type assigned to the ith word in the given string. While I no longer subscribe to this program myself, it has lately been taken up by a small number of linguists, notably by Moortgat [1988].

3 Two problems.

The program outlined above raises two problems.

(I) Is there a decision procedure for the syntactic calculus, so that we can check the sentencehood of any given string of English words by a calculation? More generally, can we decide whether $A \to B$ is provable, for given A and B?

In the original system of Ajdukiewicz and Bar-Hillel, this problem had an obvious solution, because they only accepted *contractions* such as $(C/B) \otimes B \to C$. However, the syntactic calculus also allows *expansions* such as $B \to (C/B)\backslash C$, so the existence of a decision procedure is not immediately obvious.

To lead up to another problem, note that there may be several ways of showing that a string of English words is a sentence and these may in fact amount to different interpretations of the given string. My favourite example is:

experience teaches spiders that time flies.

To see why this can be constructed as a sentence in two different ways, compare the two passives:

spiders are taught by experience that time flies;

spiders that time flies (with a stop-watch) *are taught by experience.*

(II) Assuming that we have defined the notion of equality between proofs of $A \to B$, is there a decision procedure for determining whether two such proofs are equal?

To take account of different proofs that $A \to B$, we should really work in a *category* rather than a mere deductive system. Fortunately, the syntactic calculus, or rather its proof theory, is pretty close to being a category, which I have called a "biclosed monoidal category" or "residuated category" [1968/9]. All we have to do is to introduce names for the proofs of entailments and to impose appropriate equations. Thus we have

(1) $1_A : A \to A$,

(2) $\dfrac{f : A \to B \quad g : B \to C}{gf : A \to C}$,

subject to the equations

$$f 1_A = f = 1_B f,$$
$$(hg)f = h(gf),$$

where $h : C \to D$. This is just the definition of a category.

It is also clear that (5) should be translated as asserting that the functor $- \otimes B$ has a right adjoint $-/B$ and that $A \otimes -$ has a right adjoint $A\backslash -$. We must furthermore postulate an arrow

(5) $\alpha_{ABC} : (A \otimes B) \otimes C \to A \otimes (B \otimes C)$

with an inverse in the opposite direction, and similarly for (3). Experience teaches us that we should also require certain "coherence" conditions. e.g. the *pentagonal* rule of MacLane, which asserts that the following arrow

$$((A \otimes B) \otimes C) \otimes D \to (A \otimes B) \otimes (C \otimes D) \to A \otimes (B \otimes (C \otimes D))$$

is equal to the arrow with the same source and target via

$$\cdots \to (A \otimes (B \otimes C)) \otimes D \to A \otimes ((B \otimes C) \otimes D) \to \cdots .$$

But such coherence conditions do not flow naturally from the way the syntactic calculus has been formulated so far. (If the lattice operations *inf* and *sup* are incorporated into this categorical setting, they become the *cartesian product* and *coproduct* respectively.)

4 Multicategories.

It so happens that both of the above problems (I) and (II) can be solved by reformulating the syntactic calculus as a "sequent calculus" following Gentzen (see Kleene [1952] or Szabo [1969]). As I pointed out in [1961], Gentzen's sequents are really the same as Bourbaki's [1948] multilinear operations. We shall see that such a reformulation also helps to explicate the pentagonal condition as well as the functoriality of the tensor.

By a *sequent* one understands an expression of the form

$$f : A_1 \ldots A_n \to A_{n+1},$$

where $n \geq 0$. The A_i and f go by different names in different disciplines:
in logic, f is a *deduction*, the A_i are *formulas*;
in linguistics, f is a *derivation*, the A_i are *types*;
in algebra, f is an *operation*, the A_i are *sorts*.

In the absence of Gentzen's three structural rules (see below), f is called a *multilinear* operation.

One always insists on the following axiom and rule of inference:

$$1_A : A \to A \quad \text{(identity)},$$

$$\frac{f : \Lambda \to A \quad g : \Gamma A \Delta \to B}{g \&_A f : \Gamma \Lambda \Delta \to B} \quad \text{(cut)},$$

where $\Lambda = L_1 \ldots L_k$, $\Gamma = C_1 \ldots C_m$, $\Delta = D_1 \ldots D_n$, say.

So far, we have described what linguists would call a *context-free grammar*, though from the point of view of the audience. (From the speaker's point of view, the arrow should be reversed, as is more usual among generative grammarians.)

In algebra, it is customary to introduce *variables* (or *indeterminates*) of each sort to allow the formation of polynomials. *Polynomials* of each sort are defined recursively:

(i) each variable of sort A is a polynomial of sort A;

(ii) if a_i is a polynomial of sort $A_i (i = 1, \ldots, n)$ and $f : A_1 \ldots A_n \to B$ is an operation, then $fa_1 \ldots a_n$ is a polynomial of sort B.

In logic, the variables should be thought of as *occurrences of assumptions* and the polynomials as *natural deduction proofs* in the style favoured by Prawitz [1965].

Equality of polynomials is of course a congruence relation with the usual property allowing substitution of polynomials for variables. To give meaning to the identity operation and to the cut rule, one also stipulates the identities:

$1_A a = a$, where a is of sort A;

$(g\&_A f)\overline{c\ell d} = g\overline{c}f\overline{\ell}\overline{d}$, where we have written $\overline{c} = c_1 \ldots c_m$, $\overline{d} = d_1 \ldots d_n$, $\overline{\ell} = \ell_1 \ldots \ell_k$, c_i being of type C_i, d_i of type D_i, ℓ_i of type L_i. If one so wishes, one may assume that a, c_i, d_i and ℓ_i are variables. The cut rule is thus nothing else than the rule allowing the substitution of one polynomial into another and our context-free grammar is nothing else than what might be called a *multisorted multilinear algebraic theory.*[2)]

Two operations $p, q : A_1 \ldots A_n \to B$ are said to be *equal* if one has

$$px_1 \ldots x_n = qx_1 \ldots x_n$$

with variables x_i of sort A_i. It readily follows that, for f and g as above and for $h : \Theta \to C$,

$$g\&_A 1_A = g = 1_B \&_B g,$$

$$(g\&_A f)\&_C h = \begin{cases} g\&_A(f\&_C h) & \text{if } C \text{ occurs in } \Lambda, \\ (g\&_C h)\&_A f & \text{if } C \text{ occurs in } \Gamma \text{ or } \Delta \end{cases}$$

These are precisely the equations of what I have called a *multicategory* [1969, 1989].

To see what we are missing, let me briefly mention Gentzen's three structural rules, essentially equivalent to axioms (a) to (c) of Section 1, which is absent in syntax and multilinear algebra, though present in intuitionistic logic, in semantics and in ordinary algebra (in a *multisorted algebraic theory*[2)]):

$$\frac{f : \Gamma AB\Delta \to C}{f^i : \Gamma BA\Delta \to C} \quad \text{(interchange)},$$

$$\frac{f : \Gamma AA\Delta \to C}{f^c : \Gamma A\Delta \to C} \quad \text{(contraction)},$$

$$\frac{f : \Gamma\Delta \to C}{f^w : \Gamma A\Delta \to C} \quad \text{(weakening)}.$$

These are subject to the following identities:

$f^i\bar{c}ba\bar{d} = f\bar{c}ab\bar{d}$ (interchanging two variables),
$f^c\bar{c}a\bar{d} = f\bar{c}aa\bar{d}$ (eliminating repeated variables),
$f^w\bar{c}a\bar{d} = f\bar{c}\bar{d}$ (introducing redundant variables).

5 Residuated multicategories.

Up to this point, we have not yet discussed the logical connectives $I, \otimes, /, \backslash$, $\top, \wedge, \bot, \vee$ in connection with a multicategory. For purpose of illustration, I shall pick out two, namely $\otimes$ and $/$; but the other connectives can be handled similarly [1993a].

According to Bourbaki [1948], the tensor product is determined by a bilinear operation

$$\mathbf{m}_{AB} : AB \to A \otimes B,$$

with a universal property: for every $f : \Gamma AB\Delta \to C$ (actually Bourbaki confines himself to the case when Γ and Δ are empty strings), there exists a unique $f^{\S} : \Gamma A \otimes B\Delta \to C$ such that

$$f^{\S} \&_{A\otimes B} \mathbf{m}_{AB} = f,$$

that is, identically

$$f^{\S}\bar{c}\mathbf{m}_{AB}ab\bar{d} = f\bar{c}ab\bar{d}.$$

The uniqueness of $f^{\S}$ can also be expressed as an equation; namely

$$(g\&_{A\otimes B}\mathbf{m}_{AB})^{\S} = g,$$

where $g : \Gamma A \otimes B\Delta \to C$.

This machinery can be put to use to prove the usual functoriality and coherence properties of the tensor product, as I pointed out in [1989a]. Thus, we have what logicians would call a *rule of inference*:

$$\frac{f : A \to A' \quad g : B \to B'}{f \otimes g : A \otimes B \to A' \otimes B'},$$

where $f \otimes g$ is defined by

$$(f \otimes g)\mathbf{m}_{AB}ab = \mathbf{m}_{A'B'}fagb,$$

that is,

$$f \otimes g = ((\mathbf{m}_{AB}\&_A f)\&_B g)^{\S}.$$

It is now easy to check that

$$(f' \otimes g')(f \otimes g) = (f'f) \otimes (g'g),$$

for $f' : A' \to A''$ and $g' : B' \to B''$, where we have adopted the usual categorical notation and written $f'f$ for $f'\&_{A'}f$.

We can also *define*, rather than postulate, the associativity arrow

$$\alpha_{ABC} : (A \otimes B) \otimes C \to A \otimes (B \otimes C)$$

by

$$\alpha\mathbf{mm}abc = \mathbf{m}a\mathbf{m}bc$$

and prove that it is an isomorphism. (We have omitted subscripts here, as we shall frequently do from now on.) Moreover, MacLane's pentagonal condition is checked by noting that there is a unique arrow

$$\mu_{ABCD} : ((A \otimes B) \otimes C) \otimes D \to A \otimes (B \otimes (C \otimes D))$$

such that

$$\mu\mathbf{mmm}abcd = \mathbf{m}a\mathbf{m}b\mathbf{m}cd.$$

Although Bourbaki did not point this out, the connective / is similarly determined by a bilinear operation

$$\mathbf{e}_{CB} : (C/B)B \to C \quad \text{(evaluation)},$$

with the universal property: for every $f : \Lambda B \to C$ there exists a unique $f^* : \Lambda \to C/B$ such that

$$\mathbf{e}_{CB}\&_{C/B}f^* = f,$$

that is, identically

$$\mathbf{e}f^*\bar{\ell}b = f\bar{\ell}b.$$

Again, the uniqueness can be expressed by the equation

$$(\mathbf{e}\&g)^* = g,$$

where $g : \Lambda \to C/B$.

Logicians and computer scientists are accustomed to write

$$f^*\bar{\ell} = \lambda_{y \in B} f\bar{\ell}y.$$

The two equations describing the existence and uniqueness of f^* then become:

$$\mathbf{e}(\lambda_{y\in B} f\bar{\ell}y)b == f\bar{\ell}b,$$
$$\lambda_{y\in B}(\mathbf{e}g\bar{\ell}y) == g\bar{\ell}\ .$$

The reader will recognize these as the usual β- and η-conversions respectively.

6 Decision procedures.

Gentzen had introduced his sequents for a different purpose and his method works as well, if not better, without his structural rules. Let $F(\mathcal{G})$ be the residuated multicategory, alias syntactic calculus or bilinear logic, freely generated by a multicategory, alias context-free grammar or multilinear algebraic theory, $\mathcal{G}$. Let A_i and B be formulas, alias types or sorts, in $F(\mathcal{G})$. We wish to find all deductions, alias derivations or operations, $f : A_1 \dots A_n \to B$ and later also all equations between two such deductions.

The problem is with the cut rule: if $h : \Gamma\Lambda\Delta \to B$ is to be obtained from $f : \Lambda \to A$ and $g : \Gamma A \Delta \to B$, so that $h = g\&_A f$, what is A? Gentzen's idea was to incorporate all necessary cuts into his inference rules and then to show that all other cuts are unnecessary. It was shown in [1958] that this works for the syntactic calculus even more easily than for intuitionistic logic.

First we replace the axiom $\mathbf{m}_{AB}$ by the rule of inference

$$\frac{f : \Gamma \to A \quad g : \Delta \to B}{f\mu g : \Gamma\Delta \to A \otimes B},$$

where

$$f\mu g = (\mathbf{m}\&f)\&g.$$

This has the form of an *introduction rule*: the symbol $\otimes$ is introduced on the right, just as in $f^{\S}$ the same symbol is introduced on the left.

In the same way, we replace the axiom $\mathbf{e}_{CB}$ by the rule of inference

$$\frac{f : \Lambda \to A \quad g : \Gamma B \Delta \to C}{g\varepsilon f : \Gamma B/A \Lambda\Delta \to C},$$

where

$$g\varepsilon f = g\&(\mathbf{e}\&f).$$

Again, this introduces / on the left, just as f^* introduces / on the right. Other connectives may be treated similarly [1993a].

Having adopted μ and ε in place of $\mathbf{m}$ and $\mathbf{e}$, and similarly for the other connectives, we now observe that we no longer need identity and cut, except in $\mathcal{G}$.

Thus, it is easily seen that

$$1_{A\otimes B} = \mathbf{m}_{AB}{}^{\S} = (1_A \mu 1_B)^{\S}$$

and

$$1_{C/B} = \mathbf{e}_{CB}{}^{*} = (1_C \varepsilon 1_B)^{*}.$$

If A, B and C are in $\mathcal{G}$, we are satisfied; otherwise we repeat the procedure. As to the cut, consider

$$\frac{f : \Lambda \to A \quad g : \Gamma A \Delta \to B}{g \& f : \Gamma \Lambda \Delta \to B}$$

and assume that f and g have been constructed without cut. One can then show that $g\&f$ can be replaced by, and is in fact equal to, a deduction $\Gamma\Lambda\Delta \to B$ constructed with one or two cuts of smaller degree. (The *degree* of a cut, say the above, is defined as the total number of occurrences of all the connectives in $\Lambda, A, \Gamma, \Delta$ and B. For details see my [1958, 1968/9, 1993a]).

Let us say that a deduction $f : \Lambda \to A$ in $F(\mathcal{G})$ is in *normal form* if its construction involves no cuts or identities, except those in $\mathcal{G}$. It thus follows that every deduction $f : \Lambda \to A$ is *equal* to one in normal form. In particular, this yields a solution to problem (I), because we can find all deductions $\Lambda \to A$ in normal form by working backwards.

We may now address problem (II): given two deductions $f, g : \Lambda \to A$ in $F(\mathcal{G})$, to decide whether they are equal. Without loss of generality, we may assume that f and g are in normal form. The normal form, while not quite unique, is almost so.

Recall that f and g are constructed from deductions in $\mathcal{G}$ by means of the inference rules $\mu, {}^{\S}, \varepsilon, {}^{*}$ etc. As I have shown elsewhere [1993a], f and g are equal if and only if their constructions differ only by the order in which the inference rules are applied.

More precisely, it is claimed that equality between deductions in normal form is generated by the following equations, in which $\phi(h), \phi'(h) = h^{\S}$ or h^{*} and $\chi(h, h'), \chi'(h, h') = h\mu h'$ or $h\varepsilon h'$:

$$\begin{aligned}
&\phi(\phi'(h)) = \phi'(\phi(h)),\\
&\phi(\chi(h, h')) = \chi(\phi(h), h') \text{ or } \chi(h, \phi(h')),\\
&\chi(\chi'(h_0, h_1), h) = \chi'(\chi(h_0, h)h_1) \text{ or } \chi'(h_0, \chi(h_1, h)),\\
&\chi(h, \chi'(h_0, h_1)) = \chi'(\chi(h, h_0), h_1) \text{ or } \chi'(h_0, \chi(h, h_1)).
\end{aligned}$$

There will be other generating equations of this kind, if we take the other rules of inference into account.

It follows from the existence and essential uniqueness of the normal form for deductions that the "multifunctor" $\mathcal{G} \to F(\mathcal{G})$ (categorists prefer to write $\mathcal{G} \to UF(\mathcal{G})$) is *full* and *faithful*: if Λ and A are in $\mathcal{G}$, then all deductions $\Lambda \to A$ in $F(\mathcal{G})$ are already in $\mathcal{G}$, and two such deductions can be equal in $F(\mathcal{G})$ only if they are already equal in $\mathcal{G}$.

To illustrate our method we compare two deductions of $((S/N) \otimes N)/N \to ((S/N) \otimes N)/N$ in normal form. The first may be generalized to the deduction $((S/N) \otimes N)/N' \to ((S/N') \otimes N'')/N''$ described by

$$(((1_S \varepsilon 1_N)^{\S} \varepsilon 1_{N'})^* \mu 1_{N''})^* \; .$$

The second may be generalized to the identity deduction $((S/N) \otimes N')/N'' \to ((S/N) \otimes N')/N''$, which becomes in normal form

$$(1_{N''} \varepsilon ((1_S \varepsilon 1_N)^* \mu 1_{N'}))^*.$$

Even if we identify $N'' = N' = N$, these two descriptions cannot be transformed into one another by the above generating equations. This is seen, for example, by noting that the first contains the symbol § and the latter doesn't.

7 Bimodules.

As mentioned in the introduction, the original motivation for studying the syntactic calculus came from homological algebra, which is largely concerned with bimodules. Given two rings (with unity) R and S, an $R-S$-bimodule ${}_RA_S$ is an abelian group A, at the same time a left R-module ${}_RA$ and a right S-module A_S, such that $(ra)s = r(as)$ for all $r \in R$, $a \in A$ and $s \in S$.

Given bimodules ${}_RA_S$, ${}_SB_T$ and ${}_RC_T$, one is interested in the following bimodules constructed from them:

$${}_RA_S \otimes_S B_T = A \otimes B \text{ made into an } R-T\text{- bimodule,}$$

$${}_RC_T \oslash_S B_T = \mathrm{Hom}_T(B, C) \text{ made into an } R-S\text{-bimodule,}$$

$${}_RA_S \varnothing_R C_T = \mathrm{Hom}_R(A, C) \text{ made into an } S-T\text{-bimodule.}$$

I have written $\oslash$ and $\varnothing$ rather than / and \ to avoid possible confusion with factor modules. The notation was introduced to distinguish between ${}_RC_R \oslash_R B_R$ and ${}_RB_R \varnothing_R C_R$ and to exploit the left-right symmetry. Proofs of theorems of the syntactic calculus could now be interpreted as canonical homomorphisms between bimodules.

Passing from the algebraic to the linguistic application, I made the simplifying assumption that all subscripts are equal and may therefore be omitted altogether. It now turns out that the decision to omit subscripts may have been premature. Trimble, in his thesis [1993], made crucial use of the equation $(ra)s = r(as)$ in the special case when $R = S$. But even the simplifying assumption that all subscripts are equal may have been premature. Brame, in a series of papers [1984/5/7], considered linguistic types ${}_RA_S$ provided with subscripts to allow juxtaposition of ${}_RA_S$ and ${}_{S'}B_T$ only when $S = S'$.

Let me define a *Brame grammar* as being a refined sort of context-free grammar: it consists of derivations $f : \Lambda \to A$ where Λ lives in the *free category generated by a graph*, rather than the *free monoid generated by a set* as is usual. Brame's graph consists of English words provided with two subscripts to indicate what neighbours it will admit. For example, edges of such a graph might be ${}_1the_A$, ${}_Aman_N$ and ${}_Nsnores_S$. They would allow one to form the string ${}_1$(the man snores)$_S$, but no other string made up of the same three words.

We may thus consider a variant of the syntactic calculus, call it *labeled bilinear logic*, whose formulas are of the form ${}_RA_S$, where the subscripts denote syntactic types. Recalling Bénabou's [1967] notion of "bicategory", we then have the proportion: labeled bilinear logic: Brame grammar: bicategory = bilinear logic: context-free grammar: monoidal category.

Let me take this opportunity to point out that there are three categorical ways of defining a bimodule ${}_RA_S$:

(i) as an additive functor $R \to \text{Mod}S$, where $\text{Mod}S$ is the category of right S-modules, namely the category of all additive functors $S^{\text{op}} \to \mathcal{Ab}$, the category of abelian groups (S^{op} here is the opposite of the ring S, multiplication being reversed);

(ii) as an additive functor $\text{Mod}R \to \text{Mod}S$ which has a right adjoint;

(iii) as an additive functor $R \otimes S^{\text{op}} \to \mathcal{Ab}$, where $\otimes$ denotes the tensor product, also called Kronecker product, of the two rings.

Realizing that a ring is just a small additive category with one object, we may follow Barry Mitchell [1972] in extending the word "ring" to mean any small additive category. Thus, a so-called Morita context is just a ring with two objects. Everything we said about bimodules here remains valid for rings in this extended sense.

My attention was drawn by Todd Trimble to Lawvere's [1973] generalized bimodules as also offering a very general model of labeled bilinear logic.[3)] In particular, they include a number of special cases which I had already been looking at. The following table brings out the analogous concepts, where R stands for a small additive category, a small category, a poset and a set respectively:

Add	Mod$R$	$\mathcal{A}b$	bimodule
Cat	Presh$R$	Set	profunctor
Poset	$\mathcal{P}_\downarrow(R)$	$\cdot \to \cdot$	comparison
Set	$\mathcal{P}(R)$	$\{0,1\}$	binary relation.

Here $\mathcal{P}(R)$ is the usual power set of R, $\mathcal{P}_\downarrow(R)$ the set of downward closed subsets of R, Presh R the set of presheaves on R. The word "comparison" was specially introduced for this purpose. Of course, the first three of these examples deal with $\mathcal{V}$-categories, where $\mathcal{V}$ is a symmetric monoidal closed category, namely the category listed in the third column of the table. (See Kelly [1982]). Other special cases may also be of interest.

One operation on bimodules, which had been ignored in the original syntactic calculus, is the *converse*. The converse of the bimodule ${}_RA_S$ is the bimodule ${}_{S^{op}}A^{\cup}_{R^{op}}$, where the underlying abelian group of $A^{\cup}$ is the same as that of A. The converse of a binary relation is well-known. In our Gentzen style deductive system, the converse should give rise to an inference rule:

$$\frac{f : A_1 \dots A_n \to B}{f^{\cup} : A_n^{\cup} \dots A_1^{\cup} \to B^{\cup}},$$

together with the stipulation that $f^{\cup\cup} = f$.[4)].

At the level of deductive systems, we can now obtain an arrow ${m_{AB}}^{\cup\S}$: $B^{op} \otimes A^{op} \to (A \otimes B)^{op}$, which, at the categorical level, happens to be an isomorphism. The existence of a converse will allow us to identify $B\backslash C$ with $(C^{\cup}/B^{\cup})^{\cup}$ and it will license the observed symmetry, say between $(C/B)\otimes B \to C$ and $B \otimes (B\backslash C) \to C$.

8 Abstract representation theory.

Instead of talking about $\mathcal{V}$-categories concretely, we can try to isolate those abstract properties of $\mathcal{V}$-Cat which make all this possible. We are thus led to the notion of "cosmos" studied by Bénabou and Street [1972/3]. However, I prefer a simpler description of those 2-categories which resemble $\mathcal{V}$-Cat for some $\mathcal{V}$.

The following discussion of Add (with $\mathcal{V} = \mathcal{A}b$) serves to illustrate how one may treat $\mathcal{V}$-Cat in general. As a category (actually, a closed symmetric monoidal category), Add is a non-full subcategory of Cat, the category of *locally small* categories; but as a 2-category, it inherits all natural transformations from Cat. It is closed under the operation $(\ \ )^{op}$, which sends each category onto its opposite, leaves the action of functors unchanged and reverses natural transformations.

Among the objects of Add are those which are *small*, usually called "rings", it being understood that a ring may have several objects. The full subcategory of rings is closed under $(\ )^{op}$. If R and S are rings, an $R-S$-bimodule ${}_RF_S$ is an additive functor, i.e. morphism in Add, $F : R \to \text{Mod } S$, where Mod $S = \text{Add}(S^{op}, \mathcal{A}b)$. In particular, there is the *identity* bimodule $I_R : R \to$ Mod R which sends an object $*$ of R onto Add$(-,*)$. The *converse* of F is the bimodule $F^{\cup} : S^{op} \to \text{Mod } R^{op}$, where $(\ )^{\cup}$ is the canonical mapping

$$\text{Add}(R, \text{Add}(S^{op}, \mathcal{A}b)) \to \text{Add}(S^{op}, \text{Add}(R, \mathcal{A}b))$$

such that $F^{\cup\cup} = F$.

With any bimodule $F : R \to \text{Mod } S$, there is canonically associated the additive functor $F^r : \text{Mod}S \to \text{Mod}R$ defined by $F^r(G) = \text{Nat}(F-, G)$, where $G \in \text{Mod } S$. F^r has a left adjoint F^ℓ such that $F^\ell I_R \cong F$. (Once the existence of F^ℓ has been established, we can forget about its construction with the help of colimits)[5].

Given bimodules $F : R \to \text{Mod}S$, $G : S \to \text{Mod}T$ and $H : R \to \text{Mod}T$, we may define bimodules $F \otimes G : R \to \text{Mod}T$, $H/G : R \to \text{Mod}S$ and $F\backslash H : S \to \text{Mod}T$ as follows:

$$F \otimes G = G^\ell F,\ \ H/G = G^r H,\ \ F\backslash H = (H^{\cup}/F^{\cup})^{\cup}.$$

One may now verify that

$$\text{Nat}(F \otimes G, H) \cong \text{Nat}(F, H/G) \cong \text{Nat}(G, F\backslash H)$$

and

$$(F \otimes G) \otimes H \cong F \otimes (G \otimes H),\ \ F \otimes I_S \cong F \cong I_R \otimes F.$$

Some work still has to be done to check coherence. It seems the best way to do this is to go to labeled multicategories, that is, Brame grammars. One way of interpreting a deduction $f : A_1 \ldots A_n \to B$, where A_i is an $R_{i-1} - R_i$-bimodule and B an $R_0 - R_n$-bimodule, is as a natural transformation $f : A_1^r \ldots A_n^r \to B^r$.

Time does not permit me to develop these ideas further now; but clearly the following program suggests itself:

(1) to write down all the properties a subcategory of Cat should satisfy in order to obtain a categorical model of bilinear logic;

(2) to do the same for an abstract 2-category.

REFERENCES

K. Ajdukiewicz, Die syntaktische Konnexität, Studia Philosophica I(1937), 1-27; translated in: S. McCall, Polish Logic 1920-1930, Clarendon Press, Oxford 1967.

Y. Bar-Hillel, A quasiarithmetical notation for syntactic description, Language **29**(1953), 47-58.

J. Bénabou, Introduction to bicategories, in: J. Bénabou et al., Reports of the Midwest Category Seminar, Springer LNM47(1967), 1-77.

G. Birkhoff and J.D. Lipson, Heterogeneous algebras, J. Combinatory Theory **8**(1970), 115-133.

N. Bourbaki, Algèbre multilinéaire, Hermann, Paris 1948.

M. Brame, Recursive categorical syntax and morphology I, II, III, Linguistic Analysis 14(1984), 265-287; **15**(1985), 137-176; **17**(1987), 147-185.

H. Cartan and S. Eilenberg, Homological Algebra, Princeton University Press, Princeton N.J. 1956.

G.D. Findlay and J. Lambek, Calculus of bimodules, unpublished manuscript, McGill University 1955.

V.N. Grishin, On a generalization of the Ajdukiewicz-Lambek System, in: Studies in nonclassical logics and formal systems, Nauka, Moscow 1983, 315-343.

P.J. Higgins, Algebras with a scheme of operators, Mathematische Nachrichten **27**(1963), 115-122.

G.M. Kelly, Basic concepts of enriched category theory, London Math. Soc. Lecture Notes Series 64, Cambridge University Press, Cambridge 1982.

S.C. Kleene, Introduction to metamathematics, Van Nostrand, New York N.Y. 1952.

A. Kock and G.C. Wraith, Elementary toposes, Aarhus University Lecture Notes Series 30(1971).

J. Lambek, The mathematics of sentence structure, Amer. Math. Monthly **65**(1958), 154-169.

.........., Contributions to a mathematical analysis of the English verb-phrase, J. Can. Linguistic Assoc. **5**(1959), 83-89.

.........., On the calculus of syntactic types, Amer. Math. Soc. Symposia Applied Math. **12**(1961), 166-178.

.........., Lectures on rings and modules, Ginn and Co., Waltham Mass. 1966; Chelsea Publ. Co., New York N.Y. 1986.

.........., Deductive systems and categories I, J. Math. Systems Theory **2**(1968), 278-318; II, Springer LNM 86(1969), 76-122.

.........., On the unity of algebra and logic, in: F. Borceux (ed.), Categorical algebra and its applications, Springer LNM **1348**(1988), 221-229.

.........., Multicategories revisited, Contemporary Math. **92**(1989), 217-239.

.........., On a connection between algebra, logic and linguistics, Diagrammes **22**(1989), 59-75.

.........., Logic without structural rules, in: K. Došen and P. Schroeder-Heister (eds.), Substructural logics, Oxford University Press, Oxford 1993, 179-206.

.........., From categorial grammar to bilinear logic, ibid., 207-237.

F.W. Lawvere, Metric spaces, generalized logic and closed categories, Rend. del Sem. Mat. e Fis. di Milano **43**(1973), 135-166.

S. MacLane, Natural associativity and commutativity, Rice University Studies **49**(1963), 28-46.

B. Mitchell, Rings with several objects, Advances in Math. **8**(1972), 1-161.

M. Moortgat, Categorial investigations, Foris Publications, Dordrecht-Holland/Providence RI 1988.

D. Prawitz, Natural deduction, Almquist and Wiskell, Stockholm 1965.

K.I. Rosenthal, *-autonomous categories of bimodules, manuscript, Union College 1993.

R. Street, Elementary cosmoi I, in: M. Kelly (ed.) Category Seminar, Springer LNM **420**(1974), 134-180.

S. Szabo (ed.), The collected papers of Gerhard Gentzen, Studies in Logic and the Foundations of Mathematics, North Holland, Amsterdam 1969.

T.H. Trimble, The bimodule interpretation of linear logic, manuscript, Rutgers University 1993.

FOOTNOTES

[0] This work has been supported by the Social Sciences and Humanities Research Council of Canada. The author is indebted to Djordje Čubrić and Yves Lafont for their careful reading of the manuscript.

[1] When stated thus, (a) is a consequence of (b) and (c).

[2] Multisorted algebraic theories (though not multilinear ones) were introduced by Higgins [1963] and taken up by Birkhoff and Lipson [1970]. Our notion of polynomial does not include constant polynomials, unless they are nullary operations. Note that parentheses are not needed, as long as all operations are written on the left.

[3] Quite independently, Kimmo Rosenthal [1993] has also been investigating generalized bimodules ${}_RM_R$ as providing interesting categorical models of bilinear logic, even of classical bilinear logic. The notion of "profunctor", sometimes called "distributor", is due to Bénabou. Kock and Wraith [1971] show that profunctors give rise to a biclosed monoidal category.

[4] If we look at *untyped* bilinear logic *with converse*, as illustrated by binary relations on a single set, we may replace axiom (c) of Section 1, equivalent to Gentzen's interchange rule, by the axiom

$$\text{(c}'\text{)} \qquad A \to A^{\cup};$$

then (a), (b) and (c′) together describe what might be called a *co-equivalence relation.*

[5] I_R and F^r are additive analogues of what in Cat are called *Yoneda embedding* and *Kan extension* respectively.

QUESTIONS AND ANSWERS — A CATEGORY ARISING IN LINEAR LOGIC, COMPLEXITY THEORY, AND SET THEORY

ANDREAS BLASS

ABSTRACT. A category used by de Paiva to model linear logic also occurs in Vojtáš's analysis of cardinal characteristics of the continuum. Its morphisms have been used in describing reductions between search problems in complexity theory. We describe this category and how it arises in these various contexts. We also show how these contexts suggest certain new multiplicative connectives for linear logic. Perhaps the most interesting of these is a sequential composition suggested by the set-theoretic application.

INTRODUCTION

The purpose of this paper is to discuss a category that has appeared explicitly in work of de Paiva [18] on linear logic and in work of Vojtáš [25, 26] on cardinal characteristics of the continuum. We call this category $\mathcal{PV}$ in honor of de Paiva and Vojtáš (or, more informally, in honor of Peter and Valeria). The same category is implicit in a concept of many-one reduction of search problems in complexity theory [15, 23].

The objects of $\mathcal{PV}$ are binary relations between sets; more precisely they are triples $\mathbf{A} = (A_-, A_+, A)$, where A_- and A_+ are sets and $A \subseteq A_- \times A_+$ is a binary relation between them. (We systematically use the notation of boldface capital letters for objects, the corresponding lightface letters for the relation components, and subscripts $-$ and $+$ for the two set components.) A morphism from $\mathbf{A}$ to $\mathbf{B} = (B_-, B_+, B)$ is a pair of functions $f_- : B_- \to A_-$ and $f_+ : A_+ \to B_+$ such that, for all $b \in B_-$ and all $a \in A_+$,

$$A(f_-(b), a) \implies B(b, f_+(a)).$$

(Note that the function with the minus subscript goes backward.) Composition of these morphisms is defined componentwise, with the order reversed

1991 *Mathematics Subject Classification.* 03G30 03F65 03E05 03E75 18B99 68Q25.
Partially supported by NSF grant DMS-9204276.

Typeset by $\mathcal{A}_{\mathcal{M}}\mathcal{S}$-TEX

on the minus components: $(f \circ g)_- = g_- \circ f_-$ and $(f \circ g)_+ = f_+ \circ g_+$. This clearly defines a category $\mathcal{PV}$.

The category $\mathcal{PV}$ is the special case of de Paiva's construction **GC** from [18] where **C** is the category of sets. It is also the dual of Vojtáš's category GT of generalized Galois-Tukey connections [25, 26].

Intuitively, we think of an object **A** of $\mathcal{PV}$ as representing a problem (or a type of problem). The elements of A_- are instances of the problem, i.e., specific questions of this type; the elements of A_+ are possible answers; and the relation A represents correctness, i.e., $A(x, y)$ means that y is a correct answer to the question x.

There are strong but superficial similarities between $\mathcal{PV}$ and a special case of a construction due to Chu and presented in the appendix of [2] and Section 3 of [3]. (Readers unfamiliar with the Chu construction can skip this paragraph, as it will not be mentioned later.) Specifically, Chu's construction, applied to the cartesian closed category of sets and the object 2, yields a $*$-autonomous category in which the objects are the same as those of $\mathcal{PV}$ and the morphisms differ from those of $\mathcal{PV}$ only in that they are required to satisfy $A(f_-(b), a) \iff B(b, f_+(a))$ rather than just an implication from left to right. This apparently minor difference in the definition leads to major differences in other aspects of the category, specifically in the internal hom-functor and the tensor product. But see [19] for a framework that encompasses both $\mathcal{PV}$ and the Chu construction.

In the next few sections, we shall describe how $\mathcal{PV}$ arose in various contexts. Thereafter, we indicate how ideas that arise naturally in these contexts suggest new constructions in linear logic.

I thank Valeria de Paiva for her helpful comments on an earlier version of this paper.

Reductions of Search Problems

Much of the theory of computational complexity (e.g., [10]) deals with decision problems. Such a problem is specified by giving a set of instances together with a subset called the set of positive instances; the problem is to determine, given an arbitrary instance, whether it is positive. In a typical example, the instances might be graphs and the positive instances might be the 3-colorable graphs. In another example, instances might be boolean formulas and positive instances might be the satisfiable ones. A *(many-one) reduction* from one decision problem to another is a map sending instances of the former to instances of the latter in such a way that an instance of the former is positive if and only if its image is positive. Clearly, an algorithm computing such a reduction and an algorithm solving the latter decision problem can be combined to yield an algorithm solving the former.

There are situations in complexity theory where it is useful to consider not only decision problems but also search problems. A search problem is specified by giving a set of instances, a set of witnesses, and a binary relation between them; the problem is to find, given an instance, some witness related to it. For example, the 3-colorability decision problem mentioned above (given a graph, is it 3-colorable?) can be converted into the 3-coloring search problem (given a graph, find a 3-coloring). Here the instances are graphs, the witnesses are 3-valued functions on the vertices of graphs, and the binary relation relates each graph to its (proper) 3-colorings. Similarly, there is a search version of the boolean satisfiability problem, where instances are boolean formulas, witnesses are truth assignments, and the binary relation is the satisfaction relation. Notice that a search problem is just an object $\mathbf{A}$ of $\mathcal{PV}$, the set of instances being A_- and the set of witnesses A_+.

There is a reasonable analog of many-one reducibility in the context of search problems. A reduction of $\mathbf{B}$ to $\mathbf{A}$ should first convert every instance $b \in B_-$ of $\mathbf{B}$ to an instance $a \in A_-$ of $\mathbf{A}$ (just as for decision problems), and then, if a witness w related to a is given, it should allow us, using w and remembering the original instance b, to compute a witness related to b. Again, an algorithm computing such a reduction and an algorithm solving $\mathbf{A}$ can clearly be combined to yield an algorithm solving $\mathbf{B}$. Most known many-one reductions between NP decision problems [10] implicitly involve many-one reductions of the corresponding search problems.

Formally, a *reduction* therefore consists of two functions, $f_- : B_- \to A_-$ and $f_+ : A_+ \times B_- \to B_+$ such that, for all $b \in B_-$ and $w \in A_+$,

$$A(f_-(b), w) \implies B(b, f_+(w, b)).$$

This is nearly, but not quite, the definition of a morphism from $\mathbf{A}$ to $\mathbf{B}$. The difference is that in a morphism f_+ would have only w, not b, as its argument. Thus, morphisms amount to reductions where the final witness (for b) is computed from a witness w for $a = f_-(b)$ without remembering b. This notion of reduction has been used in the literature [15, 23], but I would not argue that it is as natural as the version where one is allowed to remember b.

These observations lead to a suggestion that we record for future reference.

Suggestion 1. *Find a natural place in the theory of $\mathcal{PV}$ for reductions as described above, i.e., pairs of functions that are like morphisms except that f_+ takes an additional argument from B_- and the implication relating f_- and f_+ is amended accordingly.*

A "dual" modification of the notion of morphism, allowing f_- to have an extra argument in A_+, occurred in de Paiva's work [17] on a categorial version of Gödel's Dialectica interpretation, work that preceded the introduction of $\mathcal{PV}$ in [18].

Linear Logic

The search problems (objects of $\mathcal{PV}$) and reductions (morphisms of $\mathcal{PV}$ or generalized morphisms as in Suggestion 1) described in the preceding section are vaguely related to some of the intuitions that underlie Girard's linear logic [11]. Girard has written about linear logic as a logic of questions and answers (or actions and reactions) [11, 12], so it seems reasonable to try to model this idea in terms of $\mathcal{PV}$. Also, the fact that in a many-one reduction of **B** to **A** a witness for **B** is produced from exactly one witness for **A** is reminiscent of the central idea of linear logic that a conclusion is obtained by using each hypothesis exactly once. In this section, we attempt to make these vague intuitions precise. Our goal here is to develop de Paiva's interpretation of linear logic (at least the multiplicative and additive parts; the exponentials will be discussed briefly later) in a step by step fashion that emphasizes the naturality or necessity of the definitions used.

We intend to use objects of $\mathcal{PV}$ as the interpretations of the formulas of linear logic. This corresponds to Girard's intuition that for any formula A there are questions and answers of type A. Of course, in addition to questions and answers, objects of $\mathcal{PV}$ also have a correctness relation between them. It is reasonable to expect that one formula linearly implies another, in a particular interpretation, if and only if there is a morphism in $\mathcal{PV}$ from (the object interpreting) the former to (the object interpreting) the latter; we shall see this more precisely later.

To produce an interpretation of linear logic, we must tell how to interpret the connectives, and we must define what it means for a sequent to be true in an interpretation.

Perhaps the easiest part of this task is to interpret the additive connectives, & and $\oplus$. It seems to be universally accepted [21] that a reasonable categorial model of linear logic will interpret these as the product and coproduct of the category. Fortunately, $\mathcal{PV}$ has products and coproducts, so we adopt these as the interpretations of the additive connectives. The result is that "with" is interpreted as

$$(A_-, A_+, A)\&(B_-, B_+, B) = (A_- + B_-, A_+ \times B_+, W),$$

where

$$W(x, (a, b)) \iff \begin{cases} A(x, a), & \text{if } x \in A_- \\ B(x, b), & \text{if } x \in B_-; \end{cases}$$

"plus" is interpreted as

$$(A_-, A_+, A) \oplus (B_-, B_+, B) = (A_- \times B_-, A_+ + B_+, V)$$

where

$$V((a,b),x) \iff \begin{cases} A(a,x), & \text{if } x \in A_+ \\ B(b,x), & \text{if } x \in B_+; \end{cases}$$

and the additive units are

$$\top = (\emptyset, 1, \emptyset) \quad \text{and} \quad 0 = (1, \emptyset, \emptyset),$$

where 1 represents any one-element set.

These definitions correspond reasonably well to the intuitive meanings of the additive connectives in terms of questions and answers or in terms of Girard's "action" description of linear logic [12]. To answer a disjunction $A \oplus B$ is to provide an answer to one of A and B; correctness means that, confronted with questions of both types, we answer one of them correctly (in the sense of A or B). To answer a conjunction $A\&B$ we must give answers for both, but we are confronted with a question of only one type and only our answer to that one needs to be correct. The intuitive discussion of conjunction, in particular the fact that we must give answers of both types even though only one will be relevant to the question, might make better sense if we think of the answer as being given before the question is known. This is a rather strange way of running a dialogue, but it will arise again later in other contexts (and I've seen examples of it in real life).

There is also a natural interpretation of linear negation, since (cf. [11, 12]) questions of type A are answers of type the negation $A^\perp$ of A and vice versa. We define

$$(A_-, A_+, A)^\perp = (A_+, A_-, A^\perp),$$

where

$$A^\perp(x,y) \iff \neg A(y,x).$$

So linear negation interchanges questions with answers and replaces the correctness relation by the complement of its converse. Perhaps a few words should be said about the use of the complement of the converse rather than just the converse. There are several reasons for this, perhaps the most intuitive being that we are, after all, defining a sort of negation. Another way to look at it is to think of a contest between a questioner and an answerer, where success for the questioner is defined to mean failure for the answerer (cf. the discussion of challengers and solvers in [13]). "That's a good question" often means that I have no good answer. For another indication that the given definition of $^\perp$ is appropriate, see the section on set-theoretic applications below.

Mathematically, the strongest reason for defining $^\perp$ as we did is that it gives a contravariant involution of the category $\mathcal{PV}$. That is, the operation $^\perp$ on objects and the operation on morphisms defined by $(f_-, f_+)^\perp = (f_+, f_-)$

constitute a contravariant functor from $\mathcal{PV}$ to itself, whose square is the identity. This corresponds to the equivalences in linear logic between $A \vdash B$ and $B^\perp \vdash A^\perp$ and between $A^{\perp\perp}$ and A. (Here it is important that our underlying universe of sets is classical. In more general situations of the sort considered in [18], $A^{\perp\perp}$ and A need not be equivalent.)

We turn now to a more delicate matter, the interpretation of the multiplicative connectives. We begin with "times." Girard's intuitive explanation of the difference between the multiplicative conjunction $\otimes$ and the additive conjunction & in [12] is that the former represents an ability to perform both actions while the latter represents an ability to do either one of the two actions (chosen externally). Looking back at the interpretation of &, we would expect to modify it by allowing questions of both sorts, rather than just one, and requiring both components of the answer to be correct. This operation on objects of $\mathcal{PV}$ is quite natural, and occurs in both [18] and [25]. De Paiva uses the notation $\otimes$ for it, although it is not the interpretation of Girard's connective $\otimes$ in her interpretation of linear logic in the categories **GC** of [18]. (It is the interpretation of $\otimes$ in her earlier interpretation of intuitionistic linear logic in categories **DC**, and the notation comes from there.) Vojtáš used the notation $\times$ even though it is not the product in the category. We shall use the notation $\overline{\otimes}$ and regard it as a sort of provisional tensor product. Formally, we define

$$(A_-, A_+, A)\overline{\otimes}(B_-, B_+, B) = (A_- \times B_-, A_+ \times B_+, A \times B),$$

where the relation $A \times B$ is defined by

$$(A \times B)((x,y),(a,b)) \iff A(x,a) \text{ and } B(y,b).$$

Of course, since we have already interpreted negation, our provisional $\otimes$ gives rise to a dual connective, the provisional "par":

$$(A_-, A_+, A)\overline{\wp}(B_-, B_+, B) = (A_- \times B_-, A_+ \times B_+, P)$$

where

$$P((x,y),(a,b)) \iff A(x,a) \text{ or } B(y,b).$$

To see why these interpretations of the multiplicative connectives are only provisional and must be modified, we turn to the question of soundness of the interpretation. This requires, of course, that we define what is meant by a sequent being valid, which presumably depends on a notion of sequents being true in particular interpretations, i.e, with particular objects as values of the atomic formulas. For simplicity, we work with one-sided sequents, as in [11]. So a sequent is a finite list (or multi-set) of formulas, each interpreted as an object of $\mathcal{PV}$. Since a sequent is deductively equivalent in linear logic with the par of its members, we interpret the sequent as the (provisional) par of its members, i.e., as a certain object of $\mathcal{PV}$. So we must specify what we mean by truth of an object of $\mathcal{PV}$, and then we must try to verify the soundness of the axioms and rules of linear logic.

There are two plausible meanings for "truth" of $\mathbf{A} = (A_-, A_+, A)$, both saying intuitively that one can answer all the questions of type $\mathbf{A}$. The difference between the two is in whether the answer can depend on the question.

The first (provisional) interpretation of truth allows the answer to depend on the question, as one would probably expect intuitively.

$$\models_1 (A_-, A_+, A) \iff \forall x \in A_- \, \exists y \in A_+ \, A(x, y).$$

The second, stronger (provisional) interpretation is that one answer must uniformly answer all questions correctly.

$$\models_2 (A_-, A_+, A) \iff \exists y \in A_+ \, \forall x \in A_- \, A(x, y).$$

Before rejecting the second interpretation as unreasonably strong, one should note that the two interpretations are dual to each other in the sense that $\mathbf{A}$ is true in either sense if and only if its negation $\mathbf{A}^\perp$ is not true in the other sense. Furthermore, the second definition fits better with the idea that truth of a sequent $A \vdash B$ should mean the existence of a morphism from A to B. If we specialize to the case where A is the multiplicative unit 1, so that the sequent $A \vdash B$ becomes deductively equivalent (in linear logic) with $\vdash B$, and if we note that the unit for our provisional $\otimes$ is $(1, 1, \text{true})$, then we see that truth of B should be equivalent to existence of a morphism from $(1, 1, \text{true})$ to B. It is easily checked that existence of such a morphism is precisely the second definition of truth above. It might also be mentioned here that the second definition matches the ideas behind de Paiva's definition of the Dialectica categories in [17], for the Dialectica interpretation produces formulas of the form $\exists\forall$, not $\forall\exists$.

Finally, as we shall see in a moment, each definition has its own advantages and disadvantages when one tries to prove the soundness of linear logic, and eventually we shall need to adopt a compromise between them. The remark above about the relationship between $\models_1$, $\models_2$ and negation suggests that either version of $\models$, used alone, might have difficulties with the axioms $\vdash A, A^\perp$ (which say that linear negation is no stronger than it should be) or the cut rule (which says that linear negation is no weaker than it should be). Let us consider what happens if one tries to establish the soundness of the axioms and cut for either version of $\models$.

For the axioms, we wish to show that $\mathbf{A} \overline{\mathcal{B}} \mathbf{A}^\perp$ is true for each object $\mathbf{A}$ of $\mathcal{PV}$. In $\mathbf{A} \overline{\mathcal{B}} \mathbf{A}^\perp$, the questions are pairs (x, y) where $x \in A_-$ and

$y \in (A^{\perp})_{-} = A_{+}$, and the answers are pairs (a, b) where $a \in A_{+}$ and $b \in (A^{\perp})_{+} = A_{-}$. The answer (a, b) is correct for the question (x, y) if and only if either $A(x, a)$ or $\neg A(b, y)$ (the latter being the definition of $A^{\perp}(y, b)$). Obviously, any question (x, y) is correctly answered by (y, x). So $\models_1 \mathbf{A}\overline{\parr}\mathbf{A}^{\perp}$. On the other hand, we do not in general have $\models_2 \mathbf{A}\overline{\parr}\mathbf{A}^{\perp}$, since an easy calculation shows that this would mean that in $\mathbf{A}$ either some answer is correct for all questions or some question has no correct answer. There are, of course, easy examples of $\mathbf{A}$ where this fails; the simplest is to take $A_{-} = A_{+} = \emptyset$, and if one insists on non-empty sets then the simplest is $A_{-} = A_{+} = \{1, 2\}$ with A being the relation of equality. So, for the soundness of the axioms, $\models_1$ works properly, but $\models_2$ does not.

Now consider the cut rule. We wish to show that, if $\mathbf{B}\overline{\parr}\mathbf{A}$ and $\mathbf{C}\overline{\parr}\mathbf{A}^{\perp}$ are true, then so is $\mathbf{B}\overline{\parr}\mathbf{C}$. If we interpret truth as $\models_2$, then this is easy. Suppose (b, x) correctly answers all questions in $\mathbf{B}\overline{\parr}\mathbf{A}$ and (c, y) correctly answers all questions in $\mathbf{C}\overline{\parr}\mathbf{A}^{\perp}$; we claim that (b, c) correctly answers all questions (p, q) in $\mathbf{B}\overline{\parr}\mathbf{C}$. Indeed, if (p, q) were a counterexample, then b is not correct for p and c is not correct for q, yet (b, x) is correct for (p, y) and (c, y) is correct for (q, x) (where the four occurrences of "correct" refer to $\mathbf{B}$, $\mathbf{C}$, $\mathbf{B}\overline{\parr}\mathbf{A}$, and $\mathbf{C}\overline{\parr}\mathbf{A}^{\perp}$, respectively). But then we must have, by definition of $\overline{\parr}$, that x correctly answers y in $\mathbf{A}$ and that y correctly answers x in $\mathbf{A}^{\perp}$. That is impossible, by definition of $^{\perp}$, so the cut rule preserves $\models_2$. Unfortunately, it fails to preserve $\models_1$. The easiest counterexamples occur when both $\mathbf{B}$ and $\mathbf{C}$ have questions with no correct answers (but B_{+} and C_{+} are non-empty). Then $\mathbf{B}\overline{\parr}\mathbf{C}$ is not true, so the soundness of the cut rule would require that at least one of $\mathbf{B}\overline{\parr}\mathbf{A}$ and $\mathbf{C}\overline{\parr}\mathbf{A}^{\perp}$ also fail to be true. That means that either $\mathbf{A}$ or its negation must have a question with no correct answer, i.e., in $\mathbf{A}$ either some answer is correct for all questions or some question has no correct answer. Since that is not the case in general, we conclude that the cut rule is unsound for $\models_1$.

Summarizing the preceding discussion, we have

(1) If we define truth allowing answers to depend on questions ($\models_1$), then the axioms of linear logic are sound but the cut rule is not.
(2) If we define truth requiring the answer to be independent of the question ($\models_2$), then the cut rule is sound but the axioms are not.

Fortunately, there is a way out of this dilemma. Consider the dependence of answers on questions that was needed to obtain the soundness of the axioms. At first sight, it is an extremely strong dependence; indeed, the answer (y, x) is, except for the order of components, identical to the question (x, y). But the dependence is special in that each component of the answer depends only on the *other* component of the question.

Rather surprisingly, this sort of cross-dependence also makes the cut rule sound. To see this, suppose that both $\mathbf{B} \mathbin{⅋} \mathbf{A}$ and $\mathbf{C} \mathbin{⅋} \mathbf{A}^{\perp}$ are true in this new sense. That is, there are functions $f : B_- \to A_+$ and $g : A_- \to B_+$ such that, for all $b \in B_-$ and $x \in A_-$,

$$(1) \qquad B(b, g(x)) \quad \text{or} \quad A(x, f(b)),$$

and similarly there are $f' : C_- \to (A^{\perp})_+ = A_-$ and $g' : (A^{\perp})_- = A_+ \to C_+$ such that, for all $c \in C_-$ and all $y \in A_+$,

$$(2) \qquad C(c, g'(y)) \quad \text{or} \quad \neg A(f'(c), y).$$

Then we claim that $g' \circ f : B_- \to C_+$ and $g \circ f' : C_- \to B_+$ satisfy, for all $b \in B_-$ and $c \in C_-$,

$$B(b, g(f'(c))) \quad \text{or} \quad C(c, g'(f(b))),$$

which means that $\mathbf{B} \mathbin{⅋} \mathbf{C}$ is true in the "cross-dependence" sense. To verify the claim, let such b and c be given. If $A(f'(c), f(b))$, then (2) implies $C(c, g'(f(b)))$. If $\neg A(f'(c), f(b))$, then (1) implies $B(b, g(f'(c)))$. So the claim is true in either case, and we have verified the soundness of the cut rule.

By allowing the answer in one component of a sequent to depend on the questions in the other components but not in the same component, this "cross-dependence" notion of truth makes crucial use of the commas in a sequent, to distinguish the components. But linear logic requires (by the introduction rules for times and especially for par) that the commas in a sequent behave exactly like the connective $⅋$. So it seems necessary to build cross-dependence into the interpretation of this connective. This will lead to the correct definition of the multiplicative connectives, introduced by de Paiva [18], replacing the provisional interpretations given earlier.

The par operation on objects of $\mathcal{PV}$ is defined, as in [18], by

$$(A_-, A_+, A) ⅋ (B_-, B_+, B) = (A_- \times B_-, A_+^{B_-} \times B_+^{A_-}, P)$$

where

$$P((x, y), (f, g)) \iff A(x, f(y)) \text{ or } B(y, g(x)).$$

This operation $⅋$ is the object part of a functor, the action on morphisms being $(f ⅋ g)_- = f_- \times g_-$ and $(f ⅋ g)_+ = f_+^{g_-} \times g_+^{f_-}$. It is easy to check that $⅋$ is associative (up to natural isomorphism). In the par of several objects, questions are tuples consisting of one question from each of the objects, and answers are tuples of functions, each producing an answer in one component when given as inputs questions in all the other components.

We also interpret commas in sequents as the new $\mathbin{⅋}$ (rather than $\overline{⅋}$). This change in the interpretation of the commas makes $\models_2$ behave like the cross-dependence notion of truth described earlier. To see this, note that $\models_2$ requires the existence of a single answer correct for all questions at once, but the new $⅋$ allows that answer to consist of functions whereby each component of the answer can depend on the other components of the question. We therefore adopt $\models_2$ as the (non-provisional) definition of truth, and from now on we write it simply as $\models$. The previous discussion shows that the axioms and the cut rule are sound. We sometimes refer to an answer that is correct for all questions in an object **A** as a *solution* of the problem **A**. So truth means having a solution.

Of course, the new interpretation of par gives, by duality, a new interpretation of times.

$$(A_-, A_+, A) \otimes (B_-, B_+, B) = (A_-^{B_+} \times B_-^{A_+}, A_+ \times B_+, T),$$

where

$$T((f,g),(x,y)) \iff A(f(y),x) \text{ and } B(g(x),y).$$

(This connective was called $\oslash$ in [18].) The units for the multiplicative connectives are $1 = (1, 1, \text{true})$ and $\bot = (1, 1, \text{false})$, where true and false represent the obvious relations on a singleton. The linear implication $A \multimap B$ defined as $A^\perp ⅋ B$ is

$$(A_-, A_+, A) \multimap (B_-, B_+, B) = (A_+ \times B_-, A_-^{B_-} \times B_+^{A_+}, C),$$

where

$$C((x,y),(f,g)) \iff [A(f(y),x) \implies B(y,g(x))].$$

Notice that a solution of $\mathbf{A} \multimap \mathbf{B}$ is precisely a morphism $\mathbf{A} \to \mathbf{B}$ in $\mathcal{PV}$. This indicates that the definitions of the multiplicative connectives and of truth, though not immediately intuitive, are proper in the context of the category $\mathcal{PV}$.

The belief that these definitions are reasonable is reinforced by de Paiva's theorem [18] that the multiplicative and additive fragment of linear logic is sound for this interpretation. (Her theorem actually covers full linear logic, including the exponentials, but we have not yet discussed the interpretation of the exponentials.)

Linear logic is not complete for this interpretation. For one thing, the interpretation validates the mix rule: If **A** and **B** are both true, then so is $\mathbf{A} ⅋ \mathbf{B}$. Also, the interpretation satisfies all formulas of the form $A^\perp ⅋ (A ⅋ A)$, a special case of weakening. (General weakening, $A^\perp ⅋ (A ⅋ B)$ is not satisfied; for a counterexample, take B_+ to be empty while all of A_+, A_-, B_- are nonempty.)

The interpretations of $\otimes$ and ⅋ remain, in spite of their success at modeling linear logic, rather unintuitive. This is attested by the fact that de Paiva [18], while using $\otimes$ to interpret the multiplicative conjunction, calls it $\oslash$ and reserves the symbol $\otimes$ for the more intuitive construction that I called $\overline{\otimes}$. Vojtáš [25] also discusses $\overline{\otimes}$, calling it $\times$, but never has any use for $\otimes$. Since $\overline{\otimes}$ seems much more natural than the "correct" $\otimes$, it should have its own place in the logic.

Suggestion 2. *Find a natural place in the theory of $\mathcal{PV}$ for the operation $\overline{\otimes}$.*

DIGRESSION ON GAME SEMANTICS

At the suggestion of the editors, I give in this section a short survey of my previous work [6] on interpreting linear logic by games, and I suggest some connections between that interpretation and de Paiva's interpretation in the category $\mathcal{PV}$. This section can be skipped without loss of continuity.

The problem, which caused such difficulty in the preceding section, whether truth should require one uniform answer for all questions or (possibly) different answers for different questions ($\models_2$ versus $\models_1$) can be regarded as the problem whether the answer should be given before or after the question is known. One could regard an object of $\mathcal{PV}$ as an incomplete specification of a dialogue between the questioner and the answerer; the specification would be completed by saying who is expected to speak first. Once this idea is introduced, it is natural to consider more complicated dependences between what the two speakers say. For example, one can envision a question consisting of two components, the first of which is asked before the answer, the second after. (Cf. the intuitive description of sequential composition in the next section.) Such dialogues of length greater than two arise naturally when one interprets the additive connectives. The dialogue for **A**&**B** should begin with the questioner deciding which of **A** and **B** to discuss, and the dialogue for $\mathbf{A}\oplus\mathbf{B}$ should begin with a similar decision by the answerer. Dialogues for deeply nested iterations of the additive connectives should therefore involve long alternations between the two speakers.

Game semantics in the sense of [6] interprets the formulas of linear logic as games, i.e., as rules for conducting dialogues and for determining which of the speakers (called players in this context) "wins." (In the simple dialogues where each player speaks only once, the answerer wins if his answer is correct for his opponent's question, and the questioner wins otherwise.) Additive connectives are handled as described above, and linear negation simply interchanges the roles of the two players. Multiplicative connectives are more complicated. The dialogue for **A** $\otimes$ **B** consists of two interleaved dialogues, one for each of the constituents **A** and **B**, with the questioner allowed to decide when to switch from one constituent to the other, and with the answerer required to win both constituents in order to win the $\otimes$ compound. **A**$\parr$**B** is similar with the roles of the two players interchanged. The dialogue for !**A** consists of a potential infinity of interleaved dialogues for **A**, with the questioner deciding when to switch from one to another, with the answerer required to win all constituents in order to win !**A**, and with an additional coherence constraint on the answerer.

It is proved in [6] that affine logic, i.e., linear logic plus the rule of weakening, is sound for this semantics. There is a completeness theorem for the additive fragment (when dialogues are allowed to be infinitely long), but not for the multiplicative fragment. Subsequently, Abramsky and Jagadeesan [1] made substantial modifications in the game semantics to obtain completeness (and more) for multiplicative linear logic plus the mix rule, and further modifications by Hyland and Ong [14] eliminated the need for the mix rule. Unfortunately, these modifications no longer work as well with the additive connectives.

There is another way to connect game semantics and the question-answer approach formalized in $\mathcal{PV}$. Instead of regarding the questions and answers of the latter approach as individual moves of the players in a very short (two moves) game, one can regard the questions and answers as strategies for the two players in a longer game. This point of view may make the cross-dependence in de Paiva's interpretations of the multiplicative connectives more intuitive. If we think of A_+ as a set of strategies for the answerer in game **A** and B_- as a set of strategies for the questioner in **B**, then the elements of $A_+^{B_-}$ are strategies for the answerer in **A** that take into account what the questioner is doing in **B**. They take it into account unrealistically well, making use of the entire strategy of the questioner rather than just the moves already made by the questioner on the basis of this strategy, but at least this line of thought indicates that cross-dependence of the sort used to interpret $\otimes$ and $\parr$ in $\mathcal{PV}$ is not entirely unmotivated from the game semantical point of view.

Cardinal Characteristics of the Continuum

We begin this section by introducing a few (just enough to serve as examples later) of the many cardinal characteristics of the continuum that have been studied by set-theorists, topologists, and others. For more information about these and other characteristics, see [22] and the references cited there. All the cardinal characteristics considered here (and almost all the others) are uncountable cardinals smaller than or equal to the cardinality $\mathfrak{c} = 2^{\aleph_0}$ of the continuum. So they are of little interest if the continuum hypothesis ($\mathfrak{c} = \aleph_1$) holds, but in the absence of the continuum hypothsis there are many interesting connections, usually in the form of inequalities, between various characteristics. (There are also independence results showing that certain inequalities are not provable from the usual ZFC axioms of set theory.) Part of the work of Vojtáš [25, 26] on which this section is based can be viewed as a way to extract from the inequality proofs information which is of interest even if the continuum hypothesis holds.

Definitions. If X and Y are subsets of $\mathbb{N}$, we say that X *splits* Y if both $Y \cap X$ and $Y - X$ are infinite. The *splitting number* $\mathfrak{s}$ is the smallest cardinality of any family $\mathcal{S}$ of subsets of $\mathbb{N}$ such that every infinite subset of $\mathbb{N}$ is split by some element of $\mathcal{S}$. The *refining number* (also called the *unsplitting* or *reaping number*) $\mathfrak{r}$ is the smallest cardinality of any family $\mathcal{R}$ of infinite subsets of $\mathbb{N}$ such that no single set splits all the sets in $\mathcal{R}$. $\mathfrak{r}_\sigma$ is the smallest cardinality of any family $\mathcal{R}$ of infinite subsets of $\mathbb{N}$ such that, for any countably many subsets S_k of $\mathbb{N}$, some set in $\mathcal{R}$ is not split by any S_k.

These cardinals arise naturally in analysis, for example in connection with the Bolzano-Weierstrass theorem, which asserts that a bounded sequence of real numbers has a convergent subsequence. A straightforward diagonal argument extends this to show that, for any countably many bounded sequences of real numbers $\mathbf{x}_k = (x_{kn})_{n\in\mathbb{N}}$, there is a single infinite $A \subseteq \mathbb{N}$ such that the subsequences indexed by A, $(x_{kn})_{n\in A}$, all converge. If one tries to extend this to uncountably many sequences, then the first cardinal for which the analogous result fails is $\mathfrak{s}$. Also, $\mathfrak{r}_\sigma$ is the smallest cardinality of any family $\mathcal{R}$ of infinite subsets of $\mathbb{N}$ such that, for every bounded sequence $(x_n)_{n\in\mathbb{N}}$, there is a convergent subsequence $(x_n)_{n\in A}$ with $A \in \mathcal{R}$. There is an analogous description of $\mathfrak{r}$, where the sequences $(x_n)_{n\in\mathbb{N}}$ are required to have only finitely many distinct terms. For more information about these aspects of the cardinal characteristics, see [24].

Definitions. A function $f : \mathbb{N} \to \mathbb{N}$ *dominates* another such function g if, for all but finitely many $n \in \mathbb{N}$, $f(n) \leq g(n)$. The *dominating number* $\mathfrak{d}$ is the smallest cardinality of any family $\mathcal{D} \subseteq \mathbb{N}^\mathbb{N}$ such that every $g \in \mathbb{N}^\mathbb{N}$ is dominated by some $f \in \mathcal{D}$. The *bounding number* $\mathfrak{b}$ is the smallest cardinality of any family $\mathcal{B} \subseteq \mathbb{N}^\mathbb{N}$ such that no single g dominates all the members of $\mathcal{B}$.

The known inequalities between these cardinals (and $\aleph_1$ and $\mathfrak{c} = 2^{\aleph_0}$) are

$$\aleph_1 \leq \mathfrak{s} \leq \mathfrak{d} \leq \mathfrak{c},$$

$$\aleph_1 \leq \mathfrak{b} \leq \mathfrak{r} \leq \mathfrak{r}_\sigma \leq \mathfrak{c},$$

and

$$\mathfrak{b} \leq \mathfrak{d}.$$

It is known that any further inequalities between these cardinals are independent of ZFC, except that it is still an open problem whether $\mathfrak{r} = \mathfrak{r}_\sigma$ is provable.

The connection between the theory of these cardinals and the category $\mathcal{PV}$ discussed in previous sections becomes visible when one considers the proofs of some of these inequalities, so we shall prove the two non-trivial (but well known) ones, $\mathfrak{s} \leq \mathfrak{d}$ and $\mathfrak{b} \leq \mathfrak{r}$. (In each case, only the first of the two paragraphs in the proof is relevant to $\mathcal{PV}$, so the reader willing to take the first paragraph on faith can skip the justification in the second paragraph.)

Proof of $\mathfrak{s} \leq \mathfrak{d}$. There is a map $\alpha : \mathbb{N}^\mathbb{N} \to \mathcal{P}(\mathbb{N})$ sending every dominating family $\mathcal{D}$ (as in the definition of $\mathfrak{d}$) to a splitting family (as in the definition of $\mathfrak{s}$). In fact, one can associate to each infinite $X \subseteq \mathbb{N}$ a function $\beta(X) = f \in \mathbb{N}^\mathbb{N}$ such that, if g dominates f, then $\alpha(g)$ splits X.

Given g, to define $\alpha(g)$, partition $\mathbb{N}$ into a sequence of intervals $[0, a_1)$, $[a_1, a_2), \ldots$ such that, for each $n \in \mathbb{N}$, $g(n)$ is at most one interval beyond n (it's trivial to define such a_i's by induction), and let $\alpha(g)$ be the union of the even-numbered intervals. Define $\beta(X)$ to send each $n \in \mathbb{N}$ to the next element of X greater than n. If $f = \beta(X)$, if g dominates f, if a_i's are as in the definition of $\alpha(g)$, and if k is large enough, then the element $f(a_k - 1)$ of X lies in the interval $[a_k, a_{k+1})$. So X meets all but finitely many of the intervals $[a_k, a_{k+1})$ and is therefore split by $\alpha(g)$. □

Proof of $\mathfrak{b} \leq \mathfrak{r}$. There is a function $\beta : \mathcal{P}_\infty(\mathbb{N}) \to \mathbb{N}^\mathbb{N}$ sending every unsplittable family $\mathcal{R}$ (as in the definition of $\mathfrak{r}$) to an undominated family (as in the definition of $\mathfrak{b}$). In fact, one can associate to each $g \in \mathbb{N}^\mathbb{N}$ a set $\alpha(g) = Y \in \mathcal{P}(\mathbb{N})$ such that, if Y does not split X then g does not dominate $\beta(X)$.

The same α and β as in the preceding proof will work, as the properties required of them here are logically equivalent to the properties required there. □

In the notation of the preceding sections, the pair (β, α) in the first of these proofs is a morphism in $\mathcal{PV}$ from $(\mathbb{N}^\mathbb{N}, \mathbb{N}^\mathbb{N}, \text{is majorized by})$ to $(\mathcal{P}_\infty(\mathbb{N}), \mathcal{P}(\mathbb{N}), \text{is split by})$. In the second proof, we used the image of this under $^\perp$, namely that (α, β) is a morphism from $(\mathcal{P}(\mathbb{N}), \mathcal{P}_\infty(\mathbb{N}), \text{does not split})$ to $(\mathbb{N}^\mathbb{N}, \mathbb{N}^\mathbb{N}, \text{does not majorize})$. In both cases, the cardinal inequality follows from the following general fact. Define for each object $\mathbf{A}$ of $\mathcal{PV}$ the *norm* $\|\mathbf{A}\|$ as the smallest cardinality of any set $X \subseteq A_+$ of answers sufficient to contain at least one correct answer for every question in A_- (undefined if there is no such set, i.e., if some question has no correct answer, i.e., if $\mathbf{A}^\perp$ is true). Then the existence of a morphism $f : \mathbf{A} \to \mathbf{B}$ implies that $\|\mathbf{A}\| \geq \|\mathbf{B}\|$, because f_+ sends any set of the sort required in the definition of $\|\mathbf{A}\|$ to one as required for $\|\mathbf{B}\|$. (What I called the norm of $\mathbf{A}$ is, in Vojtáš's notation [25, 26] $\mathfrak{d}(A)$; Vojtáš's $\mathfrak{b}(A)$ is $\|\mathbf{A}^\perp\|$.)

It is an empirical fact that proofs of inequalities between cardinal characteristics of the continuum usually proceed by representing the characteristics as norms of objects in $\mathcal{PV}$ and then exhibiting explicit morphisms between those objects. This fact is explicit in Vojtáš's [25, 26] and implicit in [9]. It applies even to trivial inequalities like $\mathfrak{b} \leq \mathfrak{d}$ (where the required morphism from $(\mathbb{N}^\mathbb{N}, \mathbb{N}^\mathbb{N}, \text{is dominated by})$ to $(\mathbb{N}^\mathbb{N}, \mathbb{N}^\mathbb{N}, \text{does not dominate})$ consists of identity maps on both components) as well as to inequalities much deeper than the examples proved above; see for example the presentation in [9] of Bartoszyński's theorem [4] that the smallest number of meager sets whose union is not meager is at least as large as the corresponding number for "measure zero" in place of "meager."

It is tempting to regard the existence of a morphism $\mathbf{A} \to \mathbf{B}$ as a strong formulation of the inequality $\|\mathbf{A}\| \geq \|\mathbf{B}\|$ that is significant even in the presence of the continuum hypothesis (which makes inequalities between cardinal characteristics trivial as these cardinals lie between $\aleph_1$ and $\mathfrak{c}$ inclusive). The situation is, however, not quite so simple. My student, Olga Yiparaki, has shown that, in the presence of the continuum hypothesis (or certain weaker assumptions), there are morphisms in $\mathcal{PV}$ in both directions between any two objects that correspond (as in [25, 26]) to cardinal characteristics of the continuum. Those morphisms, however, are highly non-constructive, whereas those used in the usual proofs of cardinal inequalities are quite explicit. It therefore seems likely that a strengthening of these cardinal inequalities that retains its significance in the presence of the continuum hypothesis is to require not merely the existence of morphisms but the existence of "nice" morphisms, say ones whose components are Borel mappings.

The linear negation defined on $\mathcal{PV}$ gives a precise version of an intuitive "duality" in the theory of cardinal characteristics. In that theory, one often refers to the cardinals $\|\mathbf{A}\|$ and $\|\mathbf{A}^\perp\|$ as being dual to each other; see for example the introduction to [16]. On cardinals, this is not well defined, for two objects can have the same norm while their negations have different norms, but it is the shadow, in the world of cardinals, of the (well defined) linear negation in $\mathcal{PV}$. It may be worth noting in this connection that $(\mathcal{P}(\mathbb{N}), \mathcal{P}_\infty(\mathbb{N}), \text{does not split})$, whose norm is $\mathfrak{r}$, and $(\mathcal{P}(\mathbb{N})^\mathbb{N}, \mathcal{P}_\infty(\mathbb{N}), \text{has no component that splits})$, whose norm is $\mathfrak{r}_\sigma$, have negations both of norm $\mathfrak{s}$.

In addition to inequalities of the sort discussed above, which relate two cardinal characteristics of the continuum, there are a few theorems that relate three (occasionally even four) of them. We consider one relatively easy example here, since it leads to an idea that should connect to linear logic. The example concerns Ramsey's theorem [20], which asserts (in a simple form) that, whenever the set $[\mathbb{N}]^2$ of two-element subsets of $\mathbb{N}$ is partitioned into two pieces, then there is an infinite $H \subseteq \mathbb{N}$ that is homogeneous in the sense that all its two element subsets lie in the same piece of the partition. The cardinal $\mathfrak{hom}$ was defined in [7] as the smallest cardinality of a family $\mathcal{H}$ of infinite subsets of $\mathbb{N}$ such that, for every partition of $[\mathbb{N}]^2$ as in Ramsey's theorem, a homogeneous set can be found in $\mathcal{H}$. It was shown in [7] that this cardinal is bounded below by $\max\{\mathfrak{r}, \mathfrak{d}\}$ and above by $\max\{\mathfrak{r}_\sigma, \mathfrak{d}\}$. The lower bound amounts to two ordinary inequalities, $\mathfrak{hom} \geq \mathfrak{r}$ and $\mathfrak{hom} \geq \mathfrak{d}$, both of which were proved by exhibiting morphisms between the appropriate objects of $\mathcal{PV}$. The upper bound genuinely relates three cardinals, and we wish to make some comments about its proof, so we begin by sketching the proof.

Proof of $\mathfrak{hom} \leq \max\{\mathfrak{r}_\sigma, \mathfrak{d}\}$. Fix a family $\mathcal{R}_0$ of $\mathfrak{r}_\sigma$ subsets of $\mathbb{N}$ such that no countably many sets split all the sets in $\mathcal{R}_0$. Within each set $A \in \mathcal{R}_0$, fix a family $\mathcal{R}_1(A)$ of $\mathfrak{r}$ sets such that no single set splits them all. Also, fix a family $\mathcal{D}$ of functions dominating all functions $\mathbb{N} \to \mathbb{N}$. For each $A \in \mathcal{R}_0$, for each $B \in \mathcal{R}_1(A)$, and for each $f \in \mathcal{D}$, choose a subset $Z = Z(A, B, f)$ of B so thin that, if $x < y$ are in Z then $f(x) < y$. We claim that the family $\mathcal{H}$ of all these Z's, which clearly has cardinality $\max\{\mathfrak{r}_\sigma, \mathfrak{d}\}$ (since $\mathfrak{r} \leq \mathfrak{r}_\sigma$), contains almost homogeneous sets for all partitions of $[\mathbb{N}]^2$ into two parts. "Almost homogeneous" means that the set becomes homogeneous when finitely many of its elements are removed. Since we can close $\mathcal{H}$ under such finite changes without increasing its cardinality, the claim completes the proof.

To prove the claim, let $[\mathbb{N}]^2$ be partitioned into two parts. For each natural number n let C_n consist of those x for which $\{n, x\}$ is in the first part. By choice of $\mathcal{R}_0$, it contains a set A unsplit by any C_n. Let $g(n)$ be so large that all $x \in A$ with $x \geq g(n)$ have $\{n, x\}$ in the same piece of the partition, and let Q be the set of n for which this is the first piece. Choose $B \in \mathcal{R}_1(A)$ unsplit by Q and $f \in \mathcal{D}$ dominating g. It is then easy to check that $Z(A, B, f)$ is almost homogeneous for the given partition. $\square$

To discuss this proof in terms of $\mathcal{PV}$, we introduce the natural objects of $\mathcal{PV}$ whose norms are the cardinals under consideration. For mnemonic purposes, we name each object with the capital letter corresponding to the lower-case letter naming the cardinal.

$$\mathbf{HOM} = (\{p \mid p : [\mathbb{N}]^2 \to 2\}, \mathcal{P}_\infty(\mathbb{N}), AH),$$

where AH is the relation of almost homogeneity: $AH(p, H)$ means that H is almost homogeneous for the partition p.

$$\mathbf{D} = (\mathbb{N}^\mathbb{N}, \mathbb{N}^\mathbb{N}, \text{is dominated by}).$$

$$\mathbf{R} = (\mathcal{P}(\mathbb{N}), \mathcal{P}_\infty(\mathbb{N}), \text{does not split}).$$

$$\mathbf{R}_\sigma = (\mathcal{P}(\mathbb{N})^{\mathbb{N}}, \mathcal{P}_\infty(\mathbb{N}), \text{has no component that splits}).$$

The structure of the preceding proof is then as follows. From a question p in **HOM**, we first produced a question $(C_n)_{n\in\mathbb{N}}$ in $\mathbf{R}_\sigma$. Using an answer A to this question and also using again the original question p, we produced questions g in **D** and Q in **R**. From answers f and B to these questions, along with the previous answer A, we finally produced an answer H to the original question p in **HOM**.

This can be described as a morphism into **HOM** from a suitable combination of **D**, **R**, and $\mathbf{R}_\sigma$, but the relevant combination is a bit different from what we have considered previously. The part of the construction involving **D** and **R** is just the provisional tensor product $\overline{\otimes}$; that is, we had a question (g, Q) in $\mathbf{D}\overline{\otimes}\mathbf{R}$ and we obtained an answer (f, B) for it. (Strictly speaking, we used a version of **R** on A rather than on $\mathbb{N}$, but we shall ignore this detail.) The novelty is in how $\mathbf{D}\overline{\otimes}\mathbf{R}$ is combined with $\mathbf{R}_\sigma$. For what we produced from p was a question in $\mathbf{R}_\sigma$ together with a function converting answers to this question into questions in $\mathbf{D}\overline{\otimes}\mathbf{R}$. This thing that we produced ought to be a question in the object that is being mapped to **HOM**. An answer in that object ought to be what we used in order to get the answer H for **HOM**, namely (A, B, f).

Motivated by these considerations, we define a connective, denoted by a semi-colon (to suggest sequential composition), as follows.

$$(A_-, A_+, A); (B_-, B_+, B) = (A_- \times B_-^{A_+}, A_+ \times B_+, S),$$

where

$$S((x, f), (a, b)) \iff A(x, a) \text{ and } B(f(a), b).$$

Thus, a question of sort $\mathbf{A}; \mathbf{B}$ consists of a first question in **A**, followed by a second question in **B** that may depend on the answer to the first. A correct answer consists of correct answers to both of the constituent questions. Thus, sequential composition can be viewed as describing a dialogue in which the questioner first asks a question in **A**, is given an answer, selects on the basis of this answer a question in **B**, and is given an answer to this as well.

The proof of $\mathfrak{hom} \leq \max\{\mathfrak{r}_\sigma, \mathfrak{d}\}$ exhibits a morphism from $\mathbf{R}_\sigma; (\mathbf{D}\overline{\otimes}\mathbf{R})$ to **HOM**. The cardinal inequality follows from the existence of such a morphism, since one easily checks that the operations on infinite cardinal norms corresponding to the operations $\overline{\otimes}$ and ; are both simply max.

The sequential composition of objects of $\mathcal{PV}$ occurs repeatedly in the proofs of inequalities relating three cardinal characteristics. A typical example is the proof that the minimum number of meager sets of reals with a non-meager union is the smaller of $\mathfrak{b}$ and the minimum number of meager sets that cover the real line [16, 5, 9]. Vojtáš [25] describes the strategy for proving such three-way relations between cardinals in terms of what he calls a max-min diagram. This diagram amounts exactly to a morphism from the sequential composition of two objects to a third object. In other words, sequential composition is the reification of the max-min diagram as an object of $\mathcal{PV}$.

Sequential composition also seems a natural concept to add to linear logic from the computational point of view. Linear logic is generally viewed as a logic of parallel computation, but even parallel computations often have sequential parts, so it seems reasonable to include in the logic a way to describe sequentiality. These ideas are not yet sufficiently developed to support any claims about sequential composition, as defined in the $\mathcal{PV}$ model, being the (or a) right way to do this. In addition to semantical interpretations, one would certainly want good axioms governing any sequential composition connective that is to be added to linear logic, and one would hope that some of the pleasant proof theory of linear logic would survive the addition. Much remains to be done in this direction.

Suggestion 3. *Find a place for sequential composition (specifically for the connective called ; above) in linear logic and the theory of* $\mathcal{PV}$.

GENERALIZED MULTIPLICATIVE CONNECTIVES

The previous sections have led to three suggestions of natural connectives to add to linear logic. (Actually, Suggestion 1 concerned not a connective but a modified notion of morphism. But such a modification should correspond to a reinterpretation of $\multimap$ and therefore of $⅋$ and $\otimes$ as well.) The suggested new connectives are all analogous to the multiplicatives in that both the set of questions and the set of answers are cartesian products. (For the additive connectives, one of the two sets was a disjoint union.) The factors in these products are either sets of questions or answers from the constituent objects or else sets of functions, from questions to answers or vice versa. With these preliminary comments, it seems natural to describe general multiplicative conjunctions (cousins of $\otimes$) as follows.

A *general multiplicative conjunction* operates on n objects $\mathbf{A}_1 \dots \mathbf{A}_n$ of $\mathcal{PV}$ to produce an object $\mathbf{C}$, where $C_+ = A_{1+} \times \dots A_{n+}$ and where C_- consists of n-tuples (f_i) of functions where f_i maps some product of A_{j+}'s into A_{i-}. Which A_{j+}'s occur in the domain of which f_i's is given by the specification of the particular connective. An answer (a_i) is correct for a question (f_i) if each a_i correctly answers in $\mathbf{A}_i$ the question obtained by evaluating f_i at the relevant a_j's.

For example, $\otimes$ is a generalized multiplicative conjunction, for which

$n = 2$ and each f_i has domain A_j for the j different from i. Similarly, we obtain $\overline{\otimes}$ if the domains of the f_i's are taken to be empty products (i.e., singletons); no j is relevant to any i. Sequential composition is obtained by having f_1 depend on no arguments while f_2 has an argument in A_1.

Dual (via $^\perp$) to generalized multiplicative conjunctions are generalized multiplicative disjunctions. Here the answers are allowed to depend on some questions, rather than vice versa (exactly which dependences are allowed is the specification of a particular connective), and correctness means correctness in at least one component, rather than in all.

To avoid possible confusion, we stress that the generalization of the multiplicative connectives proposed here is quite different from that proposed by Danos and Regnier [8]. The Danos-Regnier multiplicatives can correspond to many different classical connectives, whereas mine correspond only to conjunction and disjunction. One could, of course, consider combining the two generalizations, but we do not attempt this here.

There are non-trivial *unary* conjunction and disjunction connectives. The conjunction is given by

$$\kappa(A_-, A_+, A) = (A_-^{A_+}, A_+, \kappa A),$$

where

$$\kappa A(f, a) \iff A(f(a), a).$$

The dual disjunction is

$$\alpha(A_-, A_+, A) = (A_-, A_+^{A_-}, \alpha A)$$

where

$$\alpha A(a, f) \iff A(a, f(a)).$$

These operations were called T and R in [18].

The modified concept of morphism from $\mathbf{A}$ to $\mathbf{B}$ in Suggestion 1, where f_+ maps $A_+ \times B_-$, rather than just A_+, into B_+, amounts to a morphism (in the standard $\mathcal{PV}$ sense) from $\mathbf{A}$ to $\alpha\mathbf{B}$. This concept of morphism thus gives rise to the Kleisli category of $\mathcal{PV}$ with respect to the monad α. (We have defined α only on objects, but it is routine to define it on morphisms and to describe its monad structure.)

De Paiva's Dialectica category [17] built over the category of sets has as morphisms $\mathbf{A} \to \mathbf{B}$ the $\mathcal{PV}$ morphisms $\kappa\mathbf{A} \to \mathbf{B}$. It is dual (via $^\perp$) to the category in the preceding paragraph and is the co-Kleisli category of the comonad κ (see [18, Prop. 7]).

The connective α also provides a way to reinstate the notion of truth $\models_1$ that was discarded when we replaced the provisional $\otimes$ and $⅋$ with the final versions. Indeed, $\models_1 \mathbf{A}$ holds if and only if $\models \alpha\mathbf{A}$.

EXPONENTIALS

Girard has pointed out that the exponential connectives or modalities, ! and ?, unlike the other connectives, are not determined by the axioms of linear logic. More precisely, if one added to linear logic a second pair of modalities, say !′ and ?′, subject to the same rules of inference as the original pair, then one could not deduce that the new modalities are equivalent to the old. Several versions of the exponentials could coexist in one model of linear logic.

$\mathcal{PV}$ provides an example of this phenomenon. De Paiva [18] gave an interpretation of the exponentials in which ! is a combination of the unary conjunction κ defined above and a construction S where multisets m of questions are regarded as questions and a correct answer to m is a single answer that is correct for all the questions in m. (Neither κ nor S alone can serve as an interpretation of !.) Another interpretation of the exponentials in $\mathcal{PV}$, validating the exponential rules of linear logic, is given by

$$!(A_-, A_+, A) = (1, A_+, U)$$

where 1 is a singleton, say $\{*\}$ and

$$U(*, a) \iff \forall x \in A_- \, A(x, a),$$

and its dual

$$?(A_-, A_+, A) = (A_-, 1, E)$$

where

$$E(a, *) \iff \exists x \in A_+ \, A(a, x).$$

Intuitively, a question of type **!A** (namely $*$) amounts to all questions of type **A**; a correct answer in **!A** must correctly answer all questions in **A** simultaneously.

It is easy to check that Girard's rules of inference for the exponentials are sound for this simple interpretation.

REFERENCES

1. S. Abramsky and R. Jagadeesan, *Games and full completeness for multiplicative linear logic*, J. Symbolic Logic (to appear).
2. M. Barr, **-Autonomous Categories*, Lecture Notes in Mathematics 752, Springer-Verlag, 1979.
3. M. Barr, **-Autonomous categories and linear logic*, Math. Struct. Comp. Sci. **1** (1991), 159–178.
4. T. Bartoszyński, *Additivity of measure implies additivity of category*, Trans. Amer. Math. Soc. **281** (1984), 209–213.
5. T. Bartoszyński and H. Judah, *Measure and Category — The Asymmetry* (to appear).

6. A. Blass, *A game semantics for linear logic*, Ann. Pure Appl. Logic **56** (1992), 183–220.
7. A. Blass, *Simple cardinal characteristics of the continuum*, Set Theory of the Reals (H. Judah, ed.), Israel Math. Conf. Proc. 6, 1993, pp. 63–90.
8. V. Danos and L. Regnier, *The structure of multiplicatives*, Arch. Math. Logic **28** (1989), 181–203.
9. D. H. Fremlin, *Cichoń's diagram*, Séminaire Initiation à l'Analyse (G. Choquet, M. Rogalski, and J. Saint-Raymond, eds.), Univ. Pierre et Marie Curie, 1983/84, pp. (5-01)–(5-13).
10. M. R. Garey and D. S. Johnson, *Computers and Intractability*, W. H. Freeman and Co., 1979.
11. J.-Y. Girard, *Linear logic*, Theoret. Comp. Sci. **50** (1987), 1–102.
12. J.-Y. Girard, *Toward a geometry of interaction*, Categories in Computer Science and Logic (J. W. Gray and A. Scedrov, eds.), Contemp. Math. 92, Amer. Math. Soc., 1989, pp. 69–108.
13. Y. Gurevich, *The challenger-solver game: variations on the theme of P=?NP*, Bull. Europ. Assoc. Theoret. Comp. Sci. **39** (1989), 112–121.
14. J. M. E. Hyland and C.-H. L. Ong, *Fair games and full completeness for multiplicative linear logic without the MIX-rule*, preprint (1993).
15. R. Impagliazzo and L. Levin, *No better ways to generate hard NP instances than picking uniformly at random*, Symposium on Foundations of Computer Science, IEEE Computer Society Press, 1990, pp. 812–821.
16. A. Miller, *Additivity of measure implies dominating reals*, Proc. Amer. Math. Soc. **91** (1984), 111–117.
17. V. C. V. de Paiva, *The Dialectica categories*, Categories in Computer Science and Logic (J. W. Gray and A. Scedrov, eds.), Contemp. Math. 92, Amer. Math. Soc., 1989, pp. 47–62.
18. V. C. V. de Paiva, *A Dialectica-like model of linear logic*, Category Theory and Computer Science (D. H. Pitt, D. E. Rydeheard, P. Dybjer, A. Pitts, and A. Poigné, eds.), Lecture Notes in Computer Science 389, Springer-Verlag, 1989, pp. 341–356.
19. V. C. V. de Paiva, *Categorical multirelations, linear logic, and Petri nets*, Tech. Report 225, University of Cambridge Computer Laboratory, 1991.
20. F. P. Ramsey, *On a problem of formal logic*, Proc. London Math. Soc. (2) **30** (1930), 264–286.
21. R. A. G. Seely, *Linear logic, *-autonomous categories, and cofree algebras*, Categories in Computer Science and Logic (J. W. Gray and A. Scedrov, eds.), Contemp. Math. 92, Amer. Math. Soc., 1989, pp. 371–382.
22. J. Vaughan, *Small uncountable cardinals and topology*, Open Problems in Topology (J. van Mill and G. Reed, eds.), North-Holland, 1990, pp. 195–218.
23. R. Venkatesan and L. Levin, *Random instances of a graph coloring problem are hard*, Symposium on Theory of Computing, Assoc. for Computing Machinery, 1988, pp. 217–222.
24. P. Vojtáš, *Set-theoretic characteristics of summability of sequences and convergence of series*, Comment. Math. Univ. Carolinae **28** (1987), 173–183.
25. P. Vojtáš, *Generalized Galois-Tukey connections between explicit relations on classical objects of real analysis*, preprint (1991).
26. P. Vojtáš, *Topological cardinal invariants and the Galois-Tukey category* (to appear).

MATHEMATICS DEPT., UNIVERSITY OF MICHIGAN, ANN ARBOR, MI 48109, U.S.A.
E-mail address: ablass@umich.edu

[illegible]

MATHEMATICS D[illegible] UNIVERSITY OF MICHIGAN, ANN ARBOR, MI 48109 [illegible]
E-mail [illegible]@umich.edu

Hypercoherences: a strongly stable model of linear logic

Thomas Ehrhard
Laboratoire de Mathématiques Discrètes
UPR 9016 du CNRS, 163 avenue de Luminy, case 930
F 13288 MARSEILLE CEDEX 9
ehrhard@lmd.univ-mrs.fr

Abstract

We present a model of classical linear logic based on the notion of *strong stability* that was introduced in [BE], a work about sequentiality written jointly with Antonio Bucciarelli.

Introduction

The present article is a new version of an article already published, with the same title, in *Mathematical Structures in Computer Science* (1993), vol. 3, pp. 365–385. It is identical to this previous version, except for a few minor details.

In the denotational semantics of purely functional languages (such as PCF [P, BCL]), types are interpreted as objects and programs as morphisms in a cartesian closed category (CCC for short). Usually, the objects of this category are at least Scott domains, and the morphisms are at least continuous functions. The goal of denotational semantics is to express, in terms of "abstract" properties of these functions, some interesting computational properties of the language.

One of these abstract properties is "continuity". It corresponds to the basic fact that any computation that terminates can use only a finite amount of data. The corresponding semantics of PCF is the continuous one, where objects are Scott domains, and morphisms continuous functions.

But the continuous semantics does not capture an important property of computations in PCF, namely "determinism". Vuillemin and Milner are at the origin of the first (equivalent) definitions of *sequentiality*, a semantic notion corresponding to determinism. Kahn and Plotkin ([KP]) generalized this notion of sequentiality. More precisely, they defined a category of "concrete domains" (represented by "concrete data structures") and of sequential functions.

We shall begin with an intuitive description of what sequentiality is, in the framework of concrete data structures (CDS's). A CDS D, very roughly, is a Scott domain equipped with a notion of "places" or "cells". An element of D is a partial piece of data where some cells are filled, and others are not. A cell can be filled, in general, by different values. (Think of the cartesian product of two ground types: there are two cells corresponding to the two places one can fill in a couple.) In a CDS, an element x is less than an element x' if any cell that is filled in x is also filled in x', and by the same value. If D and E are CDS's, a sequential function f from D to E is a Scott continuous function from D to E such that, if $x \in D$ (that is, x is a partial data; some cells of D may not be filled in x), for any cell d not filled in $f(x)$, there exists a cell c not filled in x and filled in any $x' \in D$ such that $x' \geq x$ and such that d is filled in $f(x')$. This definition is a bit complicated, but the idea is simple. Consider, in order to simplify a bit, the case where E has only one cell. If $f(x)$ is undefined, there is a cell c not filled in x that must be filled in any data x' more defined than x and such that $f(x')$ is defined. This means the following: if $f(x)$ is undefined, then there is some "place" in x where the computation is stuck by a lack of information. If we want the computation to go on, we *must* fill the corresponding cell in x. So sequentiality is a way of speaking about the determinism of programs, considering only their input-output behavior; the basic rule of denotational semantics is that it is forbidden to look inside programs.

The idea of sequentiality is beautiful, but the category of CDS's and sequential functions has the bad taste to not be cartesian closed. The fundamental reason for this phenomenon is that, in general, there is no reasonable notion of cell for a domain of sequential functions.

The notion of "stability", introduced by Berry [B1, B2] is a weakening of the idea of sequentiality, that allows the definition of a model of PCF (a CCC). A stable function is a continuous function which commutes to the glb's (greatest lower bounds) of finite, non-empty and bounded subsets of its domain. However, among stable maps, there are functions that are not sequential (typically the so called Berry function), and so, even if it has nice mathematical properties, the stable model is not very satisfactory with respect to the modelization of determinism. It should be noticed that stability has also been used by Girard (see [G1]) to model system F. He used a very simple kind of domains (qualitative domains), and he also observed that a subclass of these domains (coherence spaces) has very good properties with respect to stability.

This work gave rise to the first proof-theoretic model of classical linear logic.

Berry and Curien (see [BC, C]) defined a CCC where morphisms are sequential, but are not functions; they are sequential algorithms.

In [BE], a joint work with Antonio Bucciarelli, we introduced the notion of strong stability in order to build a CCC where, at first order, the morphisms are sequential functions. Our basic observation was the following: sequentiality can be expressed as a preservation property similar to stability. More precisely, let us say that a family $x_1, \ldots, x_n$ of points of a CDS is *coherent* if it has the following property: any cell that is filled in all x_i's is filled by the same value in all x_i's. Then it can be proved that a function is sequential if and only if it sends a coherent family to a coherent family, and commutes to the glb's of coherent families. (In fact, this holds only for "sequential" CDS's.) In [BE], we proved the corresponding result for coherence spaces, taking as set of cells on a coherence space a suitable set of linear open subsets of this coherence space, but the intuition is the same. The families that are coherent in the sense described above will be called "linearly coherent" in the following. In order to get a CCC (a model of PCF), we had to abandon the notion of cell (since there is no known CCC of Kahn-Plotkin sequential functions), so we kept the notion of coherence. This led us to define a category where objects are qualitative domains[1] endowed with an additional structure called "coherence of states". A coherence of states is a set of non-empty and finite subsets of the qualitative domain that has to satisfy some closure properties. A qualitative domain endowed with a coherence of states is called a qualitative domain with coherence (QDC for short), and an element of the coherence of states of a QDC is said to be a coherent subset of the qualitative domain. A morphism between two QDC's is a Scott-continuous function between the associated qualitative domains, which, furthermore, maps any coherent set to a coherent set and commutes to the intersections of coherent sets. Such a function is said to be strongly stable. It turns out that the category of QDC's and strongly stable functions is cartesian closed.

Studying more precisely the coherences of states which are generated when a model of PCF is built up starting from ground types interpreted as suitable coherence spaces endowed with a linear coherence of states, it appears that these coherences of states in fact satisfy stronger properties than the ones we required in [BE].

Let us call "coherence of atoms" of a QDC the family of all coherent subsets of the qualitative domain that are made of atoms. (So the coherence of atoms

[1] A qualitative domain is a domain whose elements are subsets of a given set called "web" of the qualitative domain. These elements are ordered under inclusion, and any subset of an element of a qualitative domain has to be an element of the qualitative domain. The singleton elements of a qualitative domain are called "atoms".

is a subset of the coherence of states.)

Essentially, for the QDC's that are obtained in the construction of a model of PCF, we observe two main phenomena:

- When the coherence of atoms of the qualitative domain is known, the whole coherence of states is known.
- When the coherence of atoms of the qualitative domain is known, the set of all states of the qualitative domain is known: the states are simply the hereditarily coherent subsets of the web. (That is, the subsets of the web, any non-empty and finite subset of which is in the coherence of atoms.)

The first of these observations is not so surprising; it simply means that the coherence of states is in some sense "prime algebraic" (that is, here, "generated by atoms"), as the qualitative domain itself. The second one is very strange, because it says that the coherence of states is actually a more primitive notion than the notion of state itself.

These observations lead to a simplification of the theory of strong stability. Instead of considering qualitative domains with coherence as objects of the category, we just have to consider a very simple kind of structure, which we call "hypercoherence". (Actually, hypercoherences are hypergraphs, this is why we choose this name.) A hypercoherence is a set of finite subsets of a given set which we call the "web" of the hypercoherence. This set of finite parts of the web is intended to represent the coherence of atoms of a QDC. There is no commitment to any primitive notion of state, since, in a qualitative domain, we certainly want any singleton to be a state.

The only difference between hypercoherences and qualitative domains is that we do not require the family of sets which defines a hypercoherence to be hereditary (i.e. down-closed under inclusion).

This difference, which at first sight could seem innocuous, is, in fact, essential, because it allows us to define the orthogonal of a hypercoherence as its complement with respect to the set of all finite parts of its web. This operation does not make any sense in the framework of qualitative domains, because the complement of a down-closed set of subsets has no reason to be down-closed (and usually, it is not).

Indeed, the category **HCohL** of hypercoherences equipped with a suitable notion of linear morphisms, gives rise to a new model of full commutative classical linear logic (with the exponential "of course" which categorically is a comonad on **HCohL**).

Formally, hypercoherences are similar to coherence spaces in the sense that the interpretations of the linear connectives in this model are similar to those of Girard in coherence spaces (see [G2]), even if there are some surprises for the "with" and for the "of course" connectives. But this model seems to authorize some considerations which were impossible with pure coherence spaces. Specifically, it seems very natural to distinguish two classes of hypercoherences that play dual roles: the hereditary ones and the antihereditary ones (many hypercoherences are in neither of these classes). These two classes might be connected to the notion of polarity that Girard introduced in his treatment of classical logic (see [G3]).

This model of linear logic is compatible with the notion of strong stability. Any hypercoherence gives rise canonically and injectively to a qualitative domain with coherence. This object is defined accordingly to the two observations stated before. So we have a notion of strongly stable maps between hypercoherences. Call the category of hypercoherences and strongly stable maps **HCohFS**. (The letters "FS" in **HCohFS** come from the french "fortement stable" which means "strongly stable".) It turns out that this category is equivalent to the co-Kleisli category of the comonad "of course" which is cartesian closed. Furthermore, **HCohFS** can be considered as a full subcategory of the category of qualitative domains with coherence and strongly stable functions, and in fact, it is a full sub-CCC. This means that the product and exponential of the co-Kleisli category of the comonad "of course" are the same as the ones we defined in [BE] for more general objects. This result can be considered as a formal statement of the two observations we started from.

The remainder of the paper consists of seven sections. Section 1 is devoted to some preliminaries. We recall the basic definitions concerning qualitative domains and stable functions, and also the results of [BE] that we use later. Section 2 gives the definition of hypercoherences and of (linear) morphisms of hypercoherences. To simplify the presentation, morphisms are presented as traces (a trace is a kind of graph) and not as functions. Section 3 presents the model of linear logic from a purely formal point of view. In section 4, we connect this model of linear logic with our previous work about strong stability. Some acquaintance with [BE] could be useful for reading this section, though all the results we need are (briefly) recalled in the preliminaries. Section 5 consists of some definitions and very simple results about a notion of polarity that seems natural in this new framework. Section 6 makes explicit a relation between this model of linear logic and Girard's model of coherence spaces. Section 7 makes explicit the connection between hypercoherences and sequentiality at first order.

We have chosen this particular presentation, although it may not be very intuitive, for two reasons: first, we hope that the above introduction has provided the reader with sufficient intuition; and second, the notion of hypercoherence is simpler than the notion of QDC and it is very easy and natural to present the category of hypercoherences and linear morphisms in a purely self-contained way.

1 Preliminaries

If E is a set, we denote its cardinality by $\#E$.

Let E and F be two sets. If $C \subseteq E \times F$, we denote the first and second projections of C by C_1 and C_2 respectively. We say that C is a *pairing* of E and F if $C_1 = E$ and $C_2 = F$.

The disjoint union of E and F will be denoted by $E + F$, and represented by $G = (E \times \{1\}) \cup (F \times \{2\})$. If $C \subseteq G$, we use $C_1 = \{a \in E \mid (a, 1) \in C\}$ for its first component and $C_2 = \{b \in F \mid (b, 2) \in C\}$ for its second component.

Definition 1 *Let E and F be sets. Let $R \subseteq E \times F$ be a binary relation. Let $A \subseteq E$ and $B \subseteq F$. We say that A and B are* paired under *R and write $A \bowtie B \bmod (R)$ if $(A \times B) \cap R$ is a pairing of A and B.*

If R is the relation "$\in$", we say that A is a multisection *of B and we write $A \lhd B$. If the relation R is "$\subseteq$", we say that A is* Egli-Milner lower *than B and we write $A \sqsubseteq B$.*

So $A \lhd B$ means that $A \subseteq \bigcup B$ and that $A \cap b$ is non empty for all $b \in B$, and $A \sqsubseteq B$ means that any element of A is a subset of an element of B and any element of B is a superset of an element of A (this corresponds to the Egli-Milner relation in power-domain theory).

Obviously, the relation $\sqsubseteq$ is a preorder on $\mathcal{P}(E)$. Furthermore, if $A \lhd B \sqsubseteq C$ then $A \lhd C$.

If E is a set, we use $\mathcal{P}^*_{\text{fin}}(E)$ to denote the set of its finite and non-empty subsets. We write $x \subseteq^*_{\text{fin}} E$ when x is a finite and non-empty subset of E.

The theory of hypercoherences is closely related to that of qualitative domains, so let us recall some basic definitions and results from [G1].

Definition 2 *A* qualitative domain *is a pair $(|Q|, Q)$ where $|Q|$ is a set (the web) and Q is a subset of $\mathcal{P}(|Q|)$ satisfying the following conditions:*

- *$\emptyset \in Q$ and, if $a \in |Q|$, then $\{a\} \in Q$.*
- *if $x \in Q$ and if $y \subseteq x$ then $y \in Q$.*
- *if $D \subseteq Q$ is directed with respect to inclusion, then $\bigcup D \in Q$.*

The elements of Q are called states *of the qualitative domain, and the qualitative domain itself will also be denoted Q (for the web can be retrieved from Q).*

Observe that a qualitative domain Q can alternatively be presented as a pair $(|Q|, Q_{\text{fin}})$ where $|Q|$ is a set and Q_{fin} is a set of finite subsets of $|Q|$ satisfying all the conditions enumerated above except the last.

If Q is a qualitative domain, we call the set of its finite states Q_{fin}.

Definition 3 *A qualitative domain Q is called* coherence space *when, for $u \subseteq |X|$, if u satisfies*

$$\forall a, b \in u \quad \{a, b\} \in Q ,$$

then $u \in Q$.

Definition 4 *Let Q and R be qualitative domains. A function $f : Q \to R$ is* stable *if it is continuous and if*

$$\forall x, x' \in Q \quad x \cup x' \in Q \Rightarrow f(x \cap x') = f(x) \cap f(x') .$$

Furthermore, f is linear *if $f(\emptyset) = \emptyset$ and*

$$\forall x, x' \in Q \quad x \cup x' \in Q \Rightarrow f(x \cup x') = f(x) \cup f(x') .$$

The adequate notion of order for stable functions is not the extensional one, but the stable one, as observed by Berry (see [B1, B2]).

Definition 5 *If $f, g : R \to Q$ are two stable functions, f is* stably less *than g (written $f \leq g$) iff*

$$\forall x, x' \in Q \quad x \subseteq x' \Rightarrow f(x) = f(x') \cap g(x) .$$

If $f : Q \to R$ is a stable function, we define its *trace* $\text{tr}(f) \subseteq Q_{\text{fin}} \times |R|$ by

$$\text{tr}(f) = \{(x, b) \mid b \in f(x) \text{ and } x \text{ minimal with this property } \},$$

giving

- $\forall x \in Q \; f(x) = \{b \mid \exists x_0 \subseteq x \; (x_0, b) \in \text{tr}(f)\}$
- if $f, g : Q \to R$ are stable, then $f \leq g$ iff $\text{tr}(f) \subseteq \text{tr}(g)$.

(See [G1] for more details about traces.)

A stable function f is linear iff all the elements of the first projection of $\text{tr}(f)$ are singletons. So the trace of a linear function $Q \to R$ will always be considered as a subset of $|Q| \times |R|$.

Definition 6 *Let Q and R be qualitative domains. A* rigid embedding *of Q into R is an injection $f : |Q| \to |R|$ such that, for all $u \subseteq |Q|$, one has $u \in Q$ iff $f(u) \in R$.*

It is the canonical notion of substructure for qualitative domains. (For more details, see [GLT].)

Now we recall some definitions and results of [BE].

Definition 7 *A* qualitative domain with coherence *(qDC) is a pair* $(Q, \mathcal{C}(Q))$ *where* Q *is a qualitative domain and* $\mathcal{C}(Q)$ *is a subset of* $\mathcal{P}^*_{\text{fin}}(Q)$ *satisfying the following properties:*

- *if* $x \in Q$ *then* $\{x\} \in \mathcal{C}(Q)$,
- *if* $A \in \mathcal{C}(Q)$ *and if* $B \in \mathcal{P}^*_{\text{fin}}(Q)$ *is such that* $B \sqsubseteq A$, *then* $B \in \mathcal{C}(Q)$,
- *if* $D_1, \ldots, D_n$ *are directed subsets of* Q *such that for any* $x_1 \in D_1, \ldots, x_n \in D_n$ *the family* $\{x_1, \ldots, x_n\}$ *is in* $\mathcal{C}(Q)$, *then the family* $\{\bigcup D_1, \ldots, \bigcup D_n\}$ *is in* $\mathcal{C}(Q)$.

An element of $\mathcal{C}(Q)$ *will be called a* coherent *set of* Q. *Such a qDC* $(Q, \mathcal{C}(Q))$ *will be denoted by* Q *simply.*

The strongly stable functions are similar to stable functions, but they have to preserve coherence as well as intersections of coherent sets of states, and not just of bounded ones:

Definition 8 *Let* Q *and* R *be qDC's. A* strongly stable map f *from* Q *to* R *is a continuous function* $f : Q \to R$ *such that, if* $A \in \mathcal{C}(Q)$, *then* $f(A) \in \mathcal{C}(R)$ *and* $f(\bigcap A) = \bigcap f(A)$.

The preservation of coherence ($A \in \mathcal{C}(Q) \Rightarrow f(A) \in \mathcal{C}(R)$) is as important as the preservations of intersections of coherent families of states. It was not present in the theory of stable functions, because a stable function has to be monotone, and thus maps a bounded set of states to a bounded set of states. Here there is no reason why the image of a coherent set of states should be coherent.

Observe that any strongly stable map is stable, because, if Q is a qDC and if $A \subseteq Q$ is finite, non-empty and bounded, then $A \in \mathcal{C}(Q)$. Actually, $A \sqsubseteq \{x\}$ for any upper bound x of A.

Definition 9 *A strongly stable function is* linear *if it is linear as a stable function.*

In [BE] we have proved that the category **QDC** of qDC's and strongly stable maps is cartesian closed. Let us just recall the characterizations of cartesian products and function spaces.

Proposition 1 *Let Q and R be qDC's. Their cartesian product $Q \times R$ in the category* **QDC** *is $(P, \mathcal{C}(P))$ where P is the usual product of the qD's Q and R (that is, up to a canonical isomorphism, P is the cartesian product of the sets Q and R, equipped with the product order) and $\mathcal{C}(P)$ is the set of non-empty and finite subsets C of P such that $C_1 \in \mathcal{C}(Q)$ and $C_2 \in \mathcal{C}(R)$.*

In the next propositions, if T is the trace of a strongly stable function, then f^T denotes this function.

Proposition 2 *Let Q and R be qDC's. The function space $\mathrm{FS}(Q,R)$ of Q and R in the category* **QDC** *is $(P, \mathcal{C}(P))$ where P is the qualitative domain of traces of strongly stable functions $Q \to R$ and $\mathcal{C}(P)$ is the set of all non-empty and finite sets $\mathcal{T}$ of traces of strongly stable functions $Q \to R$ such that, for any $A \in \mathcal{C}(Q)$ and for any pairing $\mathcal{E}$ of $\mathcal{T}$ and A, we have*

$$\{f^T(x) \mid (T,x) \in \mathcal{E}\} \in \mathcal{C}(R)$$

and

$$\bigcap\{f^T(x) \mid (T,x) \in \mathcal{E}\} = f^{\bigcap \mathcal{T}}(\bigcap A) .$$

Let us now recall how this notion of strong stability is connected to sequentiality.

Let Q be a coherence space. The orthogonal $Q^\perp$ of Q is the coherence space whose web is $|Q|$ and such that, for $a, b \in |Q|$, we have $\{a,b\} \in Q^\perp$ iff $a = b$ or $\{a,b\} \notin Q$.

In this framework, we rephrase the definition of sequentiality outlined in the introduction. The idea is to consider $Q^\perp$ as a set of cells for Q, and to say that $x \in Q$ fills $\alpha \in Q^\perp$ if $x \cap \alpha \neq \emptyset$ (observe that, in that case, $x \cap \alpha$ is a singleton).

Definition 10 *Let Q and R be coherence spaces. We say that a function $f : Q \to R$ is* sequential *iff it is continuous, and for any $x \in Q$, for any $\beta \in R^\perp$ such that $f(x) \cap \beta = \emptyset$, there exists $\alpha \in Q^\perp$ such that $x \cap \alpha = \emptyset$ and such that, for any $x' \in Q$, if $x \subseteq x'$ and $f(x') \cap \beta \neq \emptyset$, then $x' \cap \alpha \neq \emptyset$.*

Let Q be any coherence space. We endow Q with its "linear coherence" $\mathcal{C}^{\mathrm{L}}(Q)$ which is the set of non-empty and finite subsets $\{x_1, \ldots, x_n\}$ of Q such that for any $\{a_1, \ldots, a_n\} \in Q^\perp$, if $a_1 \in x_1, \ldots, a_n \in x_n$, then $a_1 = \ldots = a_n$. It is easily checked that $(Q, \mathcal{C}^{\mathrm{L}}(Q))$ is then a qDC.

Proposition 3 *Let Q and R be coherence spaces. A function $f : Q \to R$ is sequential iff it is strongly stable $(Q, \mathcal{C}^{\mathrm{L}}(Q)) \to (R, \mathcal{C}^{\mathrm{L}}(R))$.*

2 Hypercoherences: basic definitions

Definition 11 *A* hypercoherence *is a pair* $X = (|X|, \Gamma(X))$ *where* $|X|$ *is an enumerable set and* $\Gamma(X)$ *is a subset of* $\mathcal{P}^*_{\mathrm{fin}}(|X|)$ *such that for any* $a \in |X|$, $\{a\} \in \Gamma(X)$. *The set* $|X|$ *is called* web *of* X *and* $\Gamma(X)$ *is called* atomic coherence *of* X.

If X is a hypercoherence, we note $\Gamma^*(X)$ the set of all $u \in \Gamma(X)$ such that $\#u > 1$. This set is called *strict atomic coherence* of X. A hypercoherence can be described by its strict atomic coherence as well as by its atomic coherence.

Observe that the only difference between a hypercoherence and a qualitative domain is that, if $u \in \Gamma(X)$ and if $v \subseteq u$, we *do not require* that v be in $\Gamma(X)$.

Definition 12 *A hypercoherence* X *is* hereditary *if, for all* $u \in \Gamma(X)$ *and for all* $v \subseteq^*_{\mathrm{fin}} u$, *one has* $v \in \Gamma(X)$.

Not all hypercoherences are hereditary.

We explain now how to build a qDC out of a hypercoherence.

Definition 13 *Let* X *be a hypercoherence. We define* $\mathrm{qD}(X)$ *and* $\mathcal{C}(X)$ *as follows:*

$$\mathrm{qD}(X) = \{x \subseteq |X| \mid \forall u \subseteq^*_{\mathrm{fin}} |X| \quad u \subseteq x \Rightarrow u \in \Gamma(X)\}$$

and

$$\mathcal{C}(X) = \{A \subseteq^*_{\mathrm{fin}} \mathrm{qD}(X) \mid \forall u \subseteq^*_{\mathrm{fin}} |X| \quad u \lhd A \Rightarrow u \in \Gamma(X)\}\ .$$

$\mathrm{qD}(X)$ *will be called the qualitative domain generated by* X, *and* $\mathcal{C}(X)$ *will be called the state coherence generated by* X. *The couple* $(\mathrm{qD}(X), \mathcal{C}(X))$ *will be noted* $\mathrm{qDC}(X)$. *The set of finite states of* $\mathrm{qD}(X)$ *will be noted* $\mathrm{qD}_{\mathrm{fin}}(X)$.

So, $\mathrm{qD}_{\mathrm{fin}}(X)$ is the set of elements of $\Gamma(X)$ which are hereditary, that is of which any subset is either empty or in $\Gamma(X)$.

The following result justifies the terminology and notations:

Proposition 4 *If* X *is a hypercoherence, then* $(\mathrm{qD}(X), \mathcal{C}(X))$ *is a qualitative domain with coherence, and* $|\mathrm{qD}(X)| = |X|$.

The proof is straightforward. The qualitative domain with coherence $\mathrm{qDC}(X)$ will be called qualitative domain with coherence generated by X.

Observe that, for a hypercoherence X, the atomic coherence $\Gamma(X)$ can be retrieved from $\mathcal{C}(X)$ (and thus from $\mathrm{qDC}(X)$). Namely, the elements of $\Gamma(X)$ are the finite and non-empty subsets u of $|X|$ such that $\{\{a\} \mid a \in u\}$ be in $\mathcal{C}(X)$. So the hypercoherences can be seen as certain qDC's.

We give now two important classes of examples of hypercoherences:

- If Q is a qualitative domain, we can define a hereditary hypercoherence X as follows: we take $|X| = |Q|$ and $\Gamma(X) = Q_{\text{fin}} \setminus \{\emptyset\}$. Then it is easy to see that $\text{qD}(X) = Q$ and that $\mathcal{C}(X)$ is the set of all non-empty and finite *bounded* subsets of Q.

- If Q is a coherence space, we can also define a hypercoherence Y as follows: $|Y| = |Q|$ and a finite and non-empty subset of $|Q|$ is in $\Gamma(Y)$ iff it is a singleton, or it contains distinct elements a and a' of $|Q|$ such that $\{a, a'\} \in Q$. Then it is easily checked that $\text{qD}(Y) = Q$ and that $\mathcal{C}(Y) = \mathcal{C}^{\text{L}}(Q)$.

Now we have enough material to start the presentation of our model of linear logic. To avoid boring and trivial categorical calculations, we shall use the informal notion of "canonical isomorphism" between hypercoherences. An isomorphism between two hypercoherences X and Y is a bijection $f : |X| \to |Y|$ such that, for all $u \subseteq^*_{\text{fin}} |X|$, we have $f(u) \in \Gamma(Y)$ iff $u \in \Gamma(X)$. For us, a canonical isomorphism is an isomorphism which corresponds to a bijection on the webs which is standard and universal from the set-theoretic point of view. A typical example is the bijection which corresponds to the associativity of cartesian product of sets.

Definition 14 *Let X and Y be hypercoherences. We call* linear implication *of X and Y and note $X \multimap Y$ the hypercoherence defined by $|X \multimap Y| = |X| \times |Y|$ and $w \in \Gamma(X \multimap Y)$ iff w is a non-empty and finite subset of $|X \multimap Y|$ such that*

$$w_1 \in \Gamma(X) \Rightarrow (w_2 \in \Gamma(Y) \text{ and } (\#w_2 = 1 \Rightarrow \#w_1 = 1)) .$$

Equivalently, $w \in \Gamma(X \multimap Y)$ iff $w \subseteq^*_{\text{fin}} |X \multimap Y|$ and

$$w_1 \in \Gamma(X) \Rightarrow w_2 \in \Gamma(Y) \quad \text{and} \quad w_1 \in \Gamma^*(X) \Rightarrow w_2 \in \Gamma^*(Y) .$$

Of course, $X \multimap Y$ satisfies the only axiom of hypercoherences.

Definition 15 *Let X and Y be hypercoherences. A linear morphism $X \multimap Y$ is an element of* $\text{qD}(X \multimap Y)$.

We shall often write $s : X \multimap Y$ instead of $s \in \text{qD}(X \multimap Y)$.

Proposition 5 *Let X, Y and Z be hypercoherences.*

- *The set* $\text{Id}_X = \{(a, a) \mid a \in |X|\}$ *is in* $\text{qD}(X \multimap X)$.

- *Let* $s \in \mathrm{qD}(X \multimap Y)$ *and* $t \in \mathrm{qD}(Y \multimap Z)$. *Then the set*

$$t \circ s = \{(a,c) \mid \exists b\ (a,b) \in s \text{ and } (b,c) \in t\}$$

is in $\mathrm{qD}(X \multimap Z)$.

Proof: The fact that $\mathrm{Id}_X \in \mathrm{qD}(X \multimap X)$ is obvious.

Observe first that if $(a,c) \in t \circ s$ then there is exactly one b such that $(a,b) \in s$ and $(b,c) \in t$. Actually, if y is a non-empty and finite subset of $\{b \mid (a,b) \in s \text{ and } (b,c) \in t\}$, then $y \in \Gamma(Y)$ because $\{a\} \in \Gamma(X)$, and thus $\#y = 1$ because $\#\{c\} = 1$.

Let $w \subseteq t \circ s$ be finite, non-empty and such that $w_1 \in \Gamma(X)$. Let

$$u = \{(a,b) \in s \mid \exists c \in w_2\ (a,c) \in w \text{ and } (b,c) \in t\}$$

and

$$v = \{(b,c) \in t \mid \exists a \in w_1 (a,c) \in w \text{ and } (a,b) \in s\}\ .$$

Finiteness of u and v follows from the observation above. We clearly have $u_1 \subseteq w_1$. Conversely, let $a \in w_1$. Let $c \in w_2$ be such that $(a,c) \in w$. By definition, there is a b such that $(a,b) \in s$ and $(b,c) \in t$, so $a \in u_1$. Thus $u_1 = w_1 \in \Gamma(X)$. So $u_2 \in \Gamma(Y)$. But clearly $u_2 = v_1$. So $v_2 \in \Gamma(Z)$. We conclude that $w_2 \in \Gamma(Z)$ because $v_2 = w_2$.

Assume furthermore that $\#w_2 = 1$. Then $\#v_1 = 1$ and $\#u_1 = 1$ and we conclude. ∎

We have obviously defined a category, where objects are hypercoherences, composition is $\circ$ and the identity morphisms are the Id_X's. We note **HCohL** this category.

3 A model of classical linear logic

The goal of this section is to interpret in the category **HCohL** the connectives of classical linear logic. In fact, the linear implication has already been partially treated in the previous section.

We note 1 the hypercoherence whose web is a singleton.

Definition 16 *Let X and Y be hypercoherences. Their* tensor product $X \otimes Y$ *is the hypercoherence whose web is* $|X| \times |Y|$ *and whose atomic coherence is defined by:* $w \in \Gamma(X \otimes Y)$ *iff* $w_1 \in \Gamma(X)$ *and* $w_2 \in \Gamma(Y)$.

The tensor product is in fact a functor $\mathbf{HCohL} \times \mathbf{HCohL} \to \mathbf{HCohL}$. If $s : X \multimap X'$ and $t : Y \multimap Y'$, define $s \otimes t$ by

$$s \otimes t = \{((a,b),(a',b')) \mid (a,a') \in s \text{ and } (b,b') \in t\}\ .$$

Let us check that $s \otimes t : X \otimes Y \multimap X' \otimes Y'$. Let $w \subseteq^*_{\mathrm{fin}} s \otimes t$ and assume that $w_1 \in \Gamma(X \otimes Y)$, that is $w_{11} \in \Gamma(X)$ and $w_{12} \in \Gamma(Y)$. Let us prove that $w_{21} \in \Gamma(X')$. Let

$$w^1 = \{(a, a') \in s \mid \exists (b, b') \in t \; ((a, b), (a', b')) \in w\} .$$

We have $w^1 \subseteq^*_{\mathrm{fin}} s$ and $w^1_1 = w_{11}$, $w^1_2 = w_{21}$. Hence $w_{21} \in \Gamma(X')$. Similarly $w_{22} \in \Gamma(Y')$. Assume furthermore that $\# w_2 = 1$. Let (a', b') be the unique element of that set. Then $w_{11} \times \{a'\} = w^1$, so $\# w_{11} = 1$. Similarly $\# w_{12} = 1$ and we conclude.

Proposition 6 *The tensor product is, up to canonical isomorphisms, a commutative and associative operation which admits* 1 *as neutral element. Furthermore, the canonical isomorphisms associated to commutativity and associativity satisfy the axioms of symmetric monoidal categories.*

This is a purely formal verification. See [M] for details about monoidal categories.

Definition 17 *Let X be a hypercoherence. We call* orthogonal *of X and note $X^\perp$ the hypercoherence whose web is $|X|$ and whose atomic coherence is $\mathcal{P}^*_{\mathrm{fin}}(|X|) \setminus \Gamma^*(X)$.*

So that $u \in \Gamma^*\left(X^\perp\right)$ iff u is finite and non-empty and $u \notin \Gamma(X)$.

Proposition 7 *Let X be a hypercoherence. Then $(X^\perp)^\perp = X$.*

The proof is straightforward.

Proposition 8 *Let X and Y be hypercoherences. Up to a canonical isomorphism,*

$$X^\perp \multimap Y^\perp = Y \multimap X .$$

Proof: Just contrapose the definition of $X^\perp \multimap Y^\perp$. ∎

If $s : X \multimap Y$, we note ${}^t s$ the corresponding morphism $Y^\perp \multimap X^\perp$, which is simply $\{(b, a) \mid (a, b) \in s\}$. This operation on morphisms is called transposition, and turns $(\_)^\perp$ into a contravariant and involutive endofunctor on **HCohL**.

Definition 18 *Let X and Y be hypercoherences. The* par *of X and Y is the hypercoherence $X \wp Y = (X^\perp \otimes Y^\perp)^\perp$.*

Proposition 9 *Let X and Y be hypercoherences. We have $|X \wp Y| = |X| \times |Y|$, and $w \in \Gamma^*(X \wp Y)$ iff w is non-empty and finite and satisfies $w_1 \in \Gamma^*(X)$ or $w_2 \in \Gamma^*(Y)$.*

Easy calculation.

Observe that, because of propositions 8 and 6, the par is commutative, associative and admits 1 as neutral element, because clearly $1^\perp = 1$. This last phenomenon is a drawback that this semantics of linear logic shares with the coherence spaces semantics.

Proposition 10 *Let X and Y be hypercoherences, then $X \multimap Y = X^\perp \wp Y$.*

Proof: Obviously, these hypercoherences have the same web. If $w \subseteq^*_{\text{fin}} |X| \times |Y|$, then $w \in \Gamma(X \multimap Y)$ iff

$$w_1 \in \Gamma(X) \Rightarrow w_2 \in \Gamma(Y) \quad \text{and} \quad w_1 \in \Gamma^*(X) \Rightarrow w_2 \in \Gamma^*(Y)$$

that is

$$\left(w_1 \in \Gamma^*\left(X^\perp\right) \text{ or } w_2 \in \Gamma(Y)\right) \quad \text{and} \quad \left(w_1 \in \Gamma\left(X^\perp\right) \text{ or } w_2 \in \Gamma^*(Y)\right)$$

and we conclude. ∎

As a corollary, we get:

Proposition 11 *The category* **HCohL** *is monoidal closed with respect to the tensor product $\otimes$ and the function space $\multimap$. More precisely, if X, Y and Z are hypercoherences, then, up to canonical isomorphisms:*

$$(X \otimes Y) \multimap Z = X \multimap (Y \multimap Z) .$$

Proof: This results from the associativity (up to canonical isomorphisms) of par. ∎

Definition 19 *Let X and Y be hypercoherences. We call* with *of X and Y and note $X \times Y$ the hypercoherence whose web is $|X| + |Y|$ and whose atomic coherence is the set of all $w \subseteq^*_{\text{fin}} |X| + |Y|$ such that:*

$$w_1 = \emptyset \Rightarrow w_2 \in \Gamma(Y) \quad \textit{and} \quad w_2 = \emptyset \Rightarrow w_1 \in \Gamma(X) .$$

Of course, this satisfies the axiom of hypercoherence.

In that definition, the contrast with coherence spaces appears clearly: as soon as a (finite) subset of $|X| + |Y|$ is such that both of its components are non-empty, it is coherent, whereas in coherence spaces (or qualitative domains), both of its components had to be coherent. This phenomenon has important consequences. Consider for instance the hypercoherence Bool $= (\{T,F\},\{\{T\},\{F\}\})$. A subset u of $\left|\mathrm{Bool}^3\right| = |\mathrm{Bool}| \times \{1,2,3\}$ is in $\Gamma^*\left(\mathrm{Bool}^3\right)$ iff there exist $i,j \in \{1,2,3\}$ such that $i \neq j$ and $u_i \neq \emptyset$ and $u_j \neq \emptyset$. As a consequence, the set $\{\{(T,1),(F,2)\},\{(T,2),(F,3)\},\{(T,3),(F,1)\}\}$ is in $\mathcal{C}\left(\mathrm{Bool}^3\right)$, whereas it is not bounded, and not even pairwise bounded. This is why the stable but non sequential Berry's function $\mathrm{qD}\left(\mathrm{Bool}^3\right) \to \mathrm{qD}(\mathrm{Bool})$ whose trace is:

$$\{(\{(T,1),(F,2)\},T),(\{(T,2),(F,3)\},T),(\{(T,3),(F,1)\},T)\}$$

will not be in our model (see below). This definition of cartesian product is strongly related to sequentiality.

Proposition 12 *Let X and Y be hypercoherences. Then $X \times Y$ is the cartesian product of X and Y in the category* **HCohL**.

The proof is straightforward. The projection $\pi_1 : X \times Y \multimap X$ is $\{((a,1),a) \mid a \in |X|\}$, and similarly for π_2. If $s : Z \multimap X$ and $t : Z \multimap Y$ are linear morphisms, their pairing $p : Z \multimap X \times Y$ is

$$p = \{(c,(a,1)) \mid (c,a) \in s\} \cup \{(c,(b,2)) \mid (c,b) \in t\} .$$

Definition 20 *Let X and Y be hypercoherences. We call* plus *of X and Y and note $X \oplus Y$ the hypercoherence $(X^\perp \times Y^\perp)^\perp$.*

Proposition 13 *If X and Y are hypercoherences, the web of $X \oplus Y$ is $|X|+|Y|$ and its atomic coherence is the set of all $w \subseteq^*_{\mathrm{fin}} |X|+|Y|$ such that*

$$w_1 = \emptyset \text{ and } w_2 \in \Gamma(Y) \quad \text{or} \quad w_2 = \emptyset \text{ and } w_1 \in \Gamma(X) .$$

Straightforward verification.

Definition 21 *Let X be a hypercoherence. We define a hypercoherence $!X$ by setting $|!X| = \mathrm{qD_{fin}}(X)$ and by taking for $\Gamma(!X)$ the restriction of $\mathcal{C}(X)$ to $\mathrm{qD_{fin}}(X)$. In other words, if $A \subseteq^*_{\mathrm{fin}} \mathrm{qD_{fin}}(X)$, then $A \in \Gamma(!X)$ iff*

$$\forall u \subseteq^*_{\mathrm{fin}} |X| \quad u \lhd A \Rightarrow u \in \Gamma(X) .$$

Proposition 14 *An element of $\mathrm{qD}(!X)$ is a bounded subset of $\mathrm{qD_{fin}}(X)$.*

Proof: Let $A \in \mathrm{qD}(!X)$. We can assume that A is finite. Let $u \subseteq^*_{\mathrm{fin}} \bigcup A$. Let $B = \{x \in A \mid u \cap x \neq \emptyset\}$. Then $B \subseteq^*_{\mathrm{fin}} A$, so $B \in \Gamma(!X)$. So $u \in \Gamma(X)$, since $u \lhd B$ by definition of B, and thus $\bigcup A \in \mathrm{qD}(X)$. ∎

Proposition 15 *Let X and Y be hypercoherences. Up to a canonical isomorphism,*

$$!(X \times Y) = !X \otimes !Y \ .$$

Proof: It is a corollary of the forthcoming proposition 21. ∎

Definition 22 *Let X be a hypercoherence. We define the hypercoherence $?X$ by*

$$?X = (!X^{\perp})^{\perp} \ .$$

An element A of $\mathcal{P}^{*}_{\text{fin}}\left(\text{qD}_{\text{fin}}\left(X^{\perp}\right)\right)$ is in $\Gamma^{*}(?X)$ iff there exists $u \in \Gamma^{*}(X)$ such that $u \lhd A$.

We extend now the operation "!" into a functor **HCohL** → **HCohL** and we exhibit the comonad structure of this functor.

Proposition 16 *Let X and Y be hypercoherences. Let $t \in \text{qD}(X \multimap Y)$. Then the set $!t$ defined by*

$$!t = \{(x, y) \in \text{qD}_{\text{fin}}(X) \times \text{qD}_{\text{fin}}(Y) \mid x \bowtie y \bmod (t)\}$$

is an element of $\text{qD}(!X \multimap !Y)$.

Proof: Let U be any non-empty and finite subset of $!t$. Assume that $U_1 \in \Gamma(!X)$. Let $v \subseteq |Y|$ be finite, non-empty and such that $v \lhd U_2$. Let

$$w = \{(a, b) \in t \mid b \in v \text{ and } \exists (x, y) \in U \ a \in x, \ b \in y\} \ .$$

Then we have $w_2 = v$ and $w_1 \lhd U_1$. Let us just check the second of those statements. If $x \in U_1$, let $y \in U_2$ be such that $(x, y) \in U$. Let $b \in v$ be such that $b \in y$. Since $x \bowtie y \bmod (t)$, we can find some $a \in x$ such that $(a, b) \in t$. Clearly, $(a, b) \in w$, so $a \in w_1$ and we have proven one direction of the statement $w_1 \lhd U_2$, the second one being a direct consequence of the definition of w.

Since w is finite, non-empty and satisfies $w \subseteq t$, we have $w \in \Gamma(X \multimap Y)$. But $w_1 \in \Gamma(X)$ since $w_1 \lhd U_1 \in \mathcal{C}(X)$, and thus $w_2 \in \Gamma(Y)$, that is $v \in \Gamma(Y)$. This holds for any $v \lhd U_2$, so $U_2 \in \Gamma(!X)$.

Assume now that $\#U_2 = 1$, say $U_2 = \{y\}$. Take $x_0 \in U_1$. We prove that for any $x \in U_1$, we have $x_0 \subseteq x$. This clearly will entail that $\#U_1 = 1$. Let $a_0 \in x_0$. Let $b \in y$ be such that $(a_0, b) \in t$. Let

$$u = \{a \mid (a, b) \in t \text{ and } \exists x \in U_1 \ a \in x\} \ .$$

One easily checks that $u \lhd U_1$, so $u \in \Gamma(X)$. But $u \times \{b\} \subseteq t$, so $\#u = 1$, but $a_0 \in u$, so $u = \{a_0\}$. Hence, since $u \lhd U_1$, for all $x \in U_1$ one has $a_0 \in x$, and we conclude. ∎

Proposition 17 *Let X, Y and Z be hypercoherences. Then $!\mathrm{Id}_X = \mathrm{Id}_{!X}$ and if $s : X \multimap Y$ and $t : Y \multimap Z$ then $!(t \circ s) = !t \circ !s$.*

Proof: Let us check that $!(t \circ s) = !t \circ !s$. First, let $(x, z) \in !(t \circ s)$. This means that $x \bowtie z \bmod (t \circ s)$. Let

$$y = \{b \mid \exists a \in x\ \exists c \in z\ (a, b) \in s \text{ and } (b, c) \in t\} .$$

We have $(x, y) \in !s$ and $(y, z) \in !t$. Let us prove the first of these statements, the second being similar. Let $a \in x$. Let $c \in z$ be such that $(a, c) \in t \circ s$. Let b be such that $(a, b) \in s$ and $(b, c) \in t$. We have $b \in y$ by definition of y. Conversely, if $b \in y$, we can find, by definition of y, a $a \in x$ such that $(a, b) \in s$. So $x \bowtie y \bmod (s)$, that is $(x, y) \in !s$.

Next, let $(x, z) \in !t \circ !s$. Let y be such that $(x, y) \in !s$ and $(y, z) \in !t$. Let $a \in x$. Let $b \in y$ be such that $(a, b) \in s$. Since $y \bowtie z \bmod (t)$ we can find a $c \in z$ such that $(b, c) \in t$. So we have found a c such that $(a, c) \in t \circ s$. Conversely, if $c \in z$, we can similarly find a $a \in x$ such that $(a, c) \in t \circ s$, and this concludes the proof. ∎

So now we can consider the operation ! as an endofunctor on **HCohL**. We show that it has a natural structure of comonad.

Let X be a hypercoherence. Let $\varepsilon_X = \{(\{a\}, a) \mid a \in |X|\}$. It is clear that $\varepsilon_X \in \mathrm{qD}\,(!X \multimap X)$.

Let $\mu_X = \{(x, \{x_1, \ldots, x_n\}) \mid x, x_1, \ldots, x_n \in \mathrm{qD}_{\mathrm{fin}}(X) \text{ and } \bigcup_{i=1}^n x_i = x\}$. Let us check that $\mu_X \in \mathrm{qD}\,(!X \multimap !!X)$. Let $T \subseteq \mu_X$ be finite, non-empty and such that $T_1 \in \Gamma\,(!X)$. Let $A \subseteq \mathrm{qD}_{\mathrm{fin}}(X)$ be such that $A \lhd T_2$. We clearly have $A \sqsubseteq T_1$ and thus $A \in \Gamma\,(!X)$. So $T_2 \in \Gamma\,(!!X)$. If furthermore T_2 is a singleton, then T_1 is obviously also a singleton.

Proving that ε and μ are the counit and comultiplication of a comonad whose functor is ! is a straightforward verification.

For the notion of comonad, and of co-Kleisli category of a comonad, we refer to [M].

Proposition 18 *The co-Kleisli category* **coKl**(!) *of the comonad* ! *is cartesian closed.*

Proof: Remember that in this co-Kleisli category, the objects are the hypercoherences, and that a morphism $X \to Y$ is a linear morphism $!X \multimap Y$. If $S : X \to Y$ and $T : Y \to Z$ are morphisms in **coKl**(!), their composition $T \circ S : X \to Z$ is given by:

$$T \circ S = T \circ !S \circ \mu_X$$

and the identity $X \to X$ is simply ε_X. Observe that in this last equation, the symbol "$\circ$" has two different meanings: in the left-hand side, it denotes composition in **coKl**(!), whereas in the right-hand side, it denotes composition in **HCohL**.

First, this category has products, the product of X and Y being $X \times Y$.

For cartesian closedness, let X, Y and Z be hypercoherences. Up to canonical isomorphisms we have, using proposition 15:

$$\begin{aligned}(X \times Y) \to Z &= !(X \times Y) \multimap Z \\ &= (!X \otimes !Y) \multimap Z \\ &= !X \multimap (!Y \multimap Z) \\ &= X \to (Y \to Z)\,.\end{aligned}$$

To be more precise, these equalities correspond to canonical (and thus natural) isomorphisms in **HCohL** which are easily transfered to **coKl**(!) using ε. ∎

4 Hypercoherences and strong stability

The purpose of this section is to connect the model we just have presented to the model of (simply typed) λ-calculus presented in [BE]. The section is important because it contains the main intuitions at the origin of the construction of **HCohL**, and it provides an "abstract" characterization of the morphisms of this category.

Definition 23 *The category* **HCohFS** *of hypercoherences and strongly stable functions is the category where the objects are the hypercoherences, and where a morphism* $X \to Y$ *is a strongly stable function* $\mathrm{qDC}(X) \to \mathrm{qDC}(Y)$.

Proposition 19 *The categories* **coKl**(!) *and* **HCohFS** *are equivalent.*

Proof: On objects, this equivalence is simply the identity.

For morphisms, the proposition is mainly a characterization of the traces of strongly stable functions.

If X and Y are hypercoherences, we recall that $X \to Y$ denotes the hypercoherence $!X \multimap Y$.

First, let $T \in \mathrm{qD}(X \to Y)$. We prove that, by setting

$$f^T(x) = \{b \in |Y| \mid \exists x_0 \subseteq x \; (x_0, b) \in T\}$$

we define a function $\mathrm{qD}(X) \to \mathrm{qD}(Y)$ that is strongly stable.

Let us prove that if $x \in \mathrm{qD}(X)$, $y = f^T(x) \in \mathrm{qD}(Y)$. We can assume that $x \in \mathrm{qD}_{\mathrm{fin}}(X)$. Let $v \subseteq^*_{\mathrm{fin}} |Y|$ be such that $v \subseteq y$. Let $U = \{(x_0, b) \in T \mid x_0 \subseteq$

x and $b \in v\}$. We know that $U \in \Gamma(X \to Y)$, since U is non-empty and finite. But U_1 is bounded by x, and thus $U_1 \in \Gamma(!X)$. Thus $U_2 \in \Gamma(Y)$. But $U_2 = v$ and we are finished. So f^T is well defined, and Scott continuous by definition.

Let $A \in \mathcal{C}(X)$. We prove that $f^T(A) \in \mathcal{C}(Y)$. Since f^T is continuous, we can assume that any element of A is finite. Let $v \subseteq^*_{\text{fin}} |Y|$ be such that $v \lhd f^T(A)$. Let

$$U = \{(x_0, b) \in T \mid \exists x \in A \; x_0 \subseteq x \text{ and } b \in v\} .$$

Again, U is non-empty and finite, so $U \in \Gamma(X \to Y)$. We have $U_1 \sqsubseteq A$. Actually, let $x \in A$ and let $b \in v$ be such that $b \in f^T(x)$ (such a b can be found since $v \lhd f^T(A)$). Let $x_0 \subseteq x$ be such that $(x_0, b) \in T$. One has $x_0 \in U_1$. So $U_1 \in \Gamma(!X)$. So $U_2 \in \Gamma(Y)$, but $U_2 = v$ and we are finished.

Now, let $b \in \bigcap f^T(A)$. We want to prove that $b \in f^T(\bigcap A)$. We can assume again that any element of A is finite. Let

$$A_0 = \{x_0 \mid \exists x \in A \; x_0 \subseteq x \text{ and } (x_0, b) \in T\} .$$

A_0 is finite and satisfies $A_0 \sqsubseteq A$, so $A_0 \in \Gamma(!X)$, but $U = A_0 \times \{b\} \subseteq T$, thus $U \in \Gamma(X \to Y)$, and thus $\# A_0 = 1$. Let x_1 be the unique element of A_0, we have $x_1 \subseteq \bigcap A$ and $(x_1, b) \in T$, and we conclude that f^T is strongly stable.

Conversely, let $f : \mathrm{qD}(X) \to \mathrm{qD}(Y)$ be strongly stable. We shall prove now that its trace T is in $\mathrm{qD}(X \to Y)$. Let $U \subseteq T$ be finite, non-empty and such that $U_1 \in \Gamma(!X)$. We have $f(U_1) \in \mathcal{C}(Y)$ and $U_2 \lhd f(U_1)$, so $U_2 \in \Gamma(Y)$. Assume furthermore that $\# U_2 = 1$ and let b be the unique element of U_2. We have $b \in \bigcap f(U_1) = f(\bigcap U_1)$, so there exists an $x_1 \subseteq \bigcap U_1$ such that $(x_1, b) \in T$. But for $x_0 \in U_1$, we have $x_0 \supseteq \bigcap U_1 \supseteq x_1$ and thus, since f is stable, $x_0 = x_1$, so $U_1 = \{x_1\}$ and we conclude that $U \in \Gamma(X \to Y)$.

It is fairly obvious that $\mathrm{tr}\left(f^T\right) = T$ and that $f^{\mathrm{tr}(f)} = f$, since this already holds for stable functions.

It remains to prove that the correspondence we have just established is functorial.

The identity $X \to X$ in **coKl(!)** is ε_X, that is $\{(\{a\}, a) \mid a \in |X|\}$ which clearly is the trace of the identity $X \to X$ in **HCohFS**.

Let $S : X \to Y$ and $T : Y \to Z$ be morphisms in **coKl(!)**. Remember that $T \circ S = T \circ !S \circ \mu_X$, that is $T \circ S$ is the set

$$\{(x, c) \mid \exists x_1, \ldots, x_n \; \exists b_1, \ldots, b_n, \\ \textstyle\bigcup_{i=1}^n x_i = x \text{ and } \forall i \; (x_i, b_i) \in S \text{ and } (\{b_1, \ldots, b_n\}, c) \in T\} ,$$

so if $x \in \mathrm{qD}(X)$,

$$f^{T \circ S}(x) = \{c \mid \exists (x_1, b_1), \ldots, (x_n, b_n) \in S \quad \forall i \, x_i \subseteq x \text{ and } (\{b_1, \ldots, b_n\}, c) \in T\} ,$$

that is $f^{T \circ S}(x) = f^T(f^S(x))$.

For the other direction, let $f : X \to Y$ and $g : Y \to Z$ be morphisms in **HCohFS**. We have

$$\begin{aligned} \mathrm{tr}\,(g \circ f) &= \{(x,c) \mid c \in g(f(x)),\ x \text{ minimal}\} \\ &= \{(x,c) \mid \exists y\ y \subseteq f(x),\ (y,c) \in \mathrm{tr}\,(g)\,,\ x \text{ minimal}\} \\ &= \mathrm{tr}\,(g) \circ \mathrm{tr}\,(f) \quad \text{(see above the computation of this trace)} \end{aligned}$$

∎

From this result, we deduce that **HCohFS** is cartesian closed.

Observe now that the stable function $G : \mathrm{qD}\left(\mathrm{Bool}^3\right) \to \mathrm{qD}\,(\mathrm{Bool})$ whose trace is

$$\{(\{(T,1),(F,2)\},T),(\{(T,2),(F,3)\},T),(\{(T,3),(F,1)\},T)\}$$

is not in $\mathrm{qD}\left(\mathrm{Bool}^3 \to \mathrm{Bool}\right)$ since the set

$$\{\{(T,1),(F,2)\},\{(T,2),(F,3)\},\{(T,3),(F,1)\}\}$$

is in $\mathcal{C}\left(\mathrm{Bool}^3\right)$: as we have said before; the Berry's function is not a morphism in **coKl**(!).

As a corollary of the previous proposition, we get:

Proposition 20 *The category* **HCohL** *is equivalent to the category of hypercoherences and linear strongly stable functions.*

Proof: Just observe that if X is a hypercoherence, if $a_1, \ldots, a_n \in |X|$, then $\{\{a_1\}, \ldots, \{a_n\}\} \in \mathcal{C}\,(X)$ iff $\{a_1, \ldots, a_n\} \in \Gamma\,(X)$. ∎

We conclude the section with the proof that the product objects and internal arrow objects in **HCohFS** are the same as in **QDC**.

Proposition 21 *If X and Y are hypercoherences, then*

$$\mathrm{qDC}\,(X \times Y) = \mathrm{qDC}\,(X) \times \mathrm{qDC}\,(Y)\ .$$

Proof: First, let $z \in \mathrm{qD}\,(X \times Y)$, and let us prove that $z_1 \in \mathrm{qD}\,(X)$. Let $u \subseteq z_1$ be non-empty and finite. We have $u \times \{1\} \subseteq z$ and $(u \times \{1\})_2 = \emptyset$, so $u \in \Gamma\,(X)$. Similarly $z_2 \in \mathrm{qD}\,(Y)$. The inclusion $\mathrm{qD}\,(X) \times \mathrm{qD}\,(Y) \subseteq \mathrm{qD}\,(X \times Y)$ is also trivial.

Now let $C \in \mathcal{C}\,(X \times Y)$, and let us prove that $C_1 \in \mathcal{C}\,(X)$. Let u be finite and non-empty such that $u \lhd C_1$. We have $u \times \{1\} \lhd C$, and thus $u \in \Gamma\,(X)$. Similarly for C_2.

Finally, let C be in the state coherence of $\mathrm{qDC}\,(X) \times \mathrm{qDC}\,(Y)$. Let w be finite and non-empty such that $w \lhd C$. Assume that $w_2 = \emptyset$. Then certainly $w_1 \lhd C_1$, so $w_1 \in \Gamma\,(X)$, since $C_1 \in \mathcal{C}\,(X)$. Similarly if $w_1 = \emptyset$. So $C \in \mathcal{C}\,(X \times Y)$. ∎

Proposition 22 *If X and Y are hypercoherences, then*

$$\mathrm{qDC}\,(X \to Y) = \mathrm{FS}\,(\mathrm{qDC}\,(X), \mathrm{qDC}\,(Y))\ .$$

Proof: In proving proposition 19, we have shown that $\mathrm{qDC}\,(X \to Y)$ and $\mathrm{FS}\,(\mathrm{qDC}\,(X), \mathrm{qDC}\,(Y))$ have the same underlying qD. We prove now that they have the same state coherence.

First, let $\mathcal{T} \in \mathcal{C}\,(X \to Y)$. We want to prove that $\mathcal{T}$ is state coherent in $\mathrm{FS}\,(\mathrm{qDC}\,(X), \mathrm{qDC}\,(Y))$. Let $A \in \mathcal{C}\,(X)$ and let $\mathcal{E}$ be any pairing of $\mathcal{T}$ and A. Let

$$B = \{f^T(x) \mid (T, x) \in \mathcal{E}\}\ .$$

We prove that $B \in \mathcal{C}\,(Y)$. We can assume that any $T \in \mathcal{T}$ and any $x \in A$ is finite. Let $v \lhd B$. Let

$$U = \{(x_0, b) \mid \exists (T, x) \in \mathcal{E}\ x_0 \subseteq x \text{ and } (x_0, b) \in T \text{ and } b \in v\}\ .$$

It is clear that U is non-empty and finite. Let $T \in \mathcal{T}$. Let $x \in A$ be such that $(T, x) \in \mathcal{E}$. Let $b \in v$ be such that $b \in f^T(x)$. Let $x_0 \subseteq x$ be such that $(x_0, b) \in T$. We have $(x_0, b) \in U$, and thus $U \lhd \mathcal{T}$. Thus $U \in \Gamma\,(X \to Y)$. By the same kind of reasoning, we can check that $U_1 \sqsubseteq A$ and that $U_2 = v$. So $v \in \Gamma\,(Y)$.

Next we prove that $\bigcap B = f^{\bigcap \mathcal{T}}(\bigcap A)$. Let $b \in \bigcap B$. Let

$$A_0 = \{x_0 \mid \exists (T, x) \in \mathcal{E}\ x_0 \subseteq x \text{ and } (x_0, b) \in T\}\ .$$

Again we can check easily that $A_0 \times \{b\} \lhd \mathcal{T}$ and that $A_0 \sqsubseteq A$. So A_0 is a singleton $\{x_1\}$ and we get $x_1 \subseteq \bigcap A$ and $(x_1, b) \in \bigcap \mathcal{T}$, and we are finished.

Finally, let $\mathcal{T} \subseteq \mathrm{FS}\,(\mathrm{qDC}\,(X), \mathrm{qDC}\,(Y))$ be state coherent in that qDC. Let U be finite, non-empty and such that $U \lhd \mathcal{T}$. We want to prove that $U \in \Gamma\,(X \to Y)$, so assume that $U_1 \in \mathcal{C}\,(X)$ and consider the set

$$\mathcal{E} = \{(T, x_0) \in \mathcal{T} \times U_1 \mid \exists b \in U_2\ (x_0, b) \in T \cap U\}\ .$$

Clearly $\mathcal{E}$ is a pairing of $\mathcal{T}$ and U_1. Let $B = \{f^T(x_0) \mid (T, x_0) \in \mathcal{E}\}$. We know that $B \in \mathcal{C}\,(Y)$. But $U_2 \lhd B$, so $U_2 \in \Gamma\,(Y)$. Suppose furthermore that U_2 is a singleton $\{b\}$. We certainly have $b \in \bigcap B$, and thus $b \in f^{\bigcap \mathcal{T}}(\bigcap U_1)$. So there exists $x_1 \subseteq \bigcap U_1$ such that $(x_1, b) \in \bigcap \mathcal{T}$. If x_0 is an element of U_1, then $(x_0, b) \in U$ (since $U_2 = \{b\}$), and thus there is a $T \in \mathcal{T}$ such that $(x_0, b) \in T$ (since $U \lhd \mathcal{T}$). But we have seen that $(x_1, b) \in T$ and that $x_1 \subseteq x_0$, so $x_1 = x_0$, and thus $\# U_1 = 1$. This achieves the proof of the proposition. ∎

As a corollary, we get:

Proposition 23 *The category* **coKl(!)** *is (equivalent to) a full sub-cartesian-closed category of* **QDC**.

5 A notion of polarity in hypercoherences

This section contains some simple observations about two subcategories of **HCohL**. We feel intuitively that these two classes of objects could be connected to Girard's polarities (cf. [G3]). There remain, however, some mismatches and this intuition could very well be misleading.

Definition 24 *A hypercoherence X is* positive *if $\Gamma(X)$ is hereditary. It is* negative *if $X^\perp$ is positive.*

So a positive hypercoherence can simply be seen as a qualitative domain.

There is a very natural direct characterization of negative hypercoherences:

Proposition 24 *A hypercoherence X is negative iff $\Gamma^*(X)$ is antihereditary, that is, if $u \in \Gamma^*(X)$ and if $v \subseteq^*_{\mathrm{fin}} |X|$ is such that $u \subseteq v$, then $v \in \Gamma^*(X)$.*

The proof is straightforward.

The states of a negative hypercoherence have a very simple structure:

Proposition 25 *If X is a negative hypercoherence, then $\mathrm{qD}(X)$ is a coherence space.*

Proof: Let $u \subseteq |X|$ be such that for all $a, a' \in u$, $\{a, a'\} \in \Gamma(X)$. Let $v \subseteq^*_{\mathrm{fin}} u$. If $\#v = 1$, then $v \in \Gamma(X)$. Suppose $\#v > 1$. Let $a, a' \in v$ be such that $a \neq a'$. Since $a, a' \in u$, we know that $\{a, a'\} \in \Gamma^*(X)$. Since X is antihereditary, and since $\{a, a'\} \subseteq v$, we have $v \in \Gamma^*(X)$. ■

Of course, if X is a negative hypercoherence, it is impossible in general to retrieve X from $\mathrm{qD}(X)$ (in contrast with what happens for positive hypercoherences). This corresponds to the fact that, in that case, the elements of $\mathcal{C}(X)$ are far from being only the bounded elements of $\mathcal{P}^*_{\mathrm{fin}}(\mathrm{qD}(X))$.

Proposition 26 *Let X and Y be hypercoherences.*

- *If X and Y are positive, then so are $X \otimes Y$ and $X \oplus Y$, and $X^\perp$ is negative.*
- *If X and Y are negative, then so are $X \wp Y$ and $X \times Y$, and $X^\perp$ is positive.*

The proof is straightforward.

Observe that it is almost true that, when X is positive, $!X$ is positive too. However, it is false, because any $A \subseteq^*_{\mathrm{fin}} \mathrm{qD}_{\mathrm{fin}}(X)$ such that $\emptyset \in A$ belongs to $\Gamma(!X)$, and for such an A, the set $A \setminus \{\emptyset\}$ can perfectly well not be in $\Gamma(!X)$. When $A \in \Gamma(!X)$ is such that $\emptyset \notin A$, any non-empty subset B of A is in $\Gamma(!X)$

(for X positive). Here we have an important mismatch between this notion of polarity and Girard's; in his framework, $!A$ is always positive, even when A is not.

This remark also suggests that the "of course" connective could be decomposed in two operations: one operation corresponding to contraction, and the other one to weakening. We actually have, up to a canonical isomorphism:

$$!X = 1 \times C(X)$$

where $C(X)$ is the hypercoherence having $\mathrm{qD_{fin}}(X) \setminus \{\emptyset\}$ as web, and the restriction of $\mathcal{C}(X)$ to this web as coherence. This decomposition is motivated by the fact that the operation $X \mapsto C(X)$ maps positive hypercoherences to positive hypercoherences, whereas the operation $X \mapsto 1 \times X$ maps negative hypercoherences to negative hypercoherences.

Definition 25 *The full subcategory of* **HCohL** *whose objects are the positive (respectively negative) hypercoherences is denoted by* **HCohL**$^+$ *(respectively* **HCohL**$^-$*).*

Now we define two coercions.

Definition 26 *Let X be a hypercoherence. Its associated positive hypercoherence is X^+ defined by $|X^+| = |X|$ and*

$$\Gamma\left(X^+\right) = \{u \in \Gamma(X) \mid \forall v \subseteq^*_{\mathrm{fin}} u \quad v \in \Gamma(X)\}\ .$$

Its associated negative hypercoherence is $X^- = ((X^\perp)^+)^\perp$.

Clearly, X^+ is positive and X^- negative. By definition of our polarities, X is positive (respectively negative) iff it is equal to X^+ (respectively X^-).

Proposition 27 *If X is a hypercoherence, then*

$$\Gamma^*\left(X^-\right) = \{u \subseteq^*_{\mathrm{fin}} |X| \mid \exists v \subseteq u \quad v \in \Gamma^*(X)\}\ .$$

The proof is a straightforward verification.

So the operation $X \mapsto X^+$ appears as a restriction of $\Gamma(X)$, whereas, dually, the operation $X \mapsto X^-$ is a completion of $\Gamma(X)$.

Now we prove that the negative and positive coercions are functors that act trivially on morphisms.

Proposition 28 *Let X and Y be hypercoherences. If $t : X \multimap Y$, then $t : X^+ \multimap Y^+$ and $t : X^- \multimap Y^-$.*

Proof: Let $t : X \multimap Y$ and let $w \subseteq^*_{\mathrm{fin}} t$ be such that $w_1 \in \Gamma(X^+)$. If $\# w_2 = 1$, then $\# w_1 = 1$ because $w \subseteq^*_{\mathrm{fin}} t \in \mathrm{qD}(X \multimap Y)$. Let us prove that $w_2 \in \Gamma(Y^+)$. Let $v \subseteq^*_{\mathrm{fin}} w_2$ and let $w' = \{(a,b) \in w \mid b \in v\}$. Then $w'_1 \subseteq^*_{\mathrm{fin}} w_1$, and thus $w'_1 \in \Gamma(X)$, and thus, since $w' \subseteq^*_{\mathrm{fin}} t$, we have $w'_2 \in \Gamma(Y)$, that is $v \in \Gamma(Y)$. Since this holds for any $v \subseteq^*_{\mathrm{fin}} w_2$, we have $w_2 \in \Gamma(Y^+)$.

To prove that $t : X^- \multimap Y^-$, observe that ${}^t t : Y^\perp \multimap X^\perp$, thus ${}^t t : (Y^\perp)^+ \multimap (X^\perp)^+$, thus ${}^t({}^t t) : X^- \multimap Y^-$, that is $t : X^- \multimap Y^-$. ∎

We shall denote by **P** the functor $\mathbf{HCohL} \to \mathbf{HCohL}^+$ that maps X to X^+ and $t : X \multimap Y$ to $t : X^+ \multimap Y^+$, and by **N** the functor $\mathbf{HCohL} \to \mathbf{HCohL}^-$ that maps X to X^- and $t : X \multimap Y$ to $t : X^- \multimap Y^-$.

Now we prove that these functors have a universal property. Let $\mathbf{I}^+ : \mathbf{HCohL}^+ \to \mathbf{HCohL}$ and $\mathbf{I}^- : \mathbf{HCohL}^- \to \mathbf{HCohL}$ denote the inclusion functors.

Proposition 29 *The functor* **P** *is right adjoint to* $\mathbf{I}^+$ *and the functor* **N** *is left adjoint to* $\mathbf{I}^-$.

Proof: Let X be a positive hypercoherence and let Y be a hypercoherence. If $t : X \multimap Y$, we know that $t : X^+ \multimap Y^+$, that is, since X is positive, $t : X \multimap Y^+$. Conversely, if $t : X \multimap Y^+$, we have $t : X \multimap Y$ simply because $\Gamma(Y^+) \subseteq \Gamma(Y)$. So we have $\mathrm{qD}(X \multimap Y) = \mathrm{qD}(X \multimap Y^+)$, and the first adjunction holds (in a very strong sense).

Now let X be a hypercoherence and let Y be a negative hypercoherence. We have

$$\begin{array}{lll} t : X \multimap Y & \text{iff} & {}^t t : Y^\perp \multimap X^\perp \\ & \text{iff} & {}^t t : Y^\perp \multimap (X^\perp)^+ \quad \text{since } Y \text{ is negative} \\ & \text{iff} & t : X^- \multimap Y \end{array}$$

and we conclude that $\mathrm{qD}(X \multimap Y) = \mathrm{qD}(X^- \multimap Y)$, and the second adjunction holds. ∎

Observe that $\mathbf{HCohL}^+$ is equivalent to the category of qualitative domains and linear stable functions, and that $\mathbf{HCohL}^-$ is equivalent to $(\mathbf{HCohL}^+)^{\mathrm{op}}$, this equivalence being defined by the functor $(\_)^\perp$, which acts on morphisms by transposition.

6 A connection with the stable model of linear logic

We use **CS** to denote the category of coherence spaces and linear stable functions, which is the well known model of linear logic discovered by Girard (see [G2]).

Consider the functor $\mathbf{PN} = \mathbf{P} \circ \mathbf{N} : \mathbf{HCohL} \to \mathbf{HCohL}^+$. By our previous observations, we can consider this functor as having **CS** as codomain. Actually, if X is a hypercoherence, $\mathbf{PN}(X)$ is the hereditary hypercoherence whose web is $|X|$ and such that $u \in \Gamma(\mathbf{PN}(X))$ iff $u \subseteq^*_{\mathrm{fin}} |X|$ and for all $a, a' \in u$, $\{a, a'\} \in \Gamma(X)$. (See proposition 25.) So $\mathbf{PN}(X)$ can be viewed as the coherence space defined by: $\{a, a'\} \in \mathbf{PN}(X)$ iff $\{a, a'\} \in \Gamma(X)$. Furthermore, if $t : X \multimap Y$ is a linear morphism in **HCohL**, then $\mathbf{PN}(t) = t$.

Now we consider **CS** as a model of linear logic, with linear connectives interpreted as specified in [G2].

Proposition 30 *The functor* **PN** *preserves all linear connectives except the exponentials. More precisely, if X and Y are hypercoherences, then* $\mathbf{PN}(X^\perp)$ *is the orthogonal of the coherence space* $\mathbf{PN}(X)$, $\mathbf{PN}(X \otimes Y)$ *is the tensor product of the coherence spaces* $\mathbf{PN}(X)$ *and* $\mathbf{PN}(Y)$, *and so on.*

Furthermore, there is a natural rigid embedding from $\mathbf{PN}(!X)$ *into* $!\mathbf{PN}(X)$ *and from* $\mathbf{PN}(?X)$ *into* $?\mathbf{PN}(X)$

Proof: Let $\{a, a'\} \in \mathbf{PN}(X^\perp)$. This means that $\{a, a'\} \in \Gamma\left(X^\perp\right)$, that is $a = a'$ or $\{a, a'\} \notin \Gamma(X)$, but this exactly means that $\{a, a'\} \in \mathbf{PN}(X)^\perp$.

Let $\{(a, b), (a', b')\} \in \mathbf{PN}(X \otimes Y)$. This means that $\{a, a'\} \in \Gamma(X)$ and $\{b, b'\} \in \Gamma(Y)$ (because any coherent set with two elements or less is hereditary), that is $\{a, a'\} \in \mathbf{PN}(X)$ and $\{b, b'\} \in \mathbf{PN}(Y)$, that is $\{(a, b), (a', b')\} \in \mathbf{PN}(X) \otimes \mathbf{PN}(Y)$.

Let $\{(c, i), (c', j)\} \in \mathbf{PN}(X \oplus Y)$. This means that $i = j$ and that, if $i = 1$, then $\{c, c'\} \in \Gamma(X)$, and similarly for Y if $j = 2$. So we conclude. The fact that the inclusion $|\mathbf{PN}(!X)| \subseteq |!\mathbf{PN}(X)|$ defines a rigid embedding is a corollary of propositions 14.

For the remainder of the connectives, simply use the De Morgan laws. ∎

It is easy to check that **PN** is right adjoint to the inclusion functor $\mathbf{I}^+$: $\mathbf{CS} \to \mathbf{HCohL}$.

References

[B1] G. Berry. *Modèles des lambda-calculs typés.* Thèse de Doctorat, Université Paris 7, 1979.

[B2] G. Berry. *Stable Models of Typed Lambda-calculi.* Proc. ICALP 1978, Springer-Verlag LNCS **62**, 1978.

[BC] G. Berry and P.-L. Curien. *Sequential algorithms on concrete data structures.* Theor. Comp. Sci. **20**, 1982.

[BCL] G. Berry, P.-L. Curien, J.-J. Levy. *Full abstraction for sequential languages: the state of the art.* French-US Seminar on the Applications of Algebra to Language Definition and Compilation, Fontainebleau (June 1982), Cambridge University Press, 1985.

[BE] A. Bucciarelli, T. Ehrhard. *Sequentiality and strong stability.* Proc. LICS 1991.

[C] P.-L. Curien. *Categorical Combinators, Sequential Algorithms and Functional Programming.* Progress in Theoretical Computer Science, Birkhäuser, 1993.

[G1] J.-Y. Girard. *The system F of variable types fifteen years later.* Theor. Comp. Sci **45**, 1986.

[GLT] J.-Y. Girard, Y. Lafont, P. Taylor. *Proofs and Types.* vol. 7 of Cambridge Tracts in Theoretical Computer Science, Cambridge University Press, 1991.

[G2] J.-Y. Girard. *Linear Logic.* Theor. Comp. Sci. **50**, 1988.

[G3] J.-Y. Girard. *A new constructive logic: classical logic.* Math. Str. in Comp. Sci. vol. 1.3, 1991.

[KP] G. Kahn and G. Plotkin. *Domaines Concrets.* Rapport IRIA-LABORIA 336, 1978.

[M] S. MacLane. *Categories for the Working Mathematician.* Springer-Verlag Graduate Texts in Mathematics vol. 5, 1971.

[P] G. Plotkin. *LCF considered as a programming language.* Theor. Comp. Sci. **5**, 1977.

Deciding Provability of Linear Logic Formulas

Patrick Lincoln*

December 15, 1994

A great deal of insight about a logic can be derived from the study of the difficulty of deciding if a given formula is provable in that logic. Most first order logics are undecidable and are linear time decidable with no quantifiers and no propositional symbols. However, logics differ greatly in the complexity of deciding propositional formulas. For example, first-order classical logic is undecidable, propositional classical logic is NP-complete, and constant-only classical logic is decidable in linear time. Intuitionistic logic shares the same complexity characterization as classical logic except at the propositional level, where intuitionistic logic is PSPACE-complete. In this survey we review the available results characterizing various fragments of linear logic. Surprises include the fact that both propositional and constant-only linear logic are undecidable. The results of these studies can be used to guide further proof-theoretic exploration, the study of semantics, and the construction of theorem provers and logic programming languages.

1 Introduction

There are many interesting fragments of linear logic worthy of study in their own right, most described by the connectives which they employ. Full linear logic includes all the logical connectives, which come in three dual pairs: the exponentials ! and ?, the additives & and $\oplus$, and the multiplicatives $\otimes$ and $\parr$. For the most part we will consider fragments of linear logic built up using these connectives in any combination. For example, full linear logic formulas may

*SRI International Computer Science Laboratory, Menlo Park CA 94025 USA. Work supported under NSF Grant CCR-9224858. lincoln@csl.sri.com http://www.csl.sri.com/lincoln/lincoln.html

employ any connective, while multiplicative linear logic (MLL) formulas contain only the multiplicative connectives $\otimes$ and $\parr$,and multiplicative-additive linear logic (MALL) formulas contain only the multiplicative and additive connectives $\otimes$, $\parr$, &, and $\oplus$. In some cases it is easier to read a formula such as $A \parr B$ using $A^{\perp} \multimap B$ (which may be read right-to-left as the definition of the connective $\multimap$). Using the connective $\multimap$ one can define other more specific fragments such as the Horn fragment of MLL [25], but these results will be largely omitted here.

In order to gain an intuition about provability, we will usually be speaking informally of a computational process searching for a proof of a formula from the bottom up in a sequent-calculus. Thus given a conclusion sequent, we attempt to find its proof by trying each possible instance of each sequent proof rule. This point of view directly corresponds to the computational model of logic programming. Reading the sequent rules bottom-up can then lead to insights about the meanings of those rules. For example, the contraction rule can be seen as copying the principle formula, and the weakening rule can be seen as throwing it away. Not all complexity results directly flow from this viewpoint, but it is a useful starting point.

Linear logic has a great control over resources, through the elimination of weakening and contraction, and the explicit addition of a reusable (modal) operator. As will be surveyed below, the combination of these features yields a great deal of expressive power.

2 Propositional Linear Logic

The propositional fragment is considered first, since these results are central to the results for first order and constant-only logics.

2.1 Full Propositional Linear Logic

Although propositional linear logic was known to be very expressive, it was thought to be decidable for some time before a proof of undecidability surfaced [32, 31]. Briefly, the proof of undecidability goes by encoding an undecidable halting problem. A proof, read bottom up, directly corresponds to a computation. The proof of the undecidability of full linear logic proceeds by reduction of a form of alternating counter machine to propositional linear logic. An and-branching two-counter machine (ACM) is a nondeterministic machine with a finite set of states. A configuration is a triple $\langle Q_i, A, B\rangle$, where Q_i is a state, and A and B are natural numbers, the values of two counters. An ACM has a finite set of instructions of five kinds: **Increment-A**, **Increment-B**, **Decrement-A**, **Decrement-B**, and **Fork**. The **Increment** and **Decre-**

ment instructions operate as they do in standard counter machines [39]. The **Fork** instruction causes a machine to split into two independent machines: from state $\langle Q_i, A, B\rangle$ a machine taking the transition Q_i**Fork**Q_j, Q_k results in two machines, $\langle Q_j, A, B\rangle$ and $\langle Q_k, A, B\rangle$. Thus an instantaneous description is a **set of** machine configurations, which is accepting only if all machine configurations are in the final state, and all counters are zero. ACM's are essentially alternating Petri nets, and have an undecidable halting problem. It is convenient to use ACM's as opposed to standard counter machines to show undecidability, since zero-test has no natural counterpart in linear logic, but there is a natural counterpart of **Fork**: the additive conjunction &. The remaining ACM instructions may be encoded using techniques very similar to the well-studied Petri net reachability encodings [8, 18, 38, 9, 16]. The full proof of undecidability is presented in [32].

2.2 Propositional Multiplicative-Additive Linear Logic

The multiplicative-additive fragment of linear logic (MALL) excludes the reusable modals !, ?. Thus, every formula is "used" at most once in any branch of any cut-free MALL proof. Also, in every non-cut MALL rule, each hypothesis sequent has a smaller number of symbols than the conclusion sequent. This provides an immediate linear bound on the depth of cut-free MALL proofs. Since MALL enjoys a cut-elimination property [17], there is a nondeterministic PSPACE algorithm to decide MALL sequents based on simply guessing and checking the proof, recoding only the branch of the sequent proof from the root to the current point.

To show that MALL is PSPACE-Hard, one can encode classical quantified boolean formulas (QBF). For simplicity one may assume that a QBF is presented in prenex form. The quantifier-free formula may be encoded using truth tables, but the quantifiers present some difficulty. One may encode quantifiers using the additives: $\forall x$ as $(x \& x^{\perp})$, and $\exists x$ as $(x \oplus x^{\perp})$. This encoding has incorrect behavior in that it does not respect quantifier order, but using multiplicative connectives as "locks and keys" one can enforce an ordering upon the encoding of quantifiers to achieve soundness and completeness. The full proof of PSPACE-completeness is presented in [32].

2.3 Propositional Multiplicative Linear Logic

The multiplicative fragment of linear logic contains only the connectives $\otimes$ and $⅋$ (or equivalently $\otimes$ and $\multimap$), a set of propositions, and the constants 1 and $\perp$. The decision problem for this fragment is in NP, since an entire cut-free multiplicative proof may be guessed and checked in polynomial time (note that every connective is analyzed exactly once in any cut-free MLL proof). The decision problem is NP-hard by reduction from 3-Partition, a problem which requires a perfect partitioning of groups of objects in much the same way that linear logic requires a complete accounting of propositions [23, 24, 25]. The proof of correctness of the encoding makes heavy use of the 'balanced' property of MLL, which states that if a formula is provable in MLL, then the number of positive and negative occurrences of each literal are equal. This property can be used as a necessary condition to provability in MLL theorem provers or logic programming systems.

3 Constant-Only Linear Logic

3.1 Constant-Only Multiplicative Linear Logic

Some time ago, Girard developed a necessary condition for the provability of COMLL expressions based on the following definition a function M from constant multiplicative linear expressions to the integers:

$$\begin{aligned} M(1) &= 1 \\ M(\bot) &= 0 \\ M(A \parr B) &= M(A) + M(B) \\ M(A \otimes B) &= M(A) + M(B) - 1 \end{aligned}$$

Girard showed that if a constant-only MLL formula A is provable then $M(A) = 1$. There was a question about whether some similar measure might be used on constant-only MLL formulas that would be necessary and sufficient for provability. It turns out that there is no efficiently computable measure function on this class of formulas, as shown by an encoding of 3-Partition in constant-only MLL [36]. This work points out that the multiplicative constants 1 and $\bot$ have very 'propositional' behavior. The bottom line is that even for constant-only expressions of MLL deciding provability is NP-complete. This result has had an impact on the study of proof nets. Later results have generalized this result by providing general translations from arbitrary balanced MLL propositional formulas to constant-only MLL formulas [22, 26]. Together with the NP-completeness of propositional MLL, these translations provide an alternate proof of the NP-completeness of constant-only MLL.

3.2 Constant-Only Full Linear Logic

Amazingly, constant-only full linear logic is just as difficult to decide as propositional full linear logic [26]. Extending the work mentioned above, it is possible to translate any full propositional LL formula into a constant-only formula preserving provability using enumerations of constant-only formulas. Since propositional linear logic is undecidable, so then is constant-only propositional linear logic. This is remarkable, since the building blocks of expressions are so elementary. In fact, the encodings can be tuned to produce very restricted formulas containing multiple copies of only one constant (either 1 or $\perp$).

3.3 Constant-Only Multiplicative-Additive Linear Logic

The encodings mentioned above can be seen to produce only polynomial growth in the size of formulas. Thus directly translating a class of PSPACE-hard propositional MALL formulas into constant-only MALL immediately produces the result that constant-only MALL is PSPACE-complete [26].

4 First Order Linear Logic

4.1 Full First Order Linear Logic

Girard's translation of first-order classical logic into first order linear logic [17] demonstrates that first order linear logic is undecidable. One could imagine coding up a Turing machine where the instructions are exponential formulas containing implications from one state to another, and the current state of the machine are represented using first order term structure. The conclusion sequent would contain these instructions, an initial state, and a final state. The exponential nature of instructions allows them to be copied and reused arbitrarily often. The quantifier rules allow the instructions to be instantiated to the current state. Thus a Turing machine computation could be read from any cut-free proof of this conclusion sequent bottom up, the intermediate states appearing directly in the sequents all along the way.

4.2 First Order Multiplicative-Additive Linear Logic

Without the exponentials, first order MALL is decidable [35]. Intuitively, this stems from the lack of the ability to copy the instructions for reuse. However, there is no readily apparent decision procedure for this fragment since the quantifier rules allow sequents of arbitrary size to appear even in cut-free proofs. The technique showing decidability sketched here [35] provides a tight complexity bound for first order MALL and MLL.

4.2.1 Deciding first order MALL

The key problem to deciding first order MALL is the lack of control over the existential rule. Reading the rule bottom up we have no idea how to guess or bound the size of the instantiating term. However, this is a false unboundedness. In classical logic, one can apply Skolemization to remove quantifiers altogether, with changes to the proof rule for identity to require unification. If we could obtain a similar result for linear logic, we could then obtain an immediate bound on the size of instantiations of terms, and thus a bound on the size of the entire sequent proof. Unfortunately, Skolemization is unsound in linear logic, as the following example demonstrates:

$$\vdash (\exists x.p^{\perp} ⅋ q^{\perp}(x)), (\forall y.q(y)) \otimes p$$

This formula is unprovable in first order linear logic, but when Skolemized it becomes $\vdash p^{\perp} ⅋ q^{\perp}(v), q(c) \otimes p$ which unfortunately is provable in linear logic augmented with unification ($v \leftarrow c$).

One can view Skolemization in classical logic as the combination of three techniques: converting the formula to prenex form, permuting the use of quantifier rules below propositional inferences, and changing the quantifier rules to instantiate quantifiers with specific (bounded) terms. Decidability depends only on the last, which is fortunate since neither of the other techniques apply to linear logic in their full generality. In [35, 13] proof systems are developed where the quantifier rules are converted into a form without unbounded guessing. The resulting system generates cut-free proofs of at most exponential size for first-order MALL formulas. It is possible to immediately generate a standard first-order linear logic proof from a proof in this modified system. Thus this fragment can be decided in NEXPTIME by guessing and checking the entire proof in the modified system.

4.2.2 Hardness of first-order MALL

As shown above, first-order MALL is decidable, and at most NEXPTIME-hard. By the propositional result sketched above, it was known to be at least PSPACE-hard. The gap was closed by developing a direct encoding of nondeterministic exponential time Turing machines [33]. This encoding is reminiscent of the standard proof of the PSPACE-hardness of quantified boolean formula validity [43, 21], and is related to the logic programming simulation of Turing machines given in [42]. This encoding in first order MALL formulas is somewhat unique in that the computation is read 'across the top' of a completed cut-free proof, rather than 'bottom up', which is utilized in most of the above-described results. The result is that Turing machine instructions are not copied as one moves up the proof tree, but instead are shared (additively) between branches. This gives an immediate exponential time limit to the machine, since the propositional structure of first-order MALL gives rise to at most a single exponential number of leaves of the proof tree.

4.3 First Order Multiplicative Linear Logic

The same proof system used to show the decidability of first order MALL [35] can be used to show that first order MLL is decidable. In fact, this procedure generates first order MLL proofs that are at most polynomial size. Thus one can guess and check an entire first order MLL proof in polynomial time. In other words, first order MLL is in NP. The propositional hardness result for the purely propositional case can be used to show that first order MLL is NP-hard, and thus NP-complete.

5 Other Fragments

There are many related problems of interest. A few representative 'nice' fragments and some other interesting cases are sketched here.

5.1 Multiplicative-Exponential Linear Logic

The multiplicative-exponential (MELL) fragment is currently of unknown complexity. By Petri net reachability encodings [8, 18, 38, 9, 16], it must be at least EXPSPACE-hard. Although Petri net reachability is decidable, there is no known encoding of MELL formulas in Petri nets. A proof of decidability of MELL may therefore lead to a new proof of the decidability of Petri net reachability, and therefore be of independent interest. More effort has been fruitlessly expended on the decidability of MELL than any other remaining open problems in this area.

5.2 Higher-Order Linear Logic

Amiot has shown that MLL (and MALL) with first and second order quantifiers and appropriate function symbols is undecidable [1, 2].

In recent work, pure second order intuitionistic MLL (IMLL2) has been shown to be undecidable, through the encoding of second order intuitionistic logic [34]. The key point is that it is possible to encode contraction and weakening using second order formulas.

$$C \triangleq \forall X.X \multimap (X \otimes X)$$

$$W \triangleq \forall X.X \multimap 1$$

$$[\Sigma \vdash_{LJ2} \Delta] \triangleq C, C, C, W, [\Sigma] \vdash_{\mathrm{IMLL2}} [\Delta]$$

C encodes contraction and W encodes weakening. A second-order intuitionistic logic (LJ2) sequent can be translated directly into IMLL2, and by adding enough copies of C and W one can preserve provability. By the undecidability of LJ2 shown in [37, 15], IMLL2 is undecidable. This result can be extended to show the undecidability of pure second order IMALL, but has not yet been extended to pure second order MLL. The decision problem for second order MLL where quantifiers cannot be instantiated with quantified formulas is also still open. This latter fragment could correspond to the logic of a polymorphic type system for a programming language.

5.3 Intuitionistic Fragments

For all of the main fragments considered above, the complexity of the decision problem is unchanged when moving from the full two sided sequent calculus to the Intuitionistic version, where the right-hand side is restricted to a single formula (and $\parr$ is replaced by $\multimap$ in the multiplicatives, and negated propositions are disallowed). However, for the case of second order MLL and second order MALL no result is known, although IMLL2 and IMALL2 are known to be undecidable. In some of the more restricted cases the intuitionistic restriction does effect expressiveness [23]

5.4 Others

The $!, \otimes$ propositional fragment, allowing arbitrary two sided sequents of propositional formulas using only $!$ and $\otimes$, has been shown to be decidable and non-trivial (NP-hard) [14]. This fragment is of interest as it relates to full MELL.

The Horn fragment of MLL is NP-complete, and that the purely implicative fragment built from only $\multimap$ and the single constant $\perp$ is NP-complete [26].

Many fragments of linear logic with a single propositional literal (and no constants) match the complexity of the corresponding constant-only fragments, which in turn match the complexity of the propositional fragments with arbitrary numbers of propositional symbols. This has been shown for full linear logic, MALL, MLL, and MELL [26]. For example, full propositional linear logic is undecidable. Therefore, constant-only linear logic is undecidable, as is single-literal propositional linear logic. Of particular interest are the results about MELL, where these results show that the reachability problem for arbitrary Petri nets can be encoded in single-literal MELL. In fact, further, Petri net reachability can be encoded in the fragment containing only $!$, $\multimap$, and a single propositional symbol. It is remarkable that the decidability of this very small fragment is still open.

Independently, Danos and Winkler have both shown that the constant-only additive-exponential linear logic evaluates in linear time. Even if multiplicative constants are allowed this same result holds. In this fragment there are only eight 'values': $\perp$, 1, $?1$, $!\perp$, $\top$, 0, $\perp \oplus 1$, and $\perp \& 1$. It happens that all expressions involving only the multiplicative and additive constants, and the additive and exponential connectives is equivalent, up to provability, of one of these eight formulas. For example, $!\top = 1$, $!?1 = 1$, $!(\perp \oplus 1) = 1$, and $\perp \& (\perp \oplus 1) = \perp$.

A variant of MELL without unrestricted exchange (commutativity), but with the additional property that exponentials can commute (and thus exponentials enjoy all the structural rules, while other formulas exhibit none) has been studied [47, 32]. It has been found to be undecidable by encoding Turing machine tapes directly in the sequent [32]. Since the sequent comma is not commutative, the entire state of the tape including the current state of the machine and position of the read head is immediately apparent from a sequent encoding. Instructions are encoded as exponential formulas that are copied and then commute to their location of application, where they are applied to change the state of the tape. However, noncommutative variants of linear logic have some problematic aspects, and there are some seemingly arbitrary choices to be made, so these fragments are somewhat speculative.

Variants of linear logic with unrestricted weakening (sometimes called Affine Logic) have also been studied [32, 7]. Here again the logics are somewhat speculative, although there is a close relationship with direct logic [27, 11]. Some fragments of linear logic with weakening have the same complexity as the same fragment without weakening. For example, just as for linear logic, full first-order affine logic is undecidable, as can be seen by the fact that Girard's encoding of classical logic into linear logic [17] is also sound and complete as a translation into affine logic. Some fragments of affine logic are easier to decide than their linear counterpart. For example, propositional affine logic is decidable [28], where propositional linear logic is not [32]. Finally, some fragments of affine logic are harder to decide than their linear counterpart. For example, the extended Horn fragment +HL is (PSPACE-complete) in affine logic, but (NP-complete) in linear logic [23].

Finally, variants of linear logic with unrestricted contraction are very similar to relevance logic [3]. Urquhart has shown that some propositional variants are undecidable, and has studied other fragments [45, 46]. However, the results regarding relevance logic are very different in character than those described above, since they rely in an essential way on a distributivity that appears in relevance logic but does not appear in linear logic.

6 Conclusions

This survey has sketched the basic approaches used in the study of the complexity of deciding linear logic formulas. This area has led to some new understanding of the fragments in question, and has pointed out some gaps in current understanding. Of the remaining open problems, perhaps the decidability of propositional MELL and the decidability of pure second-order MALL are of the most interest.

There are some surprisingly rich fragments of linear logic, and surprisingly few differences between the complexity of many fragments at the first order, propositional, and constant-only levels. For example, even constant-only full linear logic is undecidable (as are the first-order and propositional fragments), and first-order MLL, propositional MLL, and constant-only MLL are all NP-complete. However, MALL is PSPACE-complete at the constant-only and propositional levels, but is NEXPTIME-complete at the first-order level.

This area of study is directly relevant to the logic-programming use of linear logic, where linear logic sequents are taken to be logic programs which execute by performing proof search [20, 5, 4, 19]. This area of research is also directly relevant to the construction of linear logic theorem provers [40, 44, 35, 6, 11, 10, 41]. The results here also lead into the study of semantics of linear logic, pointing to deep connections between various fragments of linear logic and familiar structures from computer science [12, 29, 30]. In particular, work has progressed in attempting to find viewpoints where the proof theory of linear logic can be viewed as a machine. For example, Kanovich's results derive from his view of fragments of linear logic as acyclic programs with stack. Turing machines can be seen as described above in special first-order encodings, but various counter or Minsky machines can be seen as somewhat more direct interpretation of the propositional fragments.

Finally, readers should not interpret the above results negatively: the fact that linear logic is expressive is an important feature. Classical logic is degenerate in its small number of well-behaved fragments of different complexity. Linear logic's rich structure simply provides more detail than many other logics. This detail negates the possibility of simple decision procedures, but can carry important information regarding computational content, where other logics record only simple binary results. That is, linear logic is not about "Truth"; it is about computation.

References

[1] Amiot. Decision Problems for Second Order Linear Logic Without Exponentials. Draft, 1990.

[2] G. Amiot. *Unification et logique du second ordre.* PhD thesis, Université Paris 7 (Denis Diderot), 1994.

[3] Anderson, A.R. and N.D. Belnap, Jr. *Entailment. Volume 1.* Princeton University Press, Princeton, New Jersey, 1975.

[4] J.-M. Andreoli. Logic programming with focusing proofs in linear logic. *Journal of Logic and Computation*, 1992.

[5] J.-M. Andreoli and R. Pareschi. Logic programming with sequent systems: a linear logic approach. In *Proc. Workshop on Extensions of Logic Programming, Tuebingen.* Lecture Notes in Artificial Intelligence, Springer-Verlag, Berlin, 1990.

[6] J.-M. Andreoli and R. Pareschi. Linear objects: Logical processes with built-in inheritance. *New Generation Computing*, 9, 1991.

[7] S. Artemov and A. Kopylov. Full propositional affine logic is decidable. Email Message 260 to Linear Mailing List, June 1994.

[8] A. Asperti. A logic for concurrency. Technical report, Dipartimento di Informatica, Universitá di Pisa, 1987.

[9] A. Asperti, G.-L. Ferrari, and R. Gorrieri. Implicative formulae in the 'proofs as computations' analogy. In *Proc. 17-th ACM Symp. on Principles of Programming Languages, San Francisco*, pages 59–71, January 1990.

[10] G. Bellin. *Mechanizing Proof Theory: Resource-Aware Logics and Proof-Transformations to Extract Implicit Information.* PhD thesis, Stanford University, 1990.

[11] G. Bellin and J. Ketonen. A decision procedure revisited: Notes on direct logic, linear logic, and its implementation. *Theoretical Computer Science*, 95:115–142, 1992.

[12] A. Blass. A game semantics for linear logic. *Annals Pure Appl. Logic*, 56:183–220, 1992. Special Volume dedicated to the memory of John Myhill.

[13] S. Cerrito. Herbrand methods in linear logic. Unpublished Draft, 1992.

[14] J. Chirimar and J. Lipton. Provability in TBLL: A Decision Procedure. In *Computer Science Logic '91. Lecture Notes in Computer Science, Springer*, 1992.

[15] D.M. Gabbay. *Semantical investigations in Heyting's intuitionistic logic.* Reidel, Dordrecht, 1981.

[16] V. Gehlot and C.A. Gunter. Normal process representatives. In *Proc. 5-th IEEE Symp. on Logic in Computer Science, Philadelphia*, June 1990.

[17] J.-Y. Girard. Linear logic. *Theoretical Computer Science*, 50:1–102, 1987.

[18] C.A. Gunter and V. Gehlot. Nets as tensor theories. In G. De Michelis, editor, *Proc. 10-th International Conference on Application and Theory of Petri Nets, Bonn*, pages 174–191, 1989.

[19] J. Harland and D. Pym. The uniform proof-theoretic foundation of linear logic programming. Technical Report ECS-LFCS-90-124, Laboratory for the Foundations of Computer Science, University of Edinburgh, November 1990.

[20] J.S. Hodas and D. Miller. Logic programming in a fragment of intuitionistic linear logic. In *Proc. 6-th Annual IEEE Symposium on Logic in Computer Science, Amsterdam*, pages 32–42. IEEE Computer Society Press, Los Alamitos, California, July 1991. Full paper to appear in *Information and Computation.* Draft available using anonymous ftp from host ftp.cis.upenn.edu and the file pub/papers/miller/ic92.dvi.Z.

[21] J. Hopcroft and J. Ullman. *Introduction to Automata Theory, Languages and Computation.* Addison-Wesley Publishing Company, 1979.

[22] H. Jervell. Eliminating variables in balanced Horn sequents. Draft.

[23] M. Kanovich. The multiplicative fragment of linear logic is NP-complete. Email Message, 1991.

[24] M. Kanovich. The multiplicative fragment of linear logic is NP-complete. Technical Report X-91-13, Institute for Language, Logic, and Information, June 1991.

[25] M. Kanovich. Horn programming in linear logic is NP-complete. In *Proc. 7-th Annual IEEE Symposium on Logic in Computer Science, Santa Cruz, California*, pages 200–210. IEEE Computer Society Press, Los Alamitos, California, June 1992.

[26] M. Kanovich. Simulating linear logic in 1-only linear logic. Technical Report 94-02, CNRS, Laboratoire de Mathematiques Discretes, January 1994.

[27] J. Ketonen and R. Weyhrauch. A decidable fragment of predicate calculus. *Theoretical Computer Science*, 32:297–307, 1984.

[28] A. Kopylov. Decidability of linear affine logic. Technical Report 94-32, CNRS, Laboratoire de Mathématiques Discrètes, October 1994. Available by ftp from `lmd.univ-mrs.fr:/pub/kopylov/fin.dvi`.

[29] Y. Lafont and T. Streicher. Game semantics for linear logic. In *Proc. 6-th Annual IEEE Symposium on Logic in Computer Science, Amsterdam*, pages 43–50. IEEE Computer Society Press, Los Alamitos, California, July 1991.

[30] P. Lincoln, J. Mitchell, and A. Scedrov. Stochastic interaction and linear logic. In This Volume. Cambridge University Press London Mathematical Society Lecture Notes Series.

[31] P. Lincoln, J. Mitchell, A. Scedrov, and N. Shankar. Decision problems for propositional linear logic. In *Proc. 31st IEEE Symp. on Foundations of Computer Science*, pages 662–671, 1990.

[32] P. Lincoln, J. Mitchell, A. Scedrov, and N. Shankar. Decision problems for propositional linear logic. *Annals Pure Appl. Logic*, 56:239–311, 1992. Special Volume dedicated to the memory of John Myhill.

[33] P. Lincoln and A. Scedrov. First order linear logic without modalities is NEXPTIME-hard. *Theoretical Computer Science*, 135(1):139–154, December 1994. Available using anonymous ftp from host ftp.cis.upenn.edu and the file pub/papers/scedrov/mall1.dvi.

[34] P. Lincoln, A. Scedrov, and N. Shankar. Decision problems for second order linear logic. Email Message 278 to Linear Mailing List, October 1994.

[35] P. Lincoln and N. Shankar. Proof search in first-order linear logic and other cut-free sequent calculi. In *Proc. 9-th IEEE Symp. on Logic in Computer Science,Paris*, July 1994.

[36] P. Lincoln and T. Winkler. Constant-Only Multiplicative Linear Logic is NP-Complete. *Theoretical Computer Science*, 135(1):155–169, December 1994.

[37] M.H. Loeb. Embedding first-order predicate logic in fragments of intuitionistic logic. *Journal of Symbolic Logic*, 41:705–718, 1976.

[38] N. Martí-Oliet and J. Meseguer. From Petri nets to linear logic. In: Springer LNCS 389, ed. by D.H. Pitt et al., 1989. 313-340.

[39] M. Minsky. Recursive unsolvability of post's problem of 'tag' and other topics in the theory of turing machines. *Annals of Mathematics*, 74:3:437–455, 1961.

[40] G. Mints. Resolution Calculus for the First Order Linear Logic. *Journal of Logic, Language and Information*, 2:59–83, 1993.

[41] H. Schellinx. Some syntactical observations on linear logic. *Journal of Logic and Computation*, 1:537–559, 1991.

[42] E.Y. Shapiro. Alternation and the computational complexity of logic programs. *Journal of Logic Programming*, 1:19–33, 1984.

[43] L. Stockmeyer. Classifying the computational complexity of problems. *Journal of Symbolic Logic*, 52:1–43, 1987.

[44] T. Tammet. Proof search strategies in linear logic. Programming Methodology Group Report 70, Chalmers University, 1993.

[45] A. Urquhart. The undecidability of entailment and relevant implication. *Journal of Symbolic Logic*, 49:1059–1073, 1984.

[46] A. Urquhart. The complexity of decision procedures in relevance logic. Technical Report 217/89, Department of Computer Science, University of Toronto, 1989.

[47] D.N. Yetter. Quantales and (noncommutative) linear logic. *Journal of Symbolic Logic*, 55:41–64, 1990.

The Direct Simulation of Minsky Machines in Linear Logic

Max I. Kanovich

Russian Humanities State University, Moscow,
and
CNRS, Laboratoire de Mathématiques Discrètes, Marseille
maxk@lmd.univ-mrs.fr

Abstract

Linear Logic was introduced by Girard [3] as a *resource-sensitive* refinement of classical logic. Lincoln, Mitchell, Scedrov, and Shankar [13] have proved the undecidability of full propositional Linear Logic. This implies that Linear Logic is more expressive than *traditional* classical or intuitionistic logic, even if we consider the modalized versions of those logics. In [9, 10] we prove that *standard* many-counter Minsky machines [17] can be simulated directly in propositional Linear Logic. Here we are going to present a more *transparent* and *fruitful* simulation of many-counter Minsky machines in Linear Logic.

Simulating one system of concepts in terms of another system is known to consist of two procedures: (A) Suggesting an *encoding* of the first system in terms of the second one, and (B) Proving that the encoding suggested is *correct* and *fair*.

Here, based on a computational interpretation of Linear Logic [9, 10], we present: (A) A direct and natural *encoding* of many-counter Minsky machines in Linear Logic, and (B) Transparent proof of the *correctness* and *fairness* of this *encoding*.

As a corollary, we prove that **all** *partial recursive relations* are **directly definable** in propositional Linear Logic.

1 Introduction and Summary

Linear Logic was introduced by Girard [3] as a *resource-sensitive* refinement of classical logic. Lincoln, Mitchell, Scedrov, and Shankar [13] have proved the undecidability of full propositional Linear Logic. In [13] the proof of undecidability of propositional Linear Logic consists of a reduction from the *Halting Problem* for *And-Branching Two Counter Machines Without Zero-Test* (specified in the same [13]) to a decision problem in Linear Logic. In [9, 10] we prove that the *Halting Problem* for many-counter *standard* Minsky machines [17] can be simulated directly in propositional Linear Logic (see [9, 10]). Here we are going to present a more *transparent* and *fruitful* simulation of many-counter Minsky machines in Linear Logic.

Simulating one system of concepts in terms of another system is known to consist of two procedures:

(A) Suggesting an *encoding* of the first system in terms of the second one, and

(B) Proving that the encoding suggested is *correct* and *fair*.

When simulating many-counter Minsky machines in Linear Logic in [9, 10], we sacrificed the simplicity of our *encoding* to achieve a simpler proof of its *fairness*.

Here, based on a computational interpretation of Linear Logic [9, 10], we present

(A) A direct and natural *encoding* of many-counter Minsky machines in Linear Logic, and

(B) A transparent proof of the *correctness* and *fairness* of this *encoding*.

Moreover, according to Theorem 5.1, we can simulate directly *computations* of *standard* n-counter Minsky machines M on inputs

$$k_1,\ k_2,\ \ldots,\ k_n$$

with the help of Linear Logic sequents of the so-called *(!,⊕)-Horn* form [1]

$$(l_1 \otimes (r_1^{k_1} \otimes r_2^{k_2} \otimes \cdots \otimes r_n^{k_n})),\ !\Gamma_M \vdash l_0$$

where multiset Γ_M consists of

(a) *Horn implications* of the form $((p \otimes q) \multimap p')$ and $(r \multimap (r' \otimes q'))$,

[1] Here, and henceforth, $G^0 = \mathbf{1}$, $G^k = (\underbrace{G \otimes G \otimes \cdots \otimes G}_{k \text{ times}})$.

(b) and $\oplus$-*Horn implications* of the form $(t \multimap (t' \oplus t''))$.

As a corollary, we prove that **all** *partial recursive relations* are **directly definable** in propositional Linear Logic, even if we use nothing but *Horn-like* formulas built up of *positive literals* with the help of *connectives*: $\{ \multimap, \otimes, \oplus, ! \}$. So we see that our exposition demonstrates directly and in a clear form higher expressive power of *propositional Linear Logic* as compared with all *traditional* propositional systems (including non-classical logics).

2 Horn-like Sequents in Linear Logic

Definition 2.1 *Formulas of propositional Linear Logic are built up of* positive literals (atomic propositions)

$$p_0, \ p_1, \ p_2, \ \ldots, \ p_n, \ \ldots$$

and constants (neutrals)

$$\bot, \ \mathbf{1}$$

with the help of connectives *from the following set:* $\{ \otimes, \ \invamp, \ \multimap, \ \&, \ \oplus, \ !, \ ? \}$.

Definition 2.2 *For a given set σ of* connectives *and* constants, *by $\mathcal{L}(\sigma)$ we mean the set of all formulas that are built up of* positive *literals and constants*[2] *with the help of connectives from the set σ.*

Here we confine ourselves to *Horn-like sequents* introduced as follows.

Definition 2.3 *The tensor product of* a positive number *of positive literals is called* ***a simple product****. A single literal q is also called* ***a simple product.***

Definition 2.4 *We will use a natural isomorphism between non-empty finite multisets over positive literals and simple products:*
A multiset $\{q_1, \ q_2, \ \ldots, \ q_k\}$ will be represented by the following simple product $(q_1 \otimes q_2 \otimes \cdots \otimes q_k)$, and vice versa.

Definition 2.5 *We will write $X \cong Y$ to indicate that X and Y represent one and the same multiset M.*

Definition 2.6 ***Horn-like formulas*** *are defined as follows:*

(a) A ***Horn implication*** *is a formula of the form*

$$(X \multimap Y).$$

[2]These constants should be taken from σ, as well.

*(b) A ⊕-**Horn implication** is a formula of the form*

$$(X \multimap (Y_1 \oplus Y_2)).$$

Definition 2.7 ***Horn-like sequents*** *are defined by the following:*

(i) For multisets Γ and Δ consisting of Horn implications only, a sequent of the form[3]

$$W, \Delta, !\Gamma \vdash Z$$

*is called an **!-Horn sequent.***

(ii) For multisets Γ and Δ consisting of Horn implications and ⊕-Horn implications, a sequent of the form

$$W, \Delta, !\Gamma \vdash Z$$

*will be called an **(!,⊕)-Horn sequent.***

Here (and henceforth) X, Y, Y_1, Y_2, W, and Z are simple products.

3 Tree-like Horn Programs

As computational counterparts of (!,⊕)-Horn sequents, we will consider *tree-like Horn programs* with the following peculiarities:

Definition 3.1 *A **tree-like Horn program** is a rooted binary tree such that*

(a) Every edge of it is labelled by a Horn implication of the form $(X \multimap Y)$.

(b) The root *of the tree is specified as the **input** vertex. A* terminal *vertex, i.e. a vertex with no outgoing edges, will be specified as an **output** one.*

*(c) A vertex v with exactly two outgoing edges (v, w_1) and (v, w_2) will be called **divergent.** These two outgoing edges should be labelled by Horn implications with one and the same antecedent, say $(X \multimap Y_1)$ and $(X \multimap Y_2)$, respectively.*

Now, we should explain how such a program P *runs* for a given input W.

[3]Where !Γ stands for the multiset resulting from putting ! before each formula in Γ.

Definition 3.2 *For a given tree-like Horn program P and any simple product W, a* ***strong computation*** *is defined by induction as follows:*
We assign a simple product [4]

$$VALUE(P, W, v)$$

to each vertex v of P in such a way that

(a) For the root v_0, $VALUE(P, W, v_0) = W$.

(b) For every non-terminal vertex v and its son w with the edge (v, w) labelled by a Horn implication $(X \multimap Y)$, if $VALUE(P, W, v)$ is defined and, for some simple product V:

$$VALUE(P, W, v) \cong (X \otimes V),$$

then

$$VALUE(P, W, w) = (Y \otimes V).$$

Otherwise, $VALUE(P, W, w)$ is declared to be undefined.

Definition 3.3 *For a tree-like Horn program P and a simple product W, we say that*

$$P(W) = Z$$

if for each terminal vertex w of P,
$VALUE(P, W, w)$ is defined and $VALUE(P, W, w) \cong Z$.

We will describe each of our program constructs by Linear Logic formulas. Namely, we will associate a certain formula A to each edge e of a given program P, and say that

"This formula A is used on the edge e."

Definition 3.4 *Let P be a tree-like Horn program.*

(a) If v is a non-divergent vertex of P with the outgoing edge e labelled by a Horn implication A, then we will say that

"Formula A itself is used on the edge e."

(b) Let v be a divergent vertex of P with two outgoing edges e_1 and e_2 labelled by Horn implications $(X \multimap Y_1)$ and $(X \multimap Y_2)$, respectively. Then we will say that

"Formula A is used on e_1."

[4]This VALUE(P, W, v) is perceived as the intermediate value of the *strong computation* performed by P, which is obtained at point v.

and

"Formula A is used on e_2."

where formula A is defined as the following $\oplus$-Horn implication:

$$A = (X \multimap (Y_1 \oplus Y_2)).$$

Definition 3.5 *A tree-like Horn program P is said to be a* ***strong solution to*** *an $(!,\oplus)$-Horn sequent of the form*

$$W, \Delta, !\Gamma \vdash Z$$

if $P(W) = Z$, *and*

(a) For every edge e in P, the formula A used on e is drawn either from Γ or from Δ.

(b) Whatever path b leading from the root to a terminal vertex we take, each formula A from Δ is used once and ***exactly*** *once on this path b.*

Theorem 3.1 *[Completeness] Any $(!,\oplus)$-Horn sequent of the form*

$$W, \Delta, !\Gamma \vdash Z,$$

is derivable in Linear Logic ***if and only if*** *one can construct a tree-like Horn program P which is a strong solution to the given sequent.*

Proof. For a given strong solution P, running from its leaves to its root, we can assemble a derivation of our sequent.

In the other direction we can apply Completeness Theorem from [9, 10][5] to transform each of the derivations of our sequent into the corresponding tree-like Horn program P. ∎

4 Many-Counter Machines

The well-known n-counter machines [17] are defined as follows.

n-Counter Minsky machines deal with n counters that can contain non-negative integers.

The current value of an m-th counter will be represented by the variable x_m. This value

(a) can be increased by 1, which is represented by the *assignment operation* $x_m := x_m + 1$;

[5]In the cited papers, Completeness Theorem has been proved for a more general case.

(b) or can be decreased by 1, which is represented by the *assignment operation* $x_m := x_m - 1$;

Definition 4.1 *The program of an n-counter machine M is a finite list of* instructions

$$I_1;\ I_2;\ \ldots;\ I_s;$$

labelled by labels

$$L_0,\ L_1,\ L_2,\ \ldots,\ L_i,\ \ldots,\ L_j,\ \ldots$$

Each of these instructions is of one of the following five types:

$$\begin{array}{lll}
(1) & L_i: & x_m := x_m + 1;\ \textbf{goto}\ L_j; \\
(2) & L_i: & x_m := x_m - 1;\ \textbf{goto}\ L_j; \\
(3) & L_i: & \textbf{\textit{if}}\,(x_m > 0)\ \textbf{then goto}\ L_j; \\
(4) & L_i: & \textbf{\textit{if}}\,(x_m = 0)\ \textbf{then goto}\ L_j; \\
(5) & L_0: & \textbf{\textit{halt;}}
\end{array}$$

where L_i and L_j are labels, *and $i \geq 1$.*

Definition 4.2 *An* instantaneous description (configuration) *is a tuple:*

$$(L,\ c_1, c_2, \ldots, c_n)$$

where L is a label,
$c_1,\ c_2,\ \ldots,\ c_n$ *are the current values of our n counters, respectively.*
A computation *of a Minsky machine M is a (finite) sequence of configurations*

$$K_1,\ K_2,\ \ldots,\ K_t,\ K_{t+1},\ \ldots,$$

such that each move *(from K_t to K_{t+1}) can be performed by applying an instruction from the program of machine M.*

5 The Main Encoding

In our encoding we will use the following *literals:*

(i) $r_1,\ r_2,\ \ldots,\ r_m,\ \ldots,\ r_n$ [6]

(ii) $l_0,\ l_1,\ l_2,\ \ldots,\ l_i,\ \ldots,\ l_j,\ \ldots$[7]

(iii) $\kappa_1,\ \kappa_2,\ \ldots,\ \kappa_m,\ \ldots,\ \kappa_n$ [8]

---

[6] Literal r_m is associated with the m-th counter.
[7] Literal l_i represents label L_i.
[8] Literal κ_m will be used to *kill* all literals except r_m.

Each instruction I from the list of instructions (1)-(4) of Definition 4.1 will be axiomatized by the corresponding Linear Logic formula φ_I as follows:

$$\begin{aligned} \varphi_{(1)} &= (l_i \multimap (l_j \otimes r_m)), \\ \varphi_{(2)} &= ((l_i \otimes r_m) \multimap l_j), \\ \varphi_{(3)} &= ((l_i \otimes r_m) \multimap (l_j \otimes r_m)), \\ \varphi_{(4)} &= (l_i \multimap (l_j \oplus \kappa_m)). \end{aligned}$$

For a given machine M, its program

$$I_1;\ I_2;\ \ldots;\ I_s;$$

is axiomatized by a multiset Φ_M in the following way:

$$\Phi_M = \varphi_{I_1}, \varphi_{I_2}, \ldots, \varphi_{I_s}.$$

In addition, for every m, by $\mathcal{K}_m$ we mean the multiset consisting of one Horn implication:

$$(\kappa_m \multimap l_0)$$

and the following $(n-1)$ *killing* implications:

$$((\kappa_m \otimes r_i) \multimap \kappa_m), \qquad (i \neq m)$$

We set that

$$\mathcal{K} = \bigcup_{m=1}^{n} \mathcal{K}_m$$

We will prove that an **exact** correspondence exists between arbitrary computations of M on inputs

$$k_1,\ k_2,\ \ldots,\ k_n$$

and derivations for a sequent of the form

$$(l_1 \otimes (r_1^{k_1} \otimes r_2^{k_2} \otimes \cdots \otimes r_n^{k_n})),\ !\Phi_M,\ !\mathcal{K} \vdash l_0.$$

More precisely, taking into account our complete computational interpretation for sequents of this kind (Theorem 3.1), we will prove an **exact** correspondence between arbitrary computations of M on inputs

$$k_1,\ k_2,\ \ldots,\ k_n$$

and *tree-like* strong solutions to this sequent

$$(l_1 \otimes (r_1^{k_1} \otimes r_2^{k_2} \otimes \cdots \otimes r_n^{k_n})),\ !\Phi_M,\ !\mathcal{K} \vdash l_0.$$

In particular, each configuration K

$$K = (L_i,\ c_1, c_2, \ldots, c_n)$$

will be represented in Linear Logic by a simple tensor product

$$\widetilde{K} = (l_i \otimes (r_1^{c_1} \otimes r_2^{c_2} \otimes \cdots \otimes r_n^{c_n})).$$

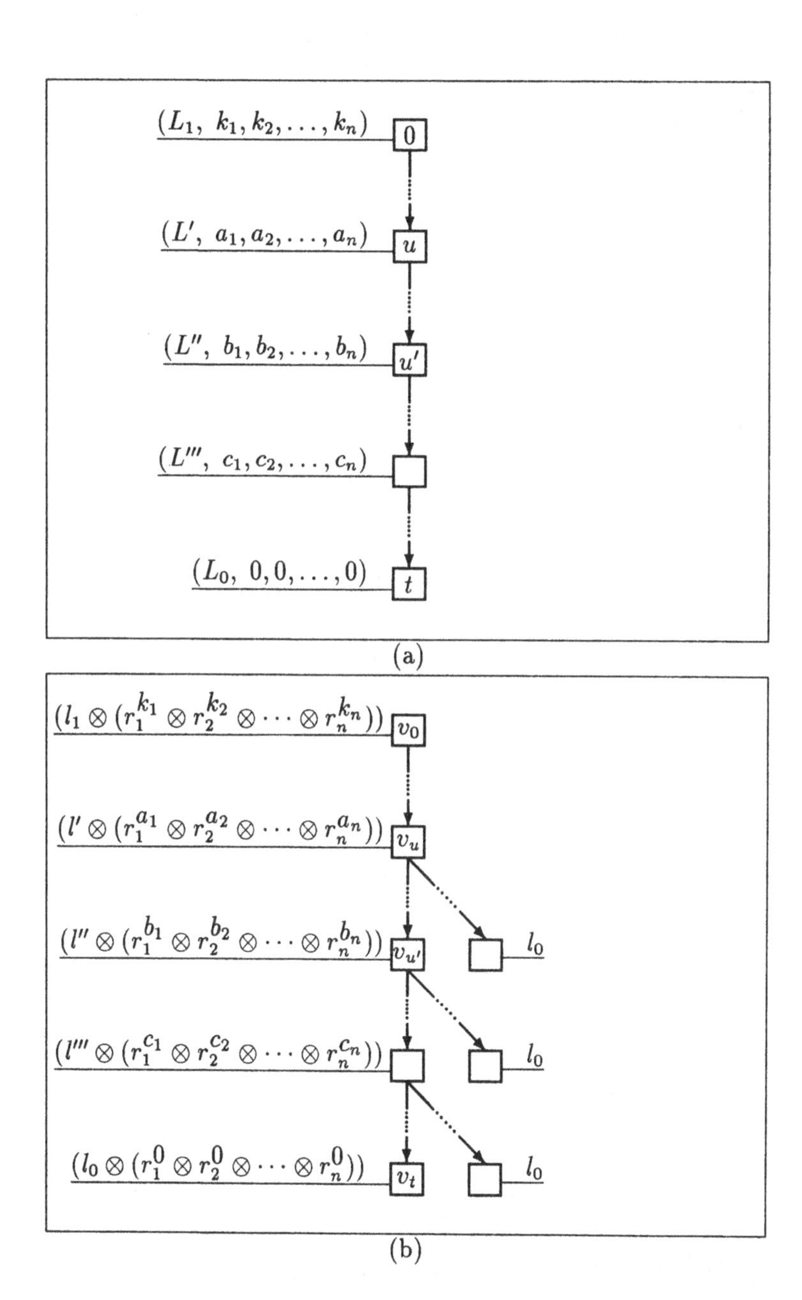

Figure 1: The correspondence: Computation (a) $\leftrightarrow$ Derivation (b).

5.1 From computations to derivations

Lemma 5.1 *For given inputs $k_1, k_2, \ldots, k_n$, let M be able to go from the* initial *configuration* $(L_1,\ k_1, k_2, \ldots, k_n)$ *to the* halting *configuration* $(L_0,\ 0, 0, \ldots, 0)$. *Then a sequent of the form*

$$(l_1 \otimes (r_1^{k_1} \otimes r_2^{k_2} \otimes \cdots \otimes r_n^{k_n})),\ !\Phi_M,\ !\mathcal{K} \vdash l_0$$

is derivable in Linear Logic.

Proof. Let (See Figure 1(a))

$$K_0,\ K_1,\ K_2,\ \ldots,\ K_u,\ K_{u+1},\ \ldots,\ K_t$$

be a computation of M leading from the initial configuration K_0:

$$K_0 = (L_1,\ k_1, k_2, \ldots, k_n),$$

to the halting configuration K_t:

$$K_t = (L_0,\ 0, 0, \ldots, 0).$$

Running from the beginning of this sequence of configurations to its end, we will construct a *tree-like* Horn program P, which is a strong solution to the sequent

$$(l_1 \otimes (r_1^{k_1} \otimes r_2^{k_2} \otimes \cdots \otimes r_n^{k_n})),\ !\Phi_M,\ !\mathcal{K} \vdash l_0,$$

and has the following peculiarities (See Figure 1(b))

(i) $P(\widetilde{K_0}) = \widetilde{K_t} = l_0$,

(ii) and, moreover, there exists a branch of P, called the *main* branch:

$$v_0,\ v_1,\ v_2,\ \ldots,\ v_u,\ v_{u+1},\ \ldots,\ v_t,$$

such that for each vertex v_u from this *main* branch:

$$\text{VALUE}(P, \widetilde{K_0}, v_u) \cong \widetilde{K_u}.$$

We construct the desired program P by induction:
The root v_0 of P is associated with the initial configuration K_0:

$$\text{VALUE}(P, \widetilde{K_0}, v_0) = (l_1 \otimes (r_1^{k_1} \otimes r_2^{k_2} \otimes \cdots \otimes r_n^{k_n})).$$

Let v_u be the terminal vertex of the fragment of P (that has already been constructed), associated with the current configuration K_u:

$$\text{VALUE}(P, \widetilde{K_0}, v_u) \cong \widetilde{K_u} = (l_i \otimes (r_1^{a_1} \otimes r_2^{a_2} \otimes \cdots \otimes r_n^{a_n})).$$

The move from K_u to K_{u+1} is simulated in the following way:

(a) If this move is performed by applying an *assignment operation* instruction I from the list of instructions (1)-(3) of Definition 4.1, then we create a new edge (v_u, v_{u+1}) and label this new edge by the corresponding Horn implication φ_I, getting for this new terminal vertex v_{u+1}:

$$\mathrm{VALUE}(P, \widetilde{K_0}, v_{u+1}) \cong \widetilde{K_{u+1}}.$$

Figure 2 shows the case where this instruction I is of the form

$$L_i : \ x_1 := x_1 - 1; \ \textbf{goto} \ L_j.$$

(b) Let the foregoing move be performed by applying a *ZERO-test* instruction I of the form (4)

$$L_i : \ \textbf{if} \ (x_m = 0) \ \textbf{then goto} \ L_j.$$

The definability conditions of this move provide that

$$a_m = 0.$$

We extend the fragment of P (that has already been constructed) as follows (See Figure 3):

First, we create two new outgoing edges (v_u, v_{u+1}) and (v_u, w_u), and label these new edges by the Horn implications

$$(l_i \multimap l_j) \ \text{ and } \ (l_i \multimap \kappa_m),$$

respectively.

It is readily seen that

$$\begin{aligned} \mathrm{VALUE}(P, \widetilde{K_0}, v_{u+1}) &\cong (l_j \otimes (r_1^{a_1} \otimes r_2^{a_2} \otimes \cdots \otimes r_m^{a_m} \otimes \cdots \otimes r_n^{a_n})) = \widetilde{K_{u+1}}, \\ \mathrm{VALUE}(P, \widetilde{K_0}, w_u) &\cong (\kappa_m \otimes (r_1^{a_1} \otimes r_2^{a_2} \otimes \cdots \otimes r_m^{a_m} \otimes \cdots \otimes r_n^{a_n})). \end{aligned}$$

Then, we create a chain of t_u new edges

$$(w_u, w_1^u), \ (w_1^u, w_2^u), \ \ldots, \ (w_{t_u-1}^u, w_{t_u}^u)$$

where

$$t_u = a_1 + a_2 + \cdots + a_{m-1} + a_{m+1} + \cdots + a_n,$$

and label these new edges by such Horn implications from $\mathcal{K}_m$ as to *kill* all occurrences of literals

$$r_1, \ r_2, \ \ldots, \ r_{m-1}, \ r_{m+1}, \ \ldots, \ r_n,$$

and ensure that

$$\text{VALUE}(P, \widetilde{K_0}, w^u_{t_u}) \cong (\kappa_m \otimes (r_1^0 \otimes r_2^0 \otimes \cdots \otimes r_m^{a_m} \otimes \cdots \otimes r_n^0)).$$

Finally, we create a new edge $(w^u_{t_u}, w^u_{t_u+1})$, and label this new edge by the Horn implication

$$(\kappa_m \multimap l_0).$$

Taking into account that $a_m = 0$, for the terminal vertex $w^u_{t_u+1}$ of the foregoing chain, we have:

$$\text{VALUE}(P, \widetilde{K_0}, w^u_{t_u+1}) = (l_0 \otimes (r_1^0 \otimes r_2^0 \otimes \cdots \otimes r_m^{a_m} \otimes \cdots \otimes r_n^0)) = l_0.$$

Hence, for all terminal vertices w, i.e. both for the terminal vertex v_t of the *main* branch and for the terminal vertices of all auxiliary chains, we obtain that

$$\text{VALUE}(P, \widetilde{K_0}, w) = l_0 = \widetilde{K_t}.$$

Thus, our inductive process results in a *tree-like* Horn program P that is a strong solution to the sequent

$$(l_1 \otimes (r_1^{k_1} \otimes r_2^{k_2} \otimes \cdots \otimes r_n^{k_n})),\ !\Phi_M,\ !\mathcal{K} \vdash l_0$$

To complete Lemma 5.1, we need only recall that, by Theorem 3.1, the latter sequent is derivable in Linear Logic. ∎

5.2 From derivations to computations

Lemma 5.2 *For given integers $k_1, k_2, \ldots, k_n$, let a sequent of the form*

$$(l_1 \otimes (r_1^{k_1} \otimes r_2^{k_2} \otimes \cdots \otimes r_n^{k_n})),\ !\Phi_M,\ !\mathcal{K} \vdash l_0$$

be derivable in Linear Logic. Then M can go from an initial *configuration of the form $(L_1,\ k_1, k_2, \ldots, k_n)$ to the* halting *configuration $(L_0,\ 0, 0, \ldots, 0)$.*

Proof. By Theorem 3.1, we can construct a *tree-like* Horn program P such that

(i) Each of the Horn implications occurring in P is drawn either from Φ_M or from $\mathcal{K}$.

(ii) For all terminal vertices w of P:

$$\text{VALUE}(P, W_0, w) = l_0$$

where

$$W_0 = (l_1 \otimes (r_1^{k_1} \otimes r_2^{k_2} \otimes \cdots \otimes r_n^{k_n})).$$

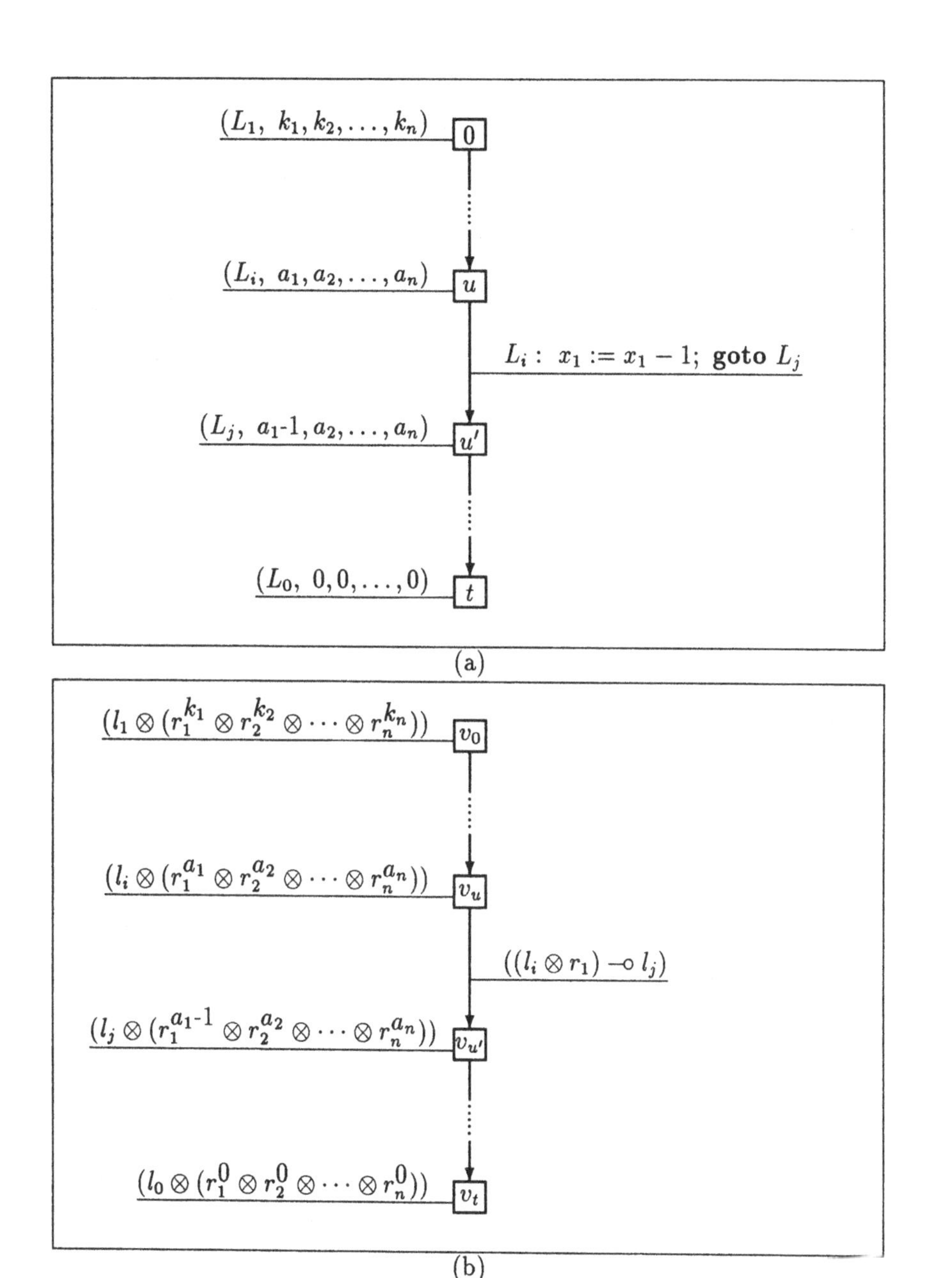

Figure 2: The *assignment operation* correspondence: (a) $\leftrightarrow$ (b).

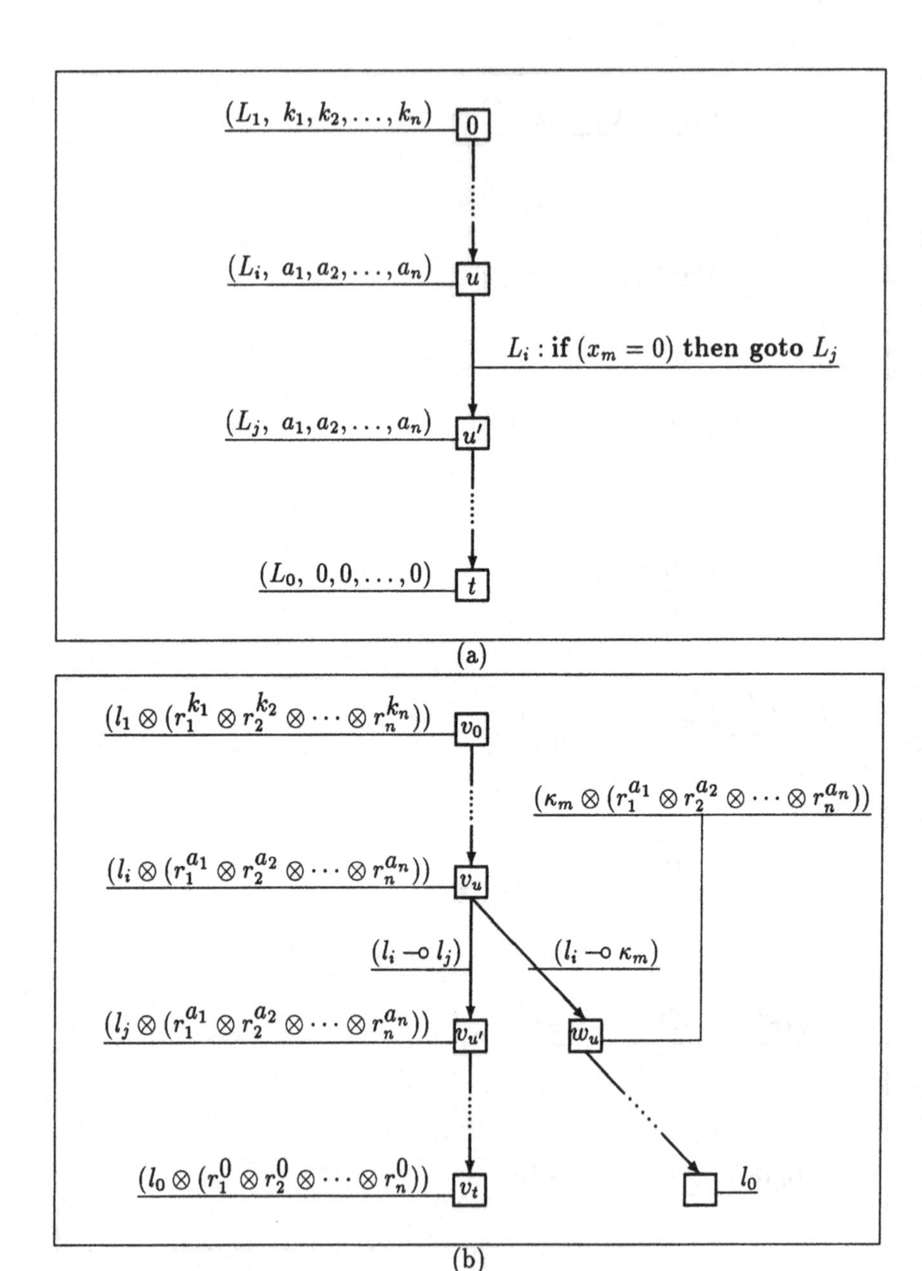

Figure 3: The *ZERO-test* correspondence: (a) $\leftrightarrow$ (b).

Lemma 5.3 *Running from the root v_0 to terminal vertices of P, we assemble the desired Minsky computation as follows:*

(a) It is proved that program P cannot be but of the form represented in Figure 1.

(b) We can identify a branch of P, called the main *branch:*

$$v_0,\ v_1,\ v_2,\ \ldots,\ v_u,\ v_{u+1},\ \ldots,\ v_t,$$

such that for all vertices v_u on this branch, $VALUE(P, W_0, v_u)$ is proved to be of the form

$$VALUE(P, W_0, v_u) \cong (l_i \otimes (r_1^{a_1} \otimes r_2^{a_2} \otimes \cdots \otimes r_n^{a_n})).$$

(c) For all non-terminal vertices w' of P that are outside the main *branch, $VALUE(P, W_0, w')$ is proved to be of the form*

$$VALUE(P, W_0, w') \cong (\kappa_m \otimes (r_1^{a_1} \otimes r_2^{a_2} \otimes \cdots \otimes r_n^{a_n})).$$

(d) Finally, the following sequence of configurations (See Figure 1)

$$K_0,\ K_1,\ K_2,\ \ldots,\ K_u,\ K_{u+1},\ \ldots,\ K_t$$

such that for every integer u

$$\widetilde{K_u} \cong VALUE(P, W_0, v_u),$$

is proved to be a successful *computation of M leading from the initial configuration K_0:*

$$K_0 = (L_1,\ k_1, k_2, \ldots, k_n),$$

to the halting configuration K_t:

$$K_t = (L_0,\ 0, 0, \ldots, 0).$$

Proof. Since

$$\widetilde{K_0} = \text{VALUE}(P, W_0, v_0) = (l_1 \otimes (r_1^{k_1} \otimes r_2^{k_2} \otimes \cdots \otimes r_n^{k_n})),$$

we have for the root v_0:

$$K_0 = (L_1,\ k_1, k_2, \ldots, k_n).$$

Let v_u be the current vertex on the *main* branch we are searching for, and, according to the inductive hypothesis, let VALUE(P, W_0, v_u) be of the form

$$\text{VALUE}(P, W_0, v_u) \cong \widetilde{K_u} = (l_i \otimes (r_1^{a_1} \otimes r_2^{a_2} \otimes \cdots \otimes r_n^{a_n})).$$

There are the following cases to be considered.

(a) Suppose that v_u is a non-divergent vertex with the only son which will be named v_{u+1}.

According to the definability conditions for our program P, the single outgoing edge (v_u, v_{u+1}) cannot be labelled but by a Horn implication A from Φ_M.

Moreover,

$$A = \varphi_I$$

for some instruction I of the form (1)-(3) from Definition 4.1.

Let this instruction I be of the form

$$L_i: \ x_1 := x_1 - 1; \ \textbf{goto} \ L_j;$$

and

$$A = ((l_i \otimes r_1) \multimap l_j).$$

Then we have (See Figure 2):

$$\text{VALUE}(P, W_0, v_{u+1}) \cong (l_j \otimes (r_1^{a_1\text{-}1} \otimes r_2^{a_2} \otimes \cdots \otimes r_n^{a_n})),$$

(and, hence, $a_1 \geq 1$).

Performing the foregoing instruction I, machine M can move from the current configuration K_u:

$$K_u = (L_i, \ a_1, a_2, \ldots, a_n),$$

to the next configuration K_{u+1}:

$$K_{u+1} = (L_j, \ a_1\text{-}1, a_2, \ldots, a_n),$$

such that

$$\widetilde{K_{u+1}} \cong \text{VALUE}(P, W_0, v_{u+1}).$$

The remaining cases are handled similarly.

(b) The **crucial point** is where v_u is a vertex with two outgoing edges, say (v_u, v_{u+1}) and (v_u, w_u), labelled by Horn implications A_1 and A_2, respectively. (See Figure 3)

It means that the implication used at this point of program P must be a $\oplus$-Horn implication A from Φ_M of the form

$$A = (l_i \multimap (l_j \oplus \kappa_m)),$$

and, in addition,

$$\begin{array}{rcl} A_1 & = & (l_i \multimap l_j) \\ A_2 & = & (l_i \multimap \kappa_m). \end{array}$$

Therefore,

$$\begin{array}{rcl} \text{VALUE}(P, W_0, v_{u+1}) & \cong & (l_j \otimes (r_1^{a_1} \otimes r_2^{a_2} \otimes \cdots \otimes r_m^{a_m} \otimes \cdots \otimes r_n^{a_n})) \\ \text{VALUE}(P, W_0, w_u) & \cong & (\kappa_m \otimes (r_1^{a_1} \otimes r_2^{a_2} \otimes \cdots \otimes r_m^{a_m} \otimes \cdots \otimes r_n^{a_n})). \end{array}$$

Let us examine the descendants of the vertex w_u.

Taking into account the definability conditions, any edge (w_1, w_2), such that w_1 is a descendant of w_u, cannot be labelled but by a Horn implication from $\mathcal{K}_m$.

So we can conclude that for all non-terminal descendants w' of the vertex w_u, VALUE(P, W_0, w') is of the form

$$\text{VALUE}(P, W_0, w') \cong (\kappa_m \otimes (r_1^{c_1} \otimes r_2^{c_2} \otimes \cdots \otimes r_m^{a_m} \otimes \cdots \otimes r_n^{c_n})).$$

For the terminal descendant w of the vertex w_u, VALUE(P, W_0, w) is to be of the form

$$\text{VALUE}(P, W_0, w) \cong (l_0 \otimes (r_1^{c_1} \otimes r_2^{c_2} \otimes \cdots \otimes r_m^{a_m} \otimes \cdots \otimes r_n^{c_n})).$$

Recalling that

$$\text{VALUE}(P, W_0, w) = l_0,$$

we get the desired

$$a_m = 0.$$

Indeed,

$$A = \varphi_I$$

for a *ZERO-test* instruction I of the form

$$L_i : \ \textbf{if } (x_m = 0) \textbf{ then goto } L_j;$$

Performing this instruction I, machine M can move from the current configuration K_u:

$$K_u = (L_i, \ a_1, a_2, \ldots, a_m, \ldots, a_n),$$

to the next configuration K_{u+1}:

$$K_{u+1} = (L_j, \ a_1, a_2, \ldots, a_m, \ldots, a_n),$$

such that

$$\widetilde{K_{u+1}} \cong \text{VALUE}(P, W_0, v_{u+1}).$$

(c) Suppose that the *main* branch we have been developing

$$v_0,\ v_1,\ v_2,\ \ldots,\ v_u,\ v_{u+1},\ \ldots,\ v_t,$$

has ended at a vertex v_t.

According to what has been said,

$$l_0 = \mathrm{VALUE}(P, W_0, v_t) \cong \widetilde{K_t} = (l_j \otimes (r_1^{c_1} \otimes r_2^{c_2} \otimes \cdots \otimes r_n^{c_n})).$$

Hence, this configuration K_t is the *halting* one:

$$K_t = (L_0,\ 0, 0, \ldots, 0).$$

Now, bringing together all the cases considered, we can complete Lemma 5.3 and, hence, Lemma 5.2. ∎

Theorem 5.1 *For given inputs* $k_1, k_2, \ldots, k_n$, *an n-counter Minsky machine M can go from the* initial *configuration* $(L_1,\ k_1, k_2, \ldots, k_n)$ *to the* halting *configuration* $(L_0,\ 0, 0, \ldots, 0)$ ***if and only if*** *a sequent of the form*

$$(l_1 \otimes (r_1^{k_1} \otimes r_2^{k_2} \otimes \cdots \otimes r_n^{k_n})),\ !\Phi_M,\ !\mathcal{K} \vdash l_0$$

is derivable in Linear Logic.

Proof. We bring together Lemma 5.1 and Lemma 5.2. ∎

6 Relations Definable in Propositional Linear Logic

For a given logical system **L**, any formula

$$F(E_1,\ E_2,\ \ldots,\ E_n)$$

can be conceived of as a *correlation* $\mathcal{P}_F$ between *propositions*

$$E_1,\ E_2,\ \ldots,\ E_n.$$

Definition 6.1 *Let*

$$F(p_1,\ p_2,\ \ldots,\ p_n)$$

be a formula in the language of logic **L** *with marked* literals [9]

$$p_1,\ p_2,\ \ldots,\ p_n.$$

[9] Non-marked *parameters* $t_1, t_2, \ldots, t_m$ are allowed to be contained in the formula.

By $\mathcal{P}_F$ *we mean the following relation between* formulas (propositions)*:*
For all formulas $E_1, E_2, \ldots, E_n$,

$$\mathcal{P}_F(E_1, E_2, \ldots, E_n) = \begin{cases} \textbf{\textit{true}}, & \textit{if} \ \vdash F(E_1, E_2, \ldots, E_n) \ \textit{is derivable in } \mathbf{L}, \\ \textbf{\textit{false}}, & \textit{otherwise.} \end{cases}$$

Within this *paradigm of definable relations,* we can give more strong reformulations of a number of complexity results related to Linear Logic, expressing the corresponding relations over integers *directly* in the language of Linear Logic.

First of all, we should make our choice among possible representations of integers in propositional Linear Logic. We will use the *simplest* representation of integers: namely, each integer k will be represented by formulas of the form

$$G^k = (\underbrace{G \otimes G \otimes \cdots \otimes G}_{k \text{ times}}).$$

In particular, the integer 0 will be represented by formulas of the form

$$G^0 = \mathbf{1}.$$

Now we introduce a strong version of the *definability* concept:

Definition 6.2 *Let*

$$Q(x_1,\ x_2,\ \ldots,\ x_n)$$

be an n*-ary predicate over non-negative integers.*

We will say that this predicate Q *is* ***directly definable in*** *Linear Logic by formulas* G_0, G_1, G_2, $\ldots$, G_n *and a multiset* Γ *if, whatever non-negative integers* k_1, k_2, $\ldots$, k_n *we take,*

$$Q(k_1, k_2, .., k_n) \ \textit{is} \ \textbf{\textit{true}}$$

if and only if the following sequent

$$G_1^{k_1}, G_2^{k_2}, \ldots, G_n^{k_n}, \Gamma \vdash G_0$$

is derivable in Linear Logic.

Theorem 6.1 *All partial recursive relations are* ***directly definable*** *in propositional Linear Logic, even if we use nothing but Horn-like formulas from* $\boldsymbol{LL}(\multimap, \otimes, \oplus, !)$.

Proof. Without loss of generality, we will consider unary partial recursive predicates where ($n = 1$).

According to [17], for a given partial recursive predicate $Q(x)$, there exists a *three*-counter Minsky machine M such that for all integers k:

$$Q(k) \;\equiv\; \text{“}M \text{ can go from } (L_1, k, 0, 0) \text{ to } (L_0, 0, 0, 0)\text{”}.$$

According to Theorem 5.1, for every integer k, our 3-counter Minsky machine M can go from the *initial* configuration $(L_1, k, 0, 0)$ to the *halting* configuration $(L_0, 0, 0, 0)$ **if and only if** a sequent of the form

$$l_1,\ r_1^k,\ !\Phi_M,\ !\mathcal{K} \vdash l_0$$

is derivable in Linear Logic.

It is readily seen that predicate Q is directly definable by these formulas l_0, r_1 and multiset

$$l_1,\ !\Phi_M,\ !\mathcal{K}.$$

■

Corollary 6.1 *We can reformulate Theorem 6.1 as follows:*

*(a) All partial recursive functions are **directly representable** in propositional Linear Logic, even if we use nothing but Horn-like formulas from* $\boldsymbol{LL}(\multimap, \otimes, \oplus, !)$.

*(b) All recursively enumerable sets are **directly definable** in Linear Logic, even if we use no more than Horn-like formulas from* $\boldsymbol{LL}(\multimap, \otimes, \oplus, !)$.

As for relations definable in the Multiplicative-Exponential Fragment of Linear Logic $\mathbf{LL}(\mathbf{1}, \multimap, \otimes, !)$ (its exact complexity level is unknown), we can demonstrate, at least, some *closure* properties for this class.

Theorem 6.2 *The class of all relations directly definable in* $\boldsymbol{LL}(\boldsymbol{1}, \multimap, \otimes, !)$ *is closed under the following operations:*
'conjunction', 'disjunction', the existential quantifier $\exists$*, and composition.*

Proof. The full proof will be presented in a forthcoming paper. ■

Example 6.1 *Each of the following relations is directly definable in the !-Horn fragment of* $\boldsymbol{LL}(\multimap, \otimes, !)$:

(a) $y = x$,

(b) $y < x$,

(c) $z = x + y$,

(d) $z \leq (x \cdot y)$,

(e) $y \leq x^2$,

(f) $y \leq 2^x$,

(g) *etc.*

Corollary 6.2 *The class of all relations directly definable in* $\boldsymbol{LL}(\boldsymbol{1}, \multimap, \otimes, !)$ *is not closed under 'negation'.*

Proof. We put a *non-constructive* proof.
Assume that the following relation

$$y > x^2$$

is directly definable in $\mathbf{LL}(\mathbf{1}, \multimap, \otimes, !)$.

Then, taking into account Theorem 6.2 and Example 6.1, we can simulate all Diophantine predicates in $\mathbf{LL}(\mathbf{1}, \multimap, \otimes, !)$, and, hence, in this case our class is not closed under *negation*.
On the other hand, if the foregoing relation

$$y > x^2$$

is not definable in $\mathbf{LL}(\mathbf{1}, \multimap, \otimes, !)$, then, according to item (e) of Example 6.1, this relation itself shows that the class of definable relations is not closed under *negation*. ∎

Acknowledgements

I owe special thanks to J.-Y.Girard, Y.Lafont, L.Regnier, and G.Amiot for providing supportive and stimulating environment to discuss all these new concepts during my stay in Laboratoire de Mathématiques Discrètes.

I am also grateful to P.Lincoln, J.Mitchell, V.Pratt, A.Scedrov, and N.Shankar for their inspiring introduction to the related problems and fruitful discussions.

I am greatly indebted to S.Artemov and his Seminar on Logical Methods in Computer Science at the Moscow State University for the opportunity of developing and clarifying my results.

References

[1] S.Abramsky, Computational Interpretations of Linear Logic, *Theoretical Computer Science,* v.111, 1992,. p.3–57. (Special Issue on the 1990 Workshop on Math. Found. Prog. Semantics.)

[2] M.R.Garey and D.S.Johnson. Computers and Intractability: A Guide to the Theory of NP-Completeness. 1979.

[3] J.-Y.Girard. Linear logic. *Theoretical Computer Science,* 50:1, 1987, pp.1–102.

[4] J.-Y.Girard. Towards a Geometry of Interaction. *Categories in Computer Science and Logic,* Contemporary Mathematics 92, AMS 1989, p.69–108.

[5] J.-Y.Girard, A.Scedrov, and P.J.Scott, Bounded Linear Logic: A Modular Approach to Polynomial Time Computability, *Theoretical Computer Science,* 97, 1992, pp.1–66.

[6] J.-Y.Girard. Geometry of interaction III : the general case, *Proceedings of the Workshop on Linear Logic,* MIT Press, (to appear)

[7] M.I.Kanovich. The multiplicative fragment of Linear Logic is NP-complete. University of Amsterdam, Institute for Logic, Language and Computation, ILLC Prepublication Series X-91-13, June 1991.

[8] M.I.Kanovich. The Horn fragment of Linear Logic is NP-complete. University of Amsterdam, Institute for Logic, Language and Computation, ILLC Prepublication Series X-91-14, August 1991.

[9] M.I.Kanovich. Horn Programming in Linear Logic is NP-complete. In *Proc. 7-th Annual IEEE Symposium on Logic in Computer Science, Santa Cruz,* June 1992, pp.200-210

[10] M.I.Kanovich. Linear logic as a logic of computations, *Annals Pure Appl. Logic,* 67 (1994) p.183–212

[11] M.I.Kanovich. The complexity of Horn Fragments of Linear Logic, *Annals Pure Appl. Logic,* 69 (1994) p.195–241. (Special Issue on LICS'92)

[12] Y.Lafont and T.Streicher. Games semantics for linear logic. In *Proc. 6-th Annual IEEE Symposium on Logic in Computer Science,* 43-51, Amsterdam, July 1991.

[13] P.Lincoln, J.Mitchell, A.Scedrov, and N.Shankar. Decision Problems for Propositional Linear Logic. Technical Report SRI-CSL-90-08, CSL, SRI International, August 1990.

[14] P.Lincoln, J.Mitchell, A.Scedrov, and N.Shankar. Decision Problems for Propositional Linear Logic. In *Proc. 31st IEEE Symp. on Foundations of Computer Science,* 662–671, 1990.

[15] P.Lincoln, J.Mitchell, A.Scedrov, and N.Shankar. Decision Problems for Propositional Linear Logic. *Annals Pure Appl. Logic,* 56 (1992) pp. 239–311.

[16] P.Lincoln and T.Winkler. Constant multiplicative Linear Logic is NP-complete. Draft, 1992.

[17] M.Minsky. Recursive unsolvability of Post's problem of 'tag' and other topics in the theory of Turing machines. *Annals of Mathematics,* 74:3:437–455, 1961.

[18] A.S.Troelstra, *Lectures on Linear Logic,* CSLI Lecture Notes No. 29, Center for the Study of Language and Information, Stanford University, 1992.

Stochastic Interaction and Linear Logic

Patrick D. Lincoln* John C. Mitchell† Andre Scedrov‡

Abstract

We present stochastic interactive semantics for propositional linear logic without modalities. The framework is based on interactive protocols considered in computational complexity theory, in which a prover with unlimited power interacts with a verifier that can only toss fair coins or perform simple tasks when presented with the given formula or with subsequent messages from the prover. The additive conjunction & is described as random choice, which reflects the intuitive idea that the verifier can perform only "random spot checks". This stochastic interactive semantic framework is shown to be sound and complete. Furthermore, the prover's winning strategies are basically proofs of the given formula. In this framework the multiplicative and additive connectives of linear logic are described by means of probabilistic operators, giving a new basis for intuitive reasoning about linear logic and a potential new tool in automated deduction.

*`lincoln@csl.sri.com` SRI International Computer Science Laboratory, Menlo Park CA 94025 USA. Work supported under NSF Grant CCR-9224858.

†`jcm@cs.stanford.edu`
WWW: `http://theory.stanford.edu/people/jcm/home.html` Department of Computer Science, Stanford University, Stanford, CA 94305. Supported in part by an NSF PYI Award, matching funds from Digital Equipment Corporation, the Powell Foundation, and Xerox Corporation; and the Wallace F. and Lucille M. Davis Faculty Scholarship.

‡`andre@cis.upenn.edu` WWW: `http://www.cis.upenn.edu/~andre` Department of Mathematics, University of Pennsylvania, Philadelphia, PA 19104-6395. Partially supported by NSF Grants CCR-91-02753 and CCR-94-00907 and by ONR Grant N00014-92-J-1916. Scedrov is an American Mathematical Society Centennial Research Fellow.

1 Introduction

Linear logic arose from the semantic study of the structure of proofs in intuitionistic logic. Girard presented the coherence space and phase space semantics of linear logic in his original work on linear logic [Gir87]. While these models provide mathematical tools for the study of several aspects of linear logic, they do not offer a simple intuitive way of reasoning about linear logic. More recently, Blass [Bla92], Abramsky and Jagadeesan [AJ94], Lamarche, and Hyland and Ong have developed semantics of linear logic by means of games and interaction. These new approaches have already proven fruitful in providing an evocative semantic paradigm for linear logic and have found a striking application to programming language theory in the work of Abramsky, Jagadeesan, and Malacaria [AJM93] and in the work of Hyland and Ong [HO93].

The game-theoretic notions used so far draw mainly on set theory and on proof theory. They do not yet fully explain linear logic formulas involving both multiplicative and additive connectives, *i.e.*, the formulas of the multiplicative-additive fragment of linear logic, MALL [Gir87, LMSS92, Sce93, Sce94]. In contrast, the direct, structural relationship between the natural fragments of linear logic and the standard computational complexity classes that has emerged from [LMSS92, LS94a, LS94b] makes it possible to draw on game-theoretic notions stemming from computational complexity theory. Here we are particularly interested in the game-theoretic notions involving stochastic (or, randomized) interaction. This setting, described in Papadimitriou [Pap94], involves interactive protocols in which a prover with unlimited power interacts with a verifier who can only toss coins or perform simple (say, polynomial time) tasks when presented with the input or with the prover's messages. Stochastic verification goes back to Solovay and Strassen [SS77].

One such setting was introduced in Papadimitriou [Pap85] as "games against Nature" or "stochastic alternating polynomial time machines" (SAPTIME) and shown to characterize polynomial space (PSPACE). More precisely, a set of finite strings L belongs to SAPTIME if some verifier V can be convinced by some prover P to accept any $x \in L$ with probability $> 1/2$, while P or any other prover can convince the verifier V to accept any $x \notin L$ only with probability $< 1/2$. In addition, if the probabilities are required to be bounded away from $1/2$, one obtains "Arthur-Merlin games" proposed by Babai and Moran [BM88] or similar "interactive proof systems" proposed by Goldwasser, Micali, and Rackoff [GMR89]. Shamir [Sha92] and Shen [She92] have shown that all these formulations describe exactly PSPACE. Since MALL is PSPACE-complete [LMSS92], it seems *a priori* possible to give a direct, structural relationship between the prover/verifier interactions just described and MALL that goes beyond an immediate Turing reduction.

In Section 2 we give an intuitive overview of the game semantics approach.

We describe a simplified 'linear game boy' where a human player plays a game essentially trying to convince a little electronic machine (which has very limited computing power) that a given MALL sequent is provable. Even if the human plays optimally, the human cannot convince the machine of the provability of an unprovable sequent very often. It happens that the game can be played to win more than half the time if and only if the starting sequent is provable.

In Section 3 we give preliminary formal definitions. These form a substrate for the formal analysis of the informal games described in Section 2. In Section 4 we present a formal semantics for MALL based on games against Nature and prove it sound and complete for provability in MALL. When the game-theoretic or the complexity-theoretic aspects are disregarded, the semantic notions involved may be distilled to a rather simple semantic characterization of MALL provability akin to the ordinary truth tables, but involving probabilistic truth values. On the other hand, the prover's winning strategies are for all practical purposes *proofs* of the given formula. In this framework the multiplicative and additive connectives of linear logic are described by means of probabilistic operators. A central feature is the description of the additive conjunction & as random choice. This approach offers a new basis for intuitive reasoning about linear logic and a potential new tool in automated deduction, and suggests a connection between linear logic and probabilistic logic such as sometimes used in artificial intelligence. This extended abstract concludes with suggestions for further research.

We would like to thank Jean-Yves Girard, Andreas Blass, Yuri Gurevich, Bill Rounds, and Larry Moss for interesting and inspiring discussions.

2 Game-Boy Semantics

Game interpretations may be used for a variety of logics. While game interpretations, or game "semantics," have a different flavor from other forms of semantics, they have certain advantages. One well-established application is the use of Ehrenfeucht-Fraissé games for characterizing the expressiveness of logics over finite structures, for instance in [Imm82].

It is easy to illustrate the general idea of game semantics using classical quantified propositional logic. The game will have two players, called *for-all* and *there-exists*, each smart enough to evaluate a quantifier-free proposition, given an assignment of truth values (*true* or *false*) to the propositional variables. The play begins with a quantified formula such as

$$\exists P\, \forall Q\, \exists R\, [(P \wedge Q \wedge \neg R) \vee (P \wedge \neg Q \wedge R)]$$

in prenex form that starts with an existential quantifier and alternates between $\forall$ and $\exists$. (Any formula can be put in this form or, alternatively, we can extend the game to more general syntactic forms.) The goal of *there-exists,* who moves first, is to choose values for the existentially quantified variables that will make the matrix true. The goal for *for-all* is to choose values for the universally quantified variables that will make the formula false. It is easy to see that if the players alternate, each choosing a truth value for the variables they are given until the quantifier prefix is exhausted, *there-exists* can win if the formula is valid, with *for-all* able to win otherwise. For example, on the valid formula above, *there-exists* may begin by letting P be *true.* Whatever value *for-all* chooses for Q, *there-exists* may win by letting R be the negation of Q. When *there-exists* has a way of winning, regardless of how *for-all* plays, we say *there-exists* has a *winning strategy,* and similarly for *for-all.* More formally, we can define a winning strategy for *there-exists,* to be a function that chooses moves in such a way that *there-exists* is guaranteed to win. A strategy is generally a function of the formula and the moves made by both players (up until the point where a next move is required).

Game interpretations for various fragments and forms of linear logic have been investigated by Blass, Abramsky and Jagadeesan, Lamarche, and Hyland and Ong [Bla92, AJ94, HO93]. Some of these games are complete, in the sense that a formula is valid iff a designated player has a winning strategy, while others yield more specialized interpretations where the proof rules are sound but not complete.

An issue we are particularly concerned with is the length of the game and the amount of "thought" (or, more technically, computational resource) that is required to play the game optimally. In the game for quantified propositional logic, for example, if *there-exists* is going to compute a winning strategy as a function of the formula and moves of the other player, this may be done by a Turing machine whose space requirement is bounded by a polynomial in the length of the formula. Moreover, if *for-all* is going to win when a formula is not valid, then poloynomial space is also required. Therefore, we can say that in this game, both players require polynomial space. This seems reasonable, since deciding whether a classical quantified propositional formula is valid is a PSPACE-complete problem.

A recent trend in complexity theory, motivated by the study of cryptography and randomness, is the use of asymmetric "games" to characterize various computational problems. Returning to classical quantified propositional formulas, there exist asymmetric games in which one player may use unbounded resources, but the other only requires polynomial time (which is less than polynomial space), and the weaker player is allowed to play randomly by "flipping coins." We will use this form of game for MALL.

2.1 Game Board

We will give several forms of proof games, all using the same board and tiles. The difference between the games is that the first versions, given primarily to illustrate the main ideas that appear in the later versions, use simple rules that require complicated strategies for both players. In the later games, we change the rules so that one player may play effectively using only randomness and very simple (polynomial time) computation. In all but the first cases, we will keep the games "short," meaning that the length of each game (the number of rounds of play) is bounded by the length of linear logic sequent that appears on the first tile played.

All games are played on a rectangular grid that looks like an uncolored checkerboard that extends infinitely up and to the right. A "home version" of this board designed for one-evening games in front of the fire can be bounded to some fixed size. In the "game boy" version we discuss below, the board is considered unbounded, with only a portion of the board displayed on the small LCD screeen at any time.

There are two players, called the prover and the verifier. The goal of the prover is to play a sequence of tiles demonstrating or giving evidence for a sequent. To do this, the prover plays tiles that contain one sequent in the lower left, and one or two shorter sequents in the top and right quadrants. The verifier tries to force the direction of the provers evidence in a way that makes it impossible for the prover to win. The verifier plays simple black (or crosshatched) tiles that may block one side of one of the provers tiles. The verifier also scores the game by blocking WIN or LOSE on a terminal prover tile, leaving the correct score exposed.

The tiles of the game are drawn in Figure 1. The main idea is that each prover tile contains an instance of a MALL proof rule, tilted to the side to fit on a rectangular grid. The prover uses these to place a proof tree, or branch of a proof tree, on the board. When a proof rule has more than one hypothesis, the verifier can block one, forcing the prover to try to complete the branch containing the other hypothesis. If the prover plays a tile with sequents Γ and Δ exposed to the top and right, then the verifier may (or may not) block one of these edges. In either case, the only legal adjacent play for the prover is to place another prover tile with sequent Γ or Δ to the left of its diagonal, Γ if this is above the previous tile and Δ if it is to the right. We consider two sequents the same, for the purposes of the game, if they are identical as multisets. In other words, the formulas may appear in any order, but the number of each type of formula must match.

The reason for the special "literal" tiles, with pseudohypotheses WIN and LOSE, is to allow the game to be "scored" by the verifier. If the prover playes until there are no connectives remaining in the sequent, the only remaining play is a matching tile with literals. At this point, the prover either wins or loses, depending (in the simple form of the game) on whether the multiset of literals is an axiom or not. Although this is not important, we establish the convention that when the prover plays a literal tile, and the verifier plays a solid tile in an adjacent square, this tile is considered *winning* if verifier plays adjacent to LOSE, and *losing* if the verifier plays adjacent to WIN. This is just to give some specific, visual representation of each play of the game.

2.2 Elementary deterministic form

An elementary form of "game" using this board and set of tiles is just to let the prover play, from any initial tile in the lower left corner of the board. The verifier checks to ensure that all tiles are legally played (that the multisets of formulas along all adjoining edges match). The verifier also checks that all exposed edge tiles are literal tiles, and that they all can be scored as a win. In this form of game, the prover exhibits a proof tree and the verifier evaluates each leaf, allowing the prover to win if each leaf of the tree is an axiom. An example play of this game is illustrated in Figure 2. Note that in some cases the prover will be required to make use of connector tiles (the last two prover tiles shown in in Figure 1) that do not change the sequent along that branch of the proof. Rather, these tiles just extend the sequent out along a branch to reach a part of the board where there may be more room. It is the prover's responsibility to plan the layout the proof so that all tiles are placed legally.

From a commonsense point of view, this is not much of a "game," amounting more to a game of solitaire than a two-person game. More technically, a play of this game may involve more prover tiles than the number of connectives in the sequent on the lower left of the lower left tile. The reason is that the tile (proof rule) for & has more symbols to the right of the diagonal (in the hypotheses) than the on the left (in the conclusion), duplicating all of the context of the &-formula. The next version of the game, which is close to the simple game described earlier for classical quantified propositional logic, we will have interaction between players and a linear bound on the number of prover (and hence verifier) tiles used in any play of the game.

2.3 Revised deterministic form

The next version of the is similar in spirit to the simple game described for classical quantified propositional logic. The prover tries to show that a sequent is provable, and the verifier tries to make this demonstration as difficult as possible. The prover is able to win if the sequent *is* provable, and the verifier is able to keep the prover from winning if the sequent is not provable. However, in order to play effectively, both players must be able to understand provability

MALL tiles for prover, with A, B arbitrary formulas, Γ an arbitrary multiset of formulas, and $L_1, \ldots, L_k$ literals.

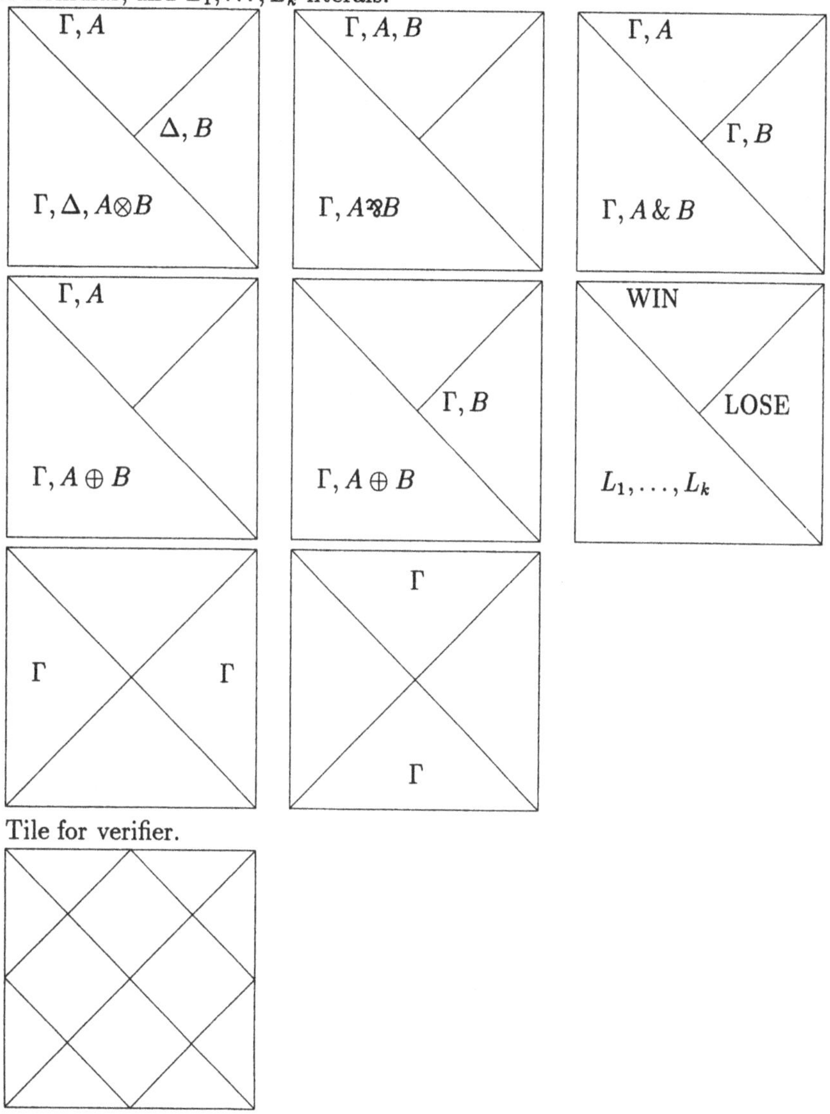

Γ, A

Δ, B

$\Gamma, \Delta, A \otimes B$

Γ, A, B

$\Gamma, A ⅋ B$

Γ, A

Γ, B

$\Gamma, A \,\&\, B$

Γ, A

$\Gamma, A \oplus B$

Γ, B

$\Gamma, A \oplus B$

WIN

LOSE

$L_1, \ldots, L_k$

Γ Γ

Γ

Γ

Tile for verifier.

Figure 1: Tiles for prover and verifier.

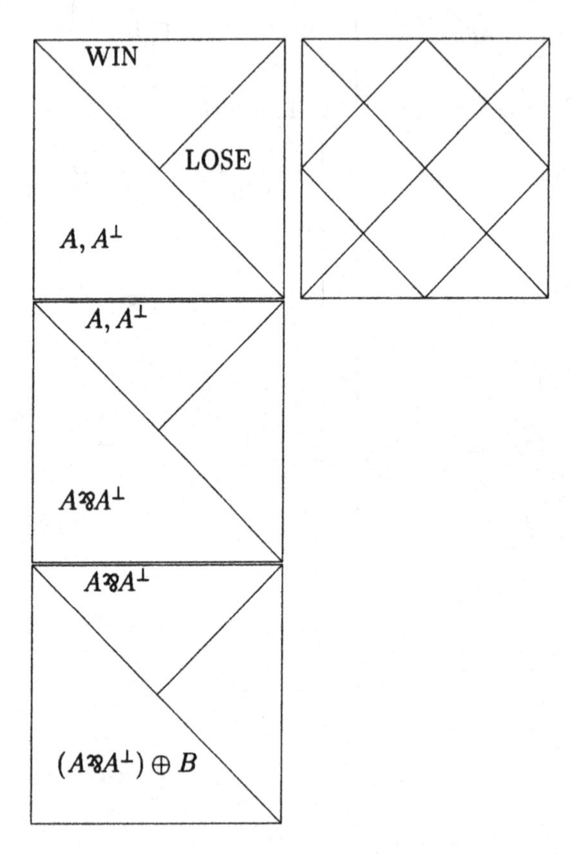

Figure 2: A winning play for $(A ⅋ A^{\perp}) \oplus B$.

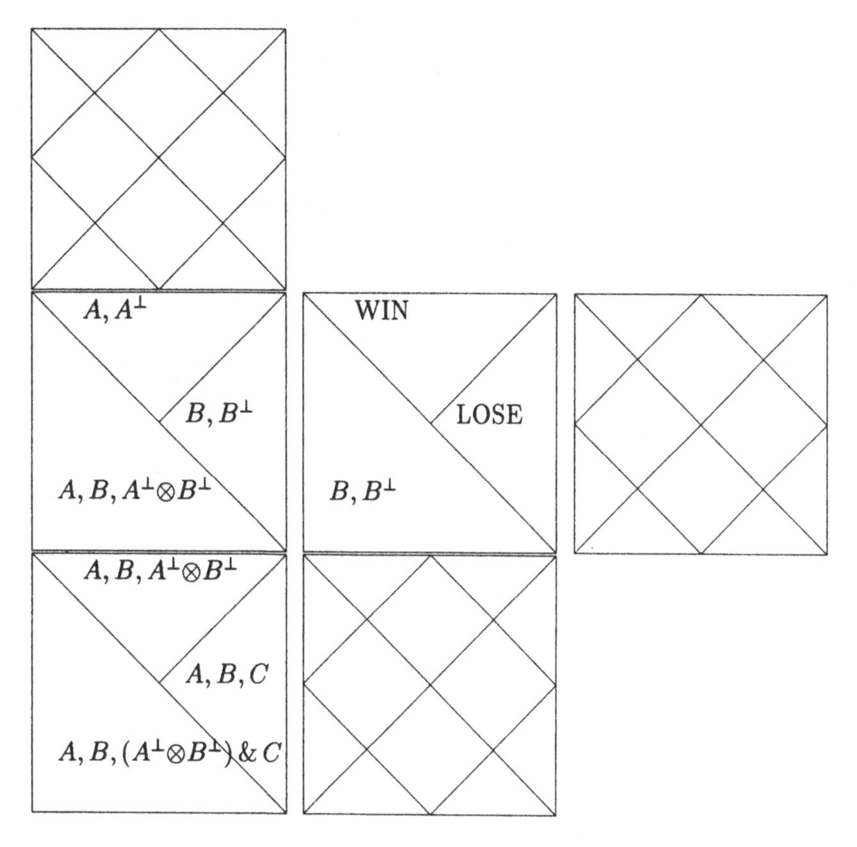

Figure 3: A winning play for $A, B, (A^{\perp} \otimes B^{\perp}) \& C$

for MALL formulas.

In this game, the two players, prover and verifier, alternate. The prover begins by placing a tile in the lower left corner of the board. The prover is not allowed to play connector tiles in this version of the game. If a tile has two sequents to the right of the diagonal, then the verifier blocks one of them and the prover must continue by placing a tile with matching sequent against the remaining, unblocked side or the tile. If there is only one sequent to the right of the diagonal, the verifier does not play and the prover may take another turn. When a literal tile is played, the verifier plays a final tile, indicating that the prover has won or lost. An example play, in which prover wins on an unprovable sequent, as the result of a legal but suboptimal play by the verifier, is shown in Figure 3.

Essentially, this game amounts to the construction of one branch of a proof tree, with the verifier allowed to select a branch for the sequent chosen by the prover. If the prover begins by playing a provable sequent (and a correct proof rule), then the prover can win, since all branches can be completed. If the prover plays an unprovable sequent, then the verifier can force the prover into a position where the branch of the proof tree exhibited on the game board cannot be completed. The length of any play is bounded by the number of connectives in the initial sequent, since each prover play reduces this number by one, and verifier keeps all sequents but one blocked.

The shortcoming of this game is that in order to play optimally, each player must have PSPACE computing power. Since the prover has to construct a proof, the prover must begin with a provable sequent and choose the correct rule at each play. If the prover tries to play an unprovable sequent, then the verifier must select the branch that cannot be completed. For example, if the prover begins with $A \,\&\, B$, where A and B are complicated formules with one provable and the other not, then the verifier can win only by determining which is provable. As a real game, this is not very satisfying, since when both players have PSPACE computing power, both can determine from the very first tile whether they have a winning strategy or not. Therefore, if this game were actually played between two capable players, it is likely that one would concede immediately after the first play, realizing that the other player had a winning strategy. This would be a bit like two adults playing tic-tac-toe.

2.4 Probabilistic form

The probabilistic game may be played by a knowledgeable prover against a limited, probabilistic opponent. A rough analogy exists with small hand-held electronic games such as Game Boy. This may help establish some intuition for the game. In our probabilistic game, the verifier has a fixed, simple strategy that is effective in a probabilistic sense. This strategy could be implemented with a minimal amount of circuitry (or simple program) in a Linear Game Boy that flips coins to decide some of its moves. This electronic game would have the properties that a knowledgeable prover could win most of the time by beginning with a provable sequent, and occasionally win with an unprovable sequent. At least in the daze of a transcontinental flight, such a game could keep a capable player occupied for some time, winning sometimes and losing others.

We outline the verifier's randomized strategy in Figures 4 and 5. A change from the previous game is that in one case, the verifier does not always block one of the sequents on a binary prover tile. This forces the prover to complete more than one branch of the proof tree. Thus the prover is allowed to play connector tiles to alleviate the inevitable board congestion (note that the number of leaves of a binary tree of depth n grows as 2^n, while our board only grows polynomially). However, the number of nonconnector prover tiles will

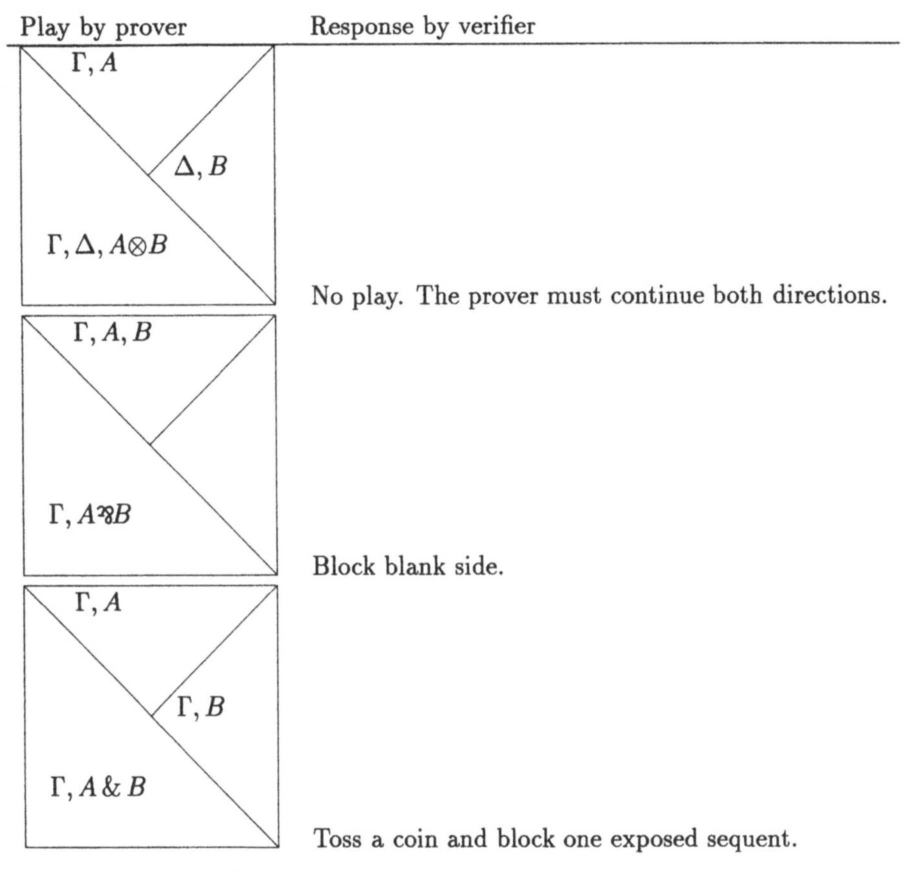

Play by prover
Response by verifier
Γ, A
Δ, B
Γ, Δ, A⊗B
No play. The prover must continue both directions.
Γ, A, B
Γ, A⅋B
Block blank side.
Γ, A
Γ, B
Γ, A & B
Toss a coin and block one exposed sequent.

Figure 4: Responses by stochastic verifier.

still be bounded by the number of connectives in the initial sequent. One may consider that at each play of any prover tile, connector tiles are automatically attached (When the sequent in the bottom left corner of the tile has n connectives, $2^{n-1}-1$ connectors could be attached in both the horizontal and vertical direction). This is unnecessarily many connectors in most cases, so we simply place the burden of planning the layout of the proof on the prover.

An initially counterintuitive aspect of this strategy is the verifier's play when the prover places a literal tile. However, this is absolutely critical to the success of the probabilistic strategy. An intuitive explanation is that if the prover begins with an unprovable sequent, then there is some branch of the proof tree of this sequent that does not lead to an axiom. However, if there are many &'s in the sequent, the proof tree will have many branches, and the verifier will choose among them randomly. Therefore, the prover has

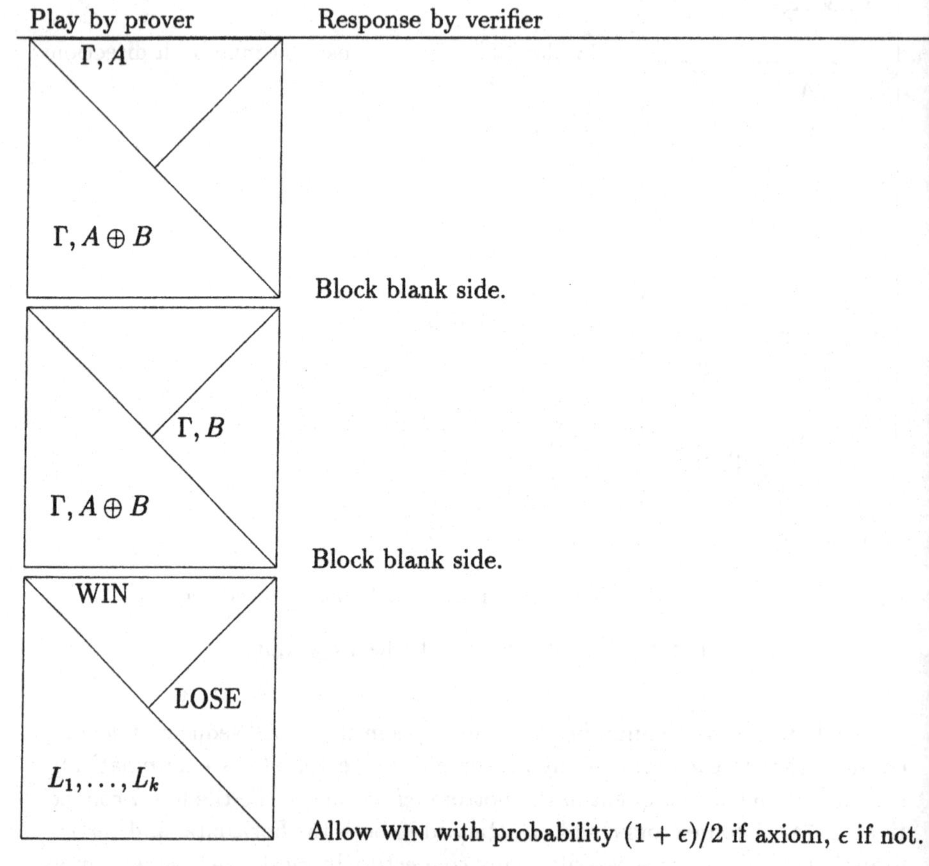

Figure 5: Responses by stochastic verifier.

a very good chance of winning, since the verifier will have only a very small probability of luckily guessing the branch that cannot be completed. Let us consider a short example illustrating this phenomenon. Say the prover begins with the sequent $A, A^{\perp} \& B$, allowing the verifier to choose between $A, A^{\perp}$ and A, B. If the verifier chooses randomly, then the prover has a 50% chance of getting $A, A^{\perp}$, which is an axiom if A is a literal, and A, B, which is not. Therefore, if the verifier were simply to mark WIN or LOSE deterministically, according to whether a sequence of literals is an axiom, the prover would have an equal chance of winning or losing from an unprovable sequent. Since the goal is to give the prover a good (over 50%) chance of winning only for provable sequents, some randomization is necessary on sequences of literals. With the strategy listed above, the prover has probability $(1 + 3\epsilon)/4$ of winning, which will be less than $1/2$ as long as ϵ is less than $1/3$.

A general analogy that may help understanding is a scenario where a class of n students are rewarded if they vote the correct answer to a true/false question. This might be a historical question, say, where there is a clear correct answer. If one student knows the answer, and the rest do not, they can vote the right answer with probability greater than one half if the student who knows the correct answer votes correctly and the remaining students flip coins and vote randomly. Of course, if they could communicate among themselves, the right answer could be passed around. But in the absence of communication, this randomized strategy works. The connection between the classroom scenario and the proof game is that when a proof branch ends in an axiom, the verifier really has very little knowledge about whether the initial sequent is really provable or not. So a randomized response is in order. On the other hand, when the proof branch ends unsucceessfully, there is more certain information, and a very low probability of winning is in order.

3 Preliminary Formalism

We now formalize the intuitive 'Game-Boy Semantics' presented in the previous section eventually making use of Papadimitrou's games against nature. We begin with a simple form of the tile-based game, and derive a min/max game semantics for MALL. This min/max game is played by two equally powerful players, the "prover" P and the "verifier" V. They are attempting to find whether the *value* of a given MALL sequent is 1 or 0, where the value v satisfies the conditions given below. Let A, B be MALL formulas, let Γ, Δ, Θ be MALL sequents, *i.e.*, finite multisets of MALL formulas, and let Ξ be a finite multiset of literals. Although we do not consider propositional constants $1, \perp, \top, 0$, the discussion here and throughout the paper may be readily extended to include them. We write $\Delta \uplus \Theta$ for the (disjoint) multiset union of Δ and Θ. As usual, we write Γ, A for the sequent obtained by adding an instance of A to Γ.

The **value** of a sequent is defined as follows:

$$v(\Gamma) = max\{v(\Gamma'; A) \mid \Gamma = \Gamma', A\},$$

$$v(\Gamma; A \otimes B) = max\{min\{v(\Delta, A), v(\Theta, B)\} \mid \Delta \uplus \Theta = \Gamma\},$$

$$v(\Gamma; A \parr B) = v(\Gamma, A, B),$$

$$v(\Gamma; A \oplus B) = max\{v(\Gamma, A), v(\Gamma, B)\},$$

$$v(\Gamma; A \& B) = min\{v(\Gamma, A), v(\Gamma, B)\},$$

$$v(\Xi) = 1 \text{ if } \Xi \text{ is an axiom, else } 0 .$$

Let us emphasize that the value of a sequent is the maximum possible satisfying these conditions. Specifically, if a sequent contains composite formulas, then several clauses regarding $v(\Gamma; A)$ might be applicable.

Player P chooses the case to be applied. In the case $\otimes$, P chooses a partition of Γ and requires both associated expresions to be evaluated. In the case $\oplus$, P chooses which of the two expressions will be evaluated. In the case $\&$, V chooses which of the two expressions will be evaluated. In the last case, V simply computes the value. Player V wins if one of the atomic sequents Ξ reached is not an axiom, else player P wins. Note that the number of moves is finite, indeed, it is polynomial in the size of a given MALL sequent. A similar game is described in Section 6 of [Bla92].

It is readily seen that:

Proposition 3.1 *The following are equivalent for any* MALL *sequent* Γ *:*

(1) Player P (prover) has a winning strategy,

(2) $v(\Gamma) = 1$,

(3) the sequent Γ is provable in MALL.

One drawback of this setting is that it *a priori* requires the verifier to have just as much insight as the prover, namely, both need to have PSPACE capability [LMSS92]. (When $A\&B$ is unprovable, the verifier must be able to choose an unprovable conjunct.) In other words, this setting is not truly interactive because both the prover and the verifier can figure out in advance who can win the game.

4 A stochastic semantics

A much more interesting interactive setting is obtained by replacing the features that involve universal branching (for instance, universal branching in the case & in the game just described) by *stochastic* features. The main idea is to replace the verifier's task of checking all the cases by "random spot checks". In such *prover/verifier interactions*, the prover has unlimited power while the verifier can only toss coins or perform simple (say, polynomial time) tasks when presented with the input or with the prover's messages [Pap85, BM88, GMR89]. Stochastic verification goes back to Solovay and Strassen [SS77].

As a first step in this direction, we consider "games against Nature" or "stochastic alternating polynomial time machines" (SAPTIME) in the style of [Pap85]. This setting consists of a polynomial time Turing machine (the *verifier*, V) and an omnipotent *prover*, P. V may toss fair coins (*i.e.*, consult a random oracle). V and P may exchange messages in polynomial time. A set of finite strings L belongs to the class SAPTIME if a single P can convince V to accept any $x \in L$ with probability $> 1/2$, while P or any other prover can convince V to accept any $x \notin L$ only with probability $< 1/2$. Note that NP is a degenerate case of a protocol in which there are no coin tosses and the probabilities are 1 and 0, respectively. It is known that SAPTIME = PSPACE [Pap85].

On the other hand, since MALL is PSPACE-complete [LMSS92] it is natural to ask whether there exists a direct, structural relationship between MALL and SAPTIME that goes beyond an immediate Turing reduction.

We seek a semantics of MALL based on randomization and interaction just described. This semantics should be compositional, sound and complete (in the sense of model theory), but it should also be efficient (in the sense of complexity theory), and it should imply that MALL is in SAPTIME.

In order to achieve this, we modify the game protocol described in Section 3 so that P can convince V to accept any provable MALL sequent Γ with probability greater than 1/2, while P or any other prover can convince V to accept any unprovable MALL sequent Γ with probability less than 1/2. The reader should keep in mind that in this setting the verifier has practically no insight: the verifier is simply a polynomial time Turing machine that can toss fair coins. More specifically, the verifier does not have the computational ability to figure out in advance who can win the game.

Let us observe that the simple two-valued semantic framework for MALL given in Section 3 does not suffice if V chooses conjuncts in the case & only at random. Indeed, if a given sequent is unprovable, a dishonest prover may concentrate all false information only on one branch of the entire game, while all other branches correctly end up with an axiom. This actually happens if the input is a single formula of the form $B_1 \& \ldots \& B_k$, where the association is to the right, B_k is unprovable, but B_i is provable for each $i < k$. Thus a more subtle semantics is needed.

Instead of two values 0 and 1, let us consider *yield functions*, linear real functions on the closed unit interval $[0,1]$. We continue the notation from Section 3. For each $0 \leq \epsilon \leq 1$ and a MALL sequent Γ the **yield** $y_\epsilon(\Gamma)$ is subject to the following conditions:

$$y_\epsilon(\Gamma) = max\{y_\epsilon(\Gamma'; A) \mid \Gamma = \Gamma', A\},$$

$$y_\epsilon(\Gamma; A \otimes B) = max\{min\{y_\epsilon(\Delta, A), y_\epsilon(\Theta, B)\} \mid \Delta \uplus \Theta = \Gamma\},$$

$$y_\epsilon(\Gamma; A ⅋ B) = y_\epsilon(\Gamma, A, B),$$

$$y_\epsilon(\Gamma; A \oplus B) = max\{y_\epsilon(\Gamma, A), y_\epsilon(\Gamma, B)\},$$

$$y_\epsilon(\Gamma; A \& B) = \frac{1}{2}[y_\epsilon(\Gamma, A) + y_\epsilon(\Gamma, B)],$$

$$y_\epsilon(\Xi) = \tfrac{1}{2}(1 + \epsilon) \text{ if } \Xi \text{ is an axiom, else } \epsilon\ .$$

For any sequent Γ let $\sigma(\Gamma) = y_0(\Gamma)$. It can be seen that $0 \leq \sigma(\Gamma) \leq 1/2$ and that the yield of a sequent Γ, as the function of ϵ, is a line segment between the points $(0, \sigma(\Gamma))$ and $(1, 1)$.

This time P is attempting to maximize the yield for a given MALL sequent. P's moves are the same as before. V, however, now chooses by a fair coin toss which of the two expressions in the & case will be evaluated. In the atomic case, with probability ϵ V accepts without looking at an atomic sequent Ξ, with probability $(1-\epsilon)/2$ V rejects without looking at Ξ, and with probability $(1-\epsilon)/2$ V first checks whether Ξ is an axiom and then accepts iff Ξ is an axiom.

Given a sequent, V wins if V ever rejects, else P wins. In other words, V accepts a given sequent iff V accepts all atomic sequents encountered.

Let us illustrate these notions on a simple example. Let A and B be purely multiplicative formulas in "conjunctive normal form", i.e., $\otimes$'s of ⅋s of literals. For such formulas it is readily seen that for any given ϵ the probabilities are: ϵ that V accepts the formula without looking at it, $(1-\epsilon)/2$ that V rejects the formula without looking at it, and $(1-\epsilon)/2$ that V first checks whether the formula is provable and then accepts it iff it is provable. Suppose A is provable and B is unprovable. When given the formula $A\&B$, the verifier tosses a fair coin and in each case proceeds as just described. Hence $y_\epsilon(A) = (1+\epsilon)/2$, $y_\epsilon(B) = \epsilon$, and thus $y_\epsilon(A\&B) = (1+3\epsilon)/4$.

Theorem 4.1 *For any sequent* Γ, $y_\epsilon(\Gamma) = \sigma(\Gamma) + (1-\sigma(\Gamma))\epsilon$. *A sequent* Γ *is provable in* MALL *iff* $\sigma(\Gamma) = 1/2$. *Furthermore:*

(1) The prover has a strategy such that if a sequent Γ *is provable in* MALL, *then for any* $0 < \epsilon \leq 1$ *the prover can convince the verifier to accept with probability* $> 1/2$. *For any such* ϵ *it is the case that* $y_\epsilon(\Gamma) > 1/2$.

(2) If Γ *is not provable in* MALL, *then there exists* $0 < \delta < 1$ *such that for*

any $0 \leq \epsilon \leq \delta$ *it is the case that* $y_\epsilon(\Gamma) < 1/2$. *For such* ϵ *any prover with any strategy can convince the verifier to accept only with probability* $< 1/2$. *It suffices to let* $\delta = 2^{-k-1}$, *where* k *is the number of occurrences of the connective* & *in the sequent* Γ.

Note that V has to be able to carry out only fair coin tosses and polynomial time tasks, because the two cases can be distinguished at $\epsilon = 2^{-k-1}$, where k is the number of occurrences of the connective & in the given sequent. For the same reason, the number of moves is polynomial in the length of the given sequent.

The protocol just described can be reformulated in such a way that the yield of a given sequent turns out to be a single real number between 0 and 1. Namely, the yield will be the probability that the verifier accepts, which will depend only on the given sequent.

Let $f(0) = f(1) = 1/2$. For $k > 1$ let

$$f(k) = \frac{1}{2} + \frac{1}{2^{2k}}.$$

In the reformulated protocol V begins by tossing a fair coin. If heads, then V tosses an unfair coin with distribution $f(k)\colon 1 - f(k)$, where k is the sum of the nesting & depths in the formulas in the given sequent, and V rejects or accepts based solely on the unfair coin toss. That is, if the first (fair) coin toss is heads, V rejects with probability $f(k)$ and accepts with probability $1 - f(k)$. If the first (fair) coin toss is tails, then P and V follow the simple protocol given in Section 3, except that in the case $A\&B$, V first computes the nesting & depths of both A and B. Say A is &-deeper than B by m. V tosses a fair coin to decide which conjunct will be kept. If the choice is B, then V replaces B by $B^+ = (\top\&\ldots(\top\&(\top\&B))\ldots)$, where there are m new &'s above B. The game now continues with a new round. The reader should observe that the unfair coin tosses may be replaced by independent subgames that involve only fair coin tosses.

It is not difficult to see that given a sequent Γ for which the sum of the nesting & depths of formulas in Γ is k, the overall probability of rejection due to the unfair coin tosses is

$$\frac{1}{2} - \frac{1}{2^{2k}} < \sum_{n=0}^{k} \frac{f(n)}{2^{1+k-n}} < \frac{1}{2}.$$

If the sequent Γ is provable, then the total probability of rejection is $< 1/2$ by the second inequality above, because P has a strategy in which the only rejections come from V's unfair coin tosses. Thus the probability of acceptance is $> 1/2$.

If Γ is not provable, then there is at least a 2^{-2k} probability that V actually discovers an atomic sequent Ξ that is not an axiom. By the first inequality above, the total probability of rejection is therefore $> 1/2$. Thus:

Theorem 4.2 *In the reformulated protocol:*

(1) The prover has a strategy such that if a sequent Γ is provable in MALL, *then the prover can convince the verifier to accept Γ with probability $> 1/2$.*
(2) If Γ is not provable in MALL, *then any prover with any strategy can convince the verifier to accept Γ only with probability $< 1/2$.*

5 Further research

In the future we plan to investigate a complete semantics of MALL based on "interactive proof systems" (IP) [BM88, GMR89]. In particular, we hope to relate MALL to random evaluation protocols for polynomial expressions [Sha92, She92].

The semantic framework for MALL discussed in this extended abstract might be just one instance of a broader relationship between linear logic and interactive protocols based on randomization. Indeed, interactive proof systems with two independent provers (MIP) are known to be NEXPTIME-complete [BFL91]. On the other hand, the first order analog of MALL, the multiplicative-additive fragment of first order linear logic MALL1 is also NEXPTIME-complete [LS94a, LS94b], thus the possibility of a direct relationship between MALL1 and MIP. The most satisfactory perspective would offer a single semantic framework for linear logic based on randomized interaction, which would have as instances the characterizations MLL = NP [Kan94, LW92], MALL = IP (=PSPACE), MALL1 = MIP (=NEXPTIME), and the undecidability of full linear logic [LMSS92].

References

[AJ94] S. Abramsky and R. Jagadeesan. Games and full completeness for multiplicative linear logic. *Journal of Symbolic Logic*, 59:543–574, 1994.

[AJM93] S. Abramsky, R. Jagadeesan, and P. Malacaria. Games and full abstraction for PCF. Manuscript, July 1993.

[BFL91] L. Babai, L. Fortnow, and C. Lund. Non-deterministic exponential time has two-prover interactive protocols. *Computational Complexity*, 1:3–40, 1991.

[Bla92] A. Blass. A game semantics for linear logic. *Annals of Pure and Applied Logic*, 56:183–220, 1992.

[BM88] L. Babai and S. Moran. Arthur-Merlin games: a randomized proof system, and a hierarchy of complexity classes. *J. Computer System Sciences*, 36:254–276, 1988.

[Gir87] J.-Y. Girard. Linear logic. *Theoretical Computer Science*, 50:1–102, 1987.

[GMR89] S. Goldwasser, S. Micali, and C. Rackoff. The knowledge complexity of interactive proof systems. *SIAM J. Computing*, 18:186–208, 1989.

[HO93] J.M.E. Hyland and L. Ong. Dialogue games and innocent strategies: an approach to intensional full abstraction for PCF. Manuscript, July 1993.

[Imm82] N. Immerman. Upper and lower bounds for first-order expressibility. *J. Computer System Sciences*, 25:76–98, 1982.

[Kan94] M.I. Kanovich. The complexity of Horn fragments of linear logic. *Annals of Pure and Applied Logic*, 69:195–241, 1994.

[LMSS92] P. Lincoln, J. Mitchell, A. Scedrov, and N. Shankar. Decision problems for propositional linear logic. *Annals of Pure and Applied Logic*, 56:239–311, 1992.

[LS94a] P. Lincoln and A. Scedrov. First order linear logic without modalities is NEXPTIME-hard. Accepted for publication in Theoretical Computer Science, 1994. Available by anonymous ftp from host ftp.cis.upenn.edu and the file pub/papers/scedrov/mall1.dvi.Z.

[LS94b] P. Lincoln and N. Shankar. Proof search in first order linear logic and other cut free sequent calculi. In *Proc. 9-th Annual IEEE Symposium on Logic in Computer Science, Paris*, pages 282–291, July 1994.

[LW92] P. Lincoln and T. Winkler. Constant-Only Multiplicative Linear Logic is NP-Complete. Manuscript, September 1992. Available using anonymous ftp from host ftp.csl.sri.com and the file pub/lincoln/comult-npc.dvi.

[Pap85] C. Papadimitriou. Games against Nature. *J. Computer System Sciences*, 31:288–301, 1985.

[Pap94] C. Papadimitriou. *Computational Complexity.* Addison-Wesley, 1994.

[Sce93] A. Scedrov. A brief guide to linear logic. In G. Rozenberg and A. Salomaa, editors, *Current Trends in Theoretical Computer Science*, pages 377–394. World Scientific Publishing Co., 1993.

[Sce94] A. Scedrov. Linear logic and computation: A survey. In H. Schwichtenberg, editor, *Proof and Computation, Proceedings Marktoberdorf Summer School 1993*, pages 281–298. NATO Advanced Science Institutes, Series F, Springer-Verlag, Berlin, 1994. To appear. Available by anonymous ftp from the host ftp.cis.upenn.edu. See pub/papers/TableOfContents.

[Sha92] A. Shamir. IP = PSPACE. *J. ACM*, 39:869–877, 1992.

[She92] A. Shen. IP = PSPACE: Simplified Proof. *J. ACM*, 39:878–880, 1992.

[SS77] R. Solovay and V. Strassen. A fast Monte-Carlo test for primality. *SIAM J. on Computing*, 6:84–85, 1977. Erratum 7:118, 1978.

Inheritance with exceptions: an attempt at formalization with linear connectives in Unified Logic

Christophe Fouqueré and Jacqueline Vauzeilles
LIPN-CNRS URA 1507
Université Paris-Nord
93430 Villetaneuse
Email: {cf,jv}@lipn.univ-paris13.fr

Abstract

The problems of inheritance reasoning in *taxonomical networks* are crucial in object-oriented languages and in artificial intelligence. A taxonomical network is a graph that enables knowledge to be represented. This paper focuses on the means linear logic offers to represent these networks and is a follow-up to the note on exceptions by Girard [Gir92a]. It is first proved that all *compatible* nodes of a taxonomical network can be deduced in the taxonomical linear theory associated to the network. Moreover, this theory can be integrated in the Unified Logic **LU** [Gir92b] and so taxonomical and classical reasoning can be combined.

1 Introduction

The problems of inheritance reasoning in taxonomical networks are crucial in object-oriented languages and in artificial intelligence. A *taxonomical network* is a graph that enables knowledge to be represented. The nodes represent concepts or properties of a set of individuals whereas the edges represent relations between concepts. The network can be viewed as a hierarchy of concepts according to levels of generality. A more specific concept is said to *inherit* informations from its subsumers. There are two kinds of edges: *default* and *exception*. A default edge between A and B means that A is generally a B or A has generally the property B. An exception edge between A and B means that there is an exception between A and B, namely A is not a B or A has not the property B. Nonmonotonic systems were developed in the last decade in order to attempt to represent defaults and exceptions in a logical way: the set of inferred grounded facts is the set of properties inherited by concepts. In this paper, we investigate the problem of formalizing inheritance in taxonomical networks with default and exception links in linear logic, or more precisely with linear connectives since the description of the taxonomical network is then immersed in the Unified Logic. This research was undertaken after a short paper by Girard [Gir92a] in which he noticed that "the consideration of weird classical models [in nomonotonic logics] definitely cuts the bridge with formal systems but also with informal reasoning" and that "the surprising ability of linear logic to encode various abstract machines indicates that much more can be done following this line". We extend his approach in three directions: first of all, we closely relate the linear formalization to intuitive reasoning on such taxonomical networks using so-called nets, second we show that we don't have to know in advance how many times facts have to be erased contrarily to Girard's assumption, finally we integrate our formalization in the Unified Logic pointing out the main properties (particularly a restriction in case of cyclicity).

In nonmonotonic systems, if T and S are two sets of formulae it is not necessarily true that the set of theorems of T is included in the set of theorems of $T \cup S$; in other words, if A, B, C are three formulae, "C is provable from A", doesn't necessarily lead to "C is provable from A and B". Of course, linear logic is monotonic; however, its linear implication behaves non-monotonically with respect to the "times" connective so, although the sequent $A \vdash C$ is provable in linear logic the sequent A "times" $B \vdash C$ is not necessarily provable in linear logic. Linear logic is therefore able to formalize problems which have so far been handled using nonmonotonic systems.

In what follows, only taxonomical networks with default and exception links are considered. An exception link inhibits in some way a concept: in figure 1[1], A is by default B and is an exception to the fact it could be C; whatever the number or the type of the paths between A and C, C is inhibited from the point of view of A.

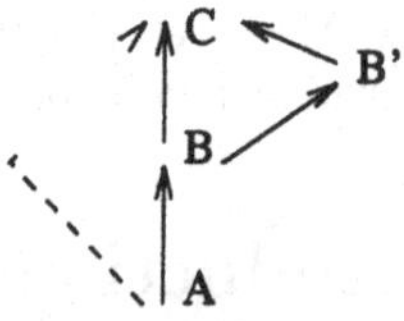

Figure 1: A simple taxonomical network

Example 1

[1]Default links (→) are drawn as undashed arrows whereas exception links (--→) are drawn as dashed arrows.

A Whale is a Cetacean; a Cetacean is a Mammal and a Mammal is generally a Land animal; but Cetaceans are an exception to be Land animals. Cetaceans are generally Plankton-eating animals. Plankton-eating animals and Land animals are Animals.
We want to conclude that a Whale is a Cetacean, a Mammal, a Plankton-eating animal, an Animal. A Mammal is generally a Land animal so an Animal. A Cetacean is a Mammal, a Plankton-eating animal so an Animal.

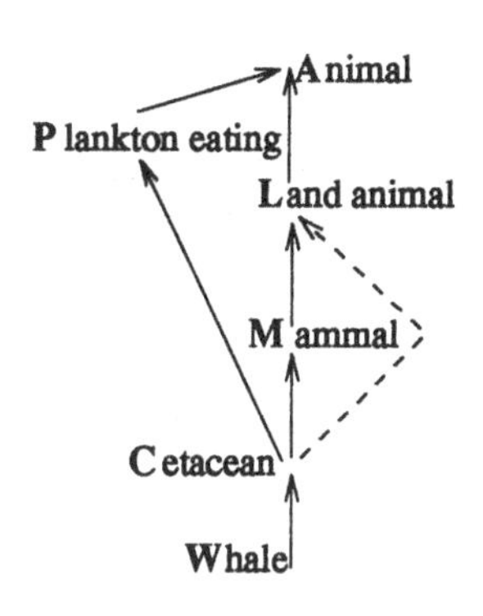

Example 2

The two exceptions (double arrow on a dashed line) define an even cycle. Let A be true; on one hand one can infer B, C; on the other hand one can infer B', C'.

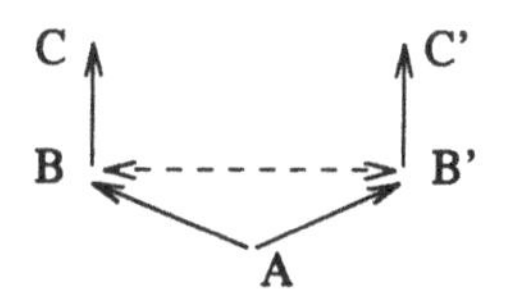

Example 3

A is generally a B, B is generally a C and C is an exception to be a B. We want to conclude that there is no solution to this problem.

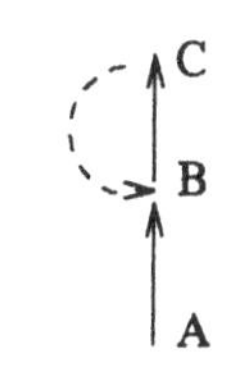

The second section concerns the definition of taxonomical networks as graphs and of compatible nodes as nodes of some (*completely correct*) subgraphs of the initial graph. These completely correct subgraphs are the solutions given in a graph theory manner. The third section is devoted to the formalization of these networks in linear logic, namely by taxonomical linear theories; we start by showing interesting properties on some provable sequents in such theories, namely simple sequents. The correspondence between completely correct graphs and simple sequents is done via nets. On one hand, nets give the essential structure of proofs of simple sequents; on the other hand it is proved that there is a one-to-one correspondence between nets and completely correct graphs, i.e. some subset of the vertices (the set of *white* vertices exactly) of a net are exactly compatible nodes of some completely correct subgraph of the taxonomical network. Then, in the section 4, the immersion of the taxonomical theories in a deductive system is performed; moreover, since the Unified Logic **LU** [Gir92b] is common to linear, intuitionistic and classical logic, taxonomical and classical reasoning can be combined.

2 Taxonomic networks with exceptions

Let us first define the taxonomic networks considered in this paper.

Definition 1 A **TNE** (Taxonomic Network with Exceptions) is defined as a quadruple $\mathcal{N} = \langle N, K, \rightarrow, \dashrightarrow \rangle$ such that:

- N is a finite set of nodes;
- $\rightarrow, \dashrightarrow$ are two irreflexive relations on $N \times N$;
- $\langle N, \rightarrow \rangle$ is an oriented acyclic graph;
- $K \subset \{a \in N /$ there is no $b \in N$ such that $(b, a) \in \rightarrow\}$; we call K the *kernel* of $\mathcal{N}$.

Definition 2 Let $\mathcal{N} = \langle N, K, \rightarrow, \dashrightarrow \rangle$ be a TNE;

- a *default path* (resp. *default path with exceptions*) in $\mathcal{N}$ is a sequence $a_1, \ldots, a_n$ of nodes of N such that:
  - $a_1 \in K$;
  - for each i $(1 \leq i \leq n-1)$ $(a_i, a_{i+1}) \in \rightarrow$; (resp. $(a_i, a_{i+1}) \in \rightarrow$ or $(a_i, a_{i+1}) \in \dashrightarrow$ and there exists at least one i $(1 \leq i \leq n-1)$ such that $(a_i, a_{i+1}) \in \dashrightarrow$);

  a_n is the *end* of the path $a_1, \ldots, a_n$;
- a *correct path* $a_1, \ldots, a_n$ in $\mathcal{N}$ is
  - either a default path;
  - either a default path with exceptions such that $a_1, \ldots, a_{n-1}$ is a default path and $(a_{n-1}, a_n) \in \dashrightarrow$, and $a_n \notin K$.

In the examples, the relations $\rightarrow$ and $\dashrightarrow$ are respectively represented by the plain line and the dashed line arrows.

Example (1 continued) The TNE $\mathcal{N}_1$ is represented by the graph of example 1: (the following abbreviations are used: W, M, C, P, L, A stand respectively for Whale, Mammal, Cetacean, Plankton-eating, Land animal, Animal)

$N_1 = \{W, C, M, P, L, A\}$,

$K_1 = \{W\}$.

For instance, W, C, M, L, A and W, C, P are default paths so are correct paths; W, C, L, A is a default path with exceptions but is not a correct path; W, C, L is a default path with exceptions and is a correct path.

Example (2 continued) The TNE $\mathcal{N}_2$ is represented by the graph of example 2:

$$N_2 = \{A, B, B', C, C'\},$$

$$K_2 = \{A\}.$$

For instance, A, B, C and A, B', C' are default paths so are correct paths; A, B, B', C' and A, B, B', B, C are default paths with exceptions but are not correct paths; A, B, B' and A, B', B are default paths with exceptions and are correct paths.

Example (3 continued) The TNE $\mathcal{N}_3$ is represented by the graph of example 3:

$$N_3 = \{A, B, C\},$$

$$K_3 = \{A\}.$$

For instance, A, B, C and A, B, C, B are correct paths.

Definition 3 Let $\mathcal{N} = \langle N, K_\mathcal{N}, \rightarrow_\mathcal{N}, \dashrightarrow_\mathcal{N}\rangle$ be a TNE;

- a subgraph $\mathcal{M} = \langle M, K_\mathcal{M}, \rightarrow_\mathcal{M}, \dashrightarrow_\mathcal{M}\rangle$ of $\mathcal{N}$ is *complete* if and only if
  - $K_\mathcal{M} = K_\mathcal{N}$;
  - if $a \in M$ then there exists a correct path in $\mathcal{M}$ (and in $\mathcal{N}$) such that its end is a;
  - if $a \in M$ then either $\{b; (a, b) \in \rightarrow_\mathcal{M}\} = \{b; (a, b) \in \rightarrow_\mathcal{N}\}$ and $\{b; (a, b) \in \dashrightarrow_\mathcal{M}\} = \{b; (a, b) \in \dashrightarrow_\mathcal{N}\}$, or $\{b; (a, b) \in \rightarrow_\mathcal{M}\} = \{b; (a, b) \in \dashrightarrow_\mathcal{M}\} = \emptyset$;
- a subgraph $\mathcal{M} = \langle M, K_\mathcal{M}, \rightarrow_\mathcal{M}, \dashrightarrow_\mathcal{M}\rangle$ of $\mathcal{N}$ is *completely correct* if and only if $\mathcal{M}$ is a complete subgraph of $\mathcal{N}$ such that:
  - each path in $\mathcal{M}$ is correct;
  - if $a \in M$ and $\{b; (a, b) \in \rightarrow_\mathcal{M}\} = \{b; (a, b) \in \dashrightarrow_\mathcal{M}\} = \emptyset$ then either $\{b; (a, b) \in \rightarrow_\mathcal{N}\} = \{b; (a, b) \in \dashrightarrow_\mathcal{N}\} = \emptyset$ or there exists a default path with exceptions in $\mathcal{M}$ such that its end is a.

Example (1 continued) This is the only completely correct subgraph of $\mathcal{N}_1$.

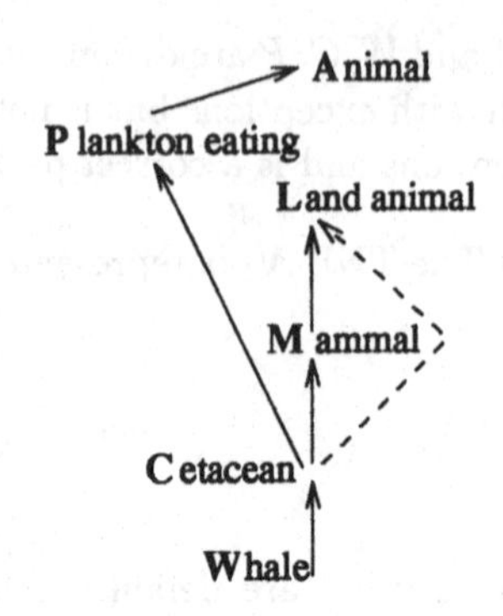

Example (2 continued) The following graphs are the two completely correct subgraphs of $\mathcal{N}_2$.

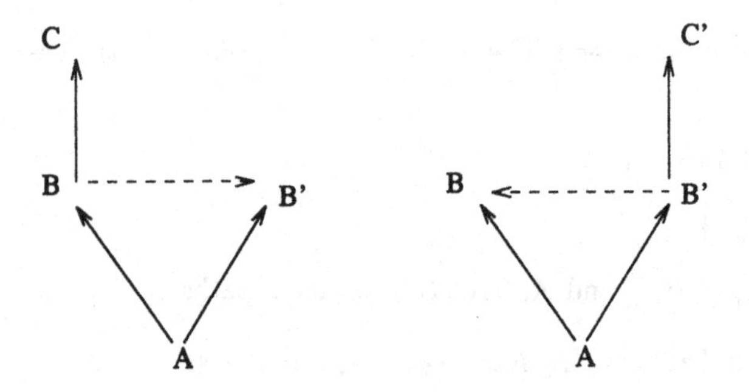

Example (3 continued) The TNE $\mathcal{N}_3$ does not admit a completely correct subgraph:

- the following subgraphs are not completely correct subgraphs of $\mathcal{N}_3$:

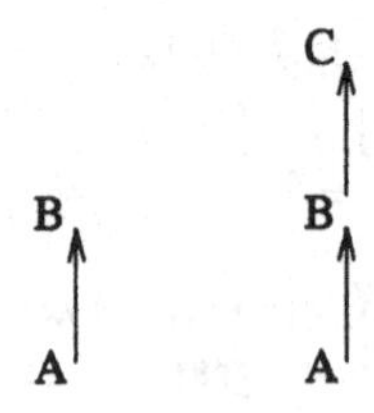

- $\mathcal{N}_3$ is not a completely correct subgraph of itself because the path A, B, C, B, C is not correct.

The three previous examples prove that a TNE can admit zero, one or many completely correct subgraphs. The lemma 22 gives a sufficient condition to the existence of one and only one completely correct subgraph of a TNE.

Definition 4 The nodes $B_1, \ldots, B_n$ of a net $\mathcal{N} = \langle N, K_{\mathcal{N}}, \rightarrow_{\mathcal{N}}, \dashrightarrow_{\mathcal{N}} \rangle$ are said to be *compatible* if and only if there exists a completely correct subgraph $\mathcal{M} = \langle M, K_{\mathcal{M}}, \rightarrow_{\mathcal{M}}, \dashrightarrow_{\mathcal{M}} \rangle$ of $\mathcal{N}$ such that $B_1, \ldots, B_n$ are nodes of $\mathcal{M}$ and for each i $(1 \leq i \leq n)$ there does not exist a default path with exceptions in $\mathcal{M}$ whose end is B_i.

Example (1 continued) For instance, Whale, Cetacean and Animal are compatible nodes of $\mathcal{N}_1$, while Whale and Land animal are not compatible nodes of $\mathcal{N}_1$.

Example (2 continued) For instance, A, B and C are compatible nodes of $\mathcal{N}_2$, while B and B' are not compatible nodes of $\mathcal{N}_2$.

Example (3 continued) Since $\mathcal{N}_3$ does not admit a completely correct subgraph, there do not exist compatible nodes of $\mathcal{N}_3$.

3 Taxonomic Linear Theories

This section is devoted to the axiomatization in linear logic of taxonomic networks with defaults and exceptions. We start by defining the fragment of linear logic in which these networks can be axiomatized. It includes mainly the multiplicative fragment of linear logic. Then we go on to formalize these networks, namely by *taxonomic linear theories.* It is then proved that inferences correspond to a subset of provable sequents, *simple sequents.* Such proofs are related to *nets* via pseudo-nets. A pseudo-net is a finite graph composed of coloured vertices and of coloured oriented edges. Pseudo-nets constructed from standard proofs of simple sequents are called nets. It is proved that the main condition for a pseudo-net to be a net is that its orientation $\prec$ is an ordering and that nets correspond exactly to completely correct subgraphs of the taxonomic network considered.

3.1 Taxonomic Linear Theories

The above problem will be formalized in the intuitionistic fragment of linear logic using the constant 1, the multiplicative connectives "times" ($\otimes$) and "linear implication" ($\multimap$), and the exponential connective "of course" (!). The reader will find a complete discussion on linear logic in [Gir87], [Tro92]. The properties of the multiplicative fragment of linear logic are essential in our representation. First of all, we have to distinguish three facts concerning a node A w.r.t. a graph: the fact that A has to be inferred (A^+), that A has to be not inferred (A^-), that A could have exiting links (A^*). In fact, only one variable A is required as the three previous ones are aliases of formulae involving A and some special variables. The set of links coming out from a node A is formalized as a (proper) axiom where A^* is the lefthand side. Suppose we have from A one by default link to B and one exception link to C. This can be interpreted as A implies by default B and an exception on C. $\multimap$ and $\otimes$ are reasonable candidates respectively for 'implies' and 'and', hence $A^* \multimap B^* \otimes C^-$. Since weakening is not available, $B^* \otimes C^-$ does not imply B^*, hence the fact that C was excepted cannot be ignored. Finally, we have to erase each occurence of C^*. This can be done using $(C^* \multimap 1)$: from $C^* \otimes (C^* \multimap 1)$, we can prove 1 that is neutral w.r.t. the connective $\otimes$. Moreover, weakening and contraction are available on ! ensuring that this removal can be done sufficiently: from $!(C^* \multimap 1)$, we can infer $(C^* \multimap 1) \otimes !(C^* \multimap 1)$ or 1. It remains to do this as soon as it is necessary. This is done by a syntactical constraint on provable sequents (see the definition of simple sequents). Remark that this constraint can be omitted modifying slightly the formalization (formula $\Sigma_{\mathcal{N}}$, see section 4); nevertheless, we let this constraint here for sake of clarity.

The language $L_{\mathcal{N}}$

- The language $L_{\mathcal{N}}$ is the linear language constructed from a set $V_N = \{A; A \in N\} \cup \{+, -, *\}$ of propositional variables and including the multiplicative connectives $\otimes$ and $\multimap$, the exponential ! and the constant 1. For each $A \in N$ we say that A is an *atom*. For each atom A, the formula $+ \multimap A$ (resp. $- \multimap A$, $* \multimap A$) is abbreviated by A^+ (resp. A^-, A^*). $A^+, B^+, \ldots, A^-, B^-, \ldots, A^*, B^*, \ldots$ are called *quasi-atomic* formulae (or *quasi-atoms*). The quasi-atomic formulae A^+ (resp. A^-, A^*) are called *plus* (resp. *minus*, *star*) quasi-atomic formulae.

- The *literals* we consider are of the following forms: quasi-atomic formulae, $(A^* \multimap 1)$ with $A \in N$, $!(A^* \multimap 1)$ with $A \in N$.

- Quasi-atomic formulae and literals of the form $!(A^* \multimap 1)$ are *strict* literals. Plus and minus quasi-atomic formulae are called *signed* literals.

- A *monomial* (resp. *signed monomial*) is a linear product ($\otimes$) of strict literals (resp. signed); a monomial has *incompatible* literals if it contains two signed literals of the same variable with opposite signs (A^+ and A^-, or B^+ and $B^-, \ldots$).

Taxonomic linear theories

Now, we can explain how taxonomic networks with exceptions are represented in linear logic.

Definition 5 Let $\mathcal{N} = \langle N, K_{\mathcal{N}}, \rightarrow, \dashrightarrow \rangle$ be a taxonomic network with exceptions, we define a taxonomic linear theory $\mathcal{T}(\mathcal{N})$ as a set of *taxonomic proper axioms* (in $L_{\mathcal{N}}$) as follows:

- $\mathcal{T}(\mathcal{N}) = \{\Gamma(A)/A \in N\}$ such that

$$\Gamma(A) \equiv A^* \vdash A^+ \otimes B_1^* \otimes \ldots \otimes B_m^* \otimes C_1^- \otimes !(C_1^* \multimap 1) \otimes \ldots \otimes C_p^- \otimes !(C_p^* \multimap 1)$$

$$\text{with} \qquad \begin{array}{l} \{B_i/i \in [1,m]\} = \{B/(A,B) \in \rightarrow\}, \\ \{C_j/j \in [1,p]\} = \{C/(A,C) \in \dashrightarrow\}, \end{array}$$

- $A \in K_{\mathcal{N}}$ iff A^* belongs to the set of *prerequisites* of simple sequents (see below).

Note that each taxonomic proper axiom could be written $A^* \vdash M$ where A is an atomic formula, M is a monomial such that A^+ occurs exactly once; no other plus quasi-atomic formula B^+ occurs in M; a strict literal $!(B^* \multimap 1)$ occurs in M if and only if B^- occurs in M; A^* does not occur in M.

A sequent is *simple* iff it is of the following form: $A_1^*, \ldots, A_r^* \vdash X$ where X is a signed monomial without incompatible literals and $K_{\mathcal{N}} = \{A_1, \ldots, A_r\}$ is the kernel of $\mathcal{N}$. We say that $A_1^*, \ldots, A_r^*$ are the *prerequisites* of the simple sequent $A_1^*, \ldots, A_r^* \vdash X$.

Example (1 continued)

$$\begin{array}{lll} \mathcal{T}(\mathcal{N}_1) \equiv \{ & W^* \vdash W^+ \otimes C^*, & C^* \vdash C^+ \otimes M^* \otimes P^* \otimes L^- \otimes !(L^* \multimap 1), \\ & M^* \vdash M^+ \otimes L^*, & L^* \vdash L^+ \otimes A^*, \\ & P^* \vdash P^+ \otimes A^*, & A^* \vdash A^+\} \end{array}$$

The only provable simple sequent of $\mathcal{T}(\mathcal{N}_1)$ is:

$$W^* \vdash W^+ \otimes C^+ \otimes M^+ \otimes P^+ \otimes A^+ \otimes L^-.$$

Example (2 continued)

$$\begin{array}{lll} \mathcal{T}(\mathcal{N}_2) \equiv \{ & A^* \vdash A^+ \otimes B^* \otimes B'^*, & B^* \vdash B^+ \otimes B'^- \otimes !(B'^* \multimap 1) \otimes C^*, \\ & C^* \vdash C^+, \quad C'^* \vdash C'^+, & B'^* \vdash B'^+ \otimes B^- \otimes !(B^* \multimap 1) \otimes C'^*\} \end{array}$$

The two provable simple sequents of $\mathcal{T}(\mathcal{N}_2)$ are:

$$A^* \vdash A^+ \otimes B^+ \otimes B'^- \otimes C^+$$
$$A^* \vdash A^+ \otimes B^- \otimes B'^+ \otimes C'^+$$

We show here an informal proof of the first provable simple sequent, while a formal (standard) proof is given in the figure 2. If we replace B^* by $B^+ \otimes B'^- \otimes !(B'^* \multimap 1) \otimes C^*$ in the sequent $A^* \vdash A^+ \otimes B^* \otimes B'^*$ we get the provable sequent (1) $A^* \vdash A^+ \otimes B^+ \otimes B'^- \otimes !(B'^* \multimap 1) \otimes C^* \otimes B'^*$; now, if we replace B'^* by $B'^+ \otimes B^- \otimes !(B^* \multimap 1) \otimes C'^*$ in the previous sequent we obtain a provable sequent $A^* \vdash A^+ \otimes B^+ \otimes B'^- \otimes !(B'^* \multimap 1) \otimes C^* \otimes B'^+ \otimes B^- \otimes !(B^* \multimap 1) \otimes C'^*$ but the consequent of this sequent contains incompatible literals (B^+ and B^-, B'^+ and B'^-). To obtain a provable simple sequent from the sequent (1), the only possibility is to delete in the sequent (1) B'^* using $!(B'^* \multimap 1)$ (since $(B'^* \otimes !(B'^* \multimap 1)) \multimap 1$), and then to replace C^* by C^+. The second provable simple sequent is obtained by replacing first B'^* by $B'^+ \otimes B^- \otimes !(B^* \multimap 1) \otimes C'^*$ in the sequent $A^* \vdash A^+ \otimes B^* \otimes B'^*$, then deleting B^*.

Example (3 continued)

$$\mathcal{T}(\mathcal{N}_3) \equiv \{ \quad A^* \vdash A^+ \otimes B^*, \qquad B^* \vdash B^+ \otimes C^*, \qquad C^* \vdash C^+ \otimes B^- \otimes !(B^* \multimap 1)\}$$

If we replace B^* by $B^+ \otimes C^*$ in the sequent $A^* \vdash A^+ \otimes B^*$ we get the provable sequent (1) $A^* \vdash A^+ \otimes B^+ \otimes C^*$; then, to obtain a provable simple sequent, C^* must be replaced by $C^+ \otimes B^- \otimes !(B^* \multimap 1)$; but the obtained sequent $A^* \vdash A^+ \otimes B^+ \otimes C^+ \otimes B^- \otimes !(B^* \multimap 1)$ contains incompatible literals. We conclude that there does not exist a simple sequent provable in $\mathcal{T}(\mathcal{N}_3)$.

Remark As it can be seen in the previous example 2, the formula $!(B'^* \multimap 1)$ is used to delete an occurrence of B'^*; moreover the formula $!(B'^* \multimap 1)$ can *'erase attribute B'^* wherever it occurs'* [Gir92a], and if the attribute B'^* does not occur the formula $!(B'^* \multimap 1)$ can also erase itself (since $!(B'^* \multimap 1)$ implies 1). To know in advance how many occurrences of B'^* must be destroyed is no longer necessary, in opposite with what Girard claimed [Gir92a].

It is reminded to the reader that Gentzen's cut elimination theorem holds for linear logic: *Let $\mathcal{T}$ be a linear theory and S a sequent (written in the language of $\mathcal{T}$), then for each proof Π of S, one can construct a proof Π' of S, all the cut-formulae of which belong to proper axioms of $\mathcal{T}$ (such cuts are called proper cuts).*

As consequence of the above theorem we have the well-known subformula property: in a proof Π of a sequent S (written in the language of $\mathcal{T}$), only occur subformulae of the proper axioms used in Π, and of formulae of S.

Definition 6 A *standard* proof of a simple sequent is a proof (with only proper cuts) in which the identity axioms $B \vdash B$ are restricted to the case where B is a quasi-atomic formula, every left premise of a $\multimap$-rule is an identity axiom, all $\otimes$-r, $\multimap$-l, !-l rules are used before cut-rules, and such that each proper axiom occurring in the proof is a premise of a cut-rule.

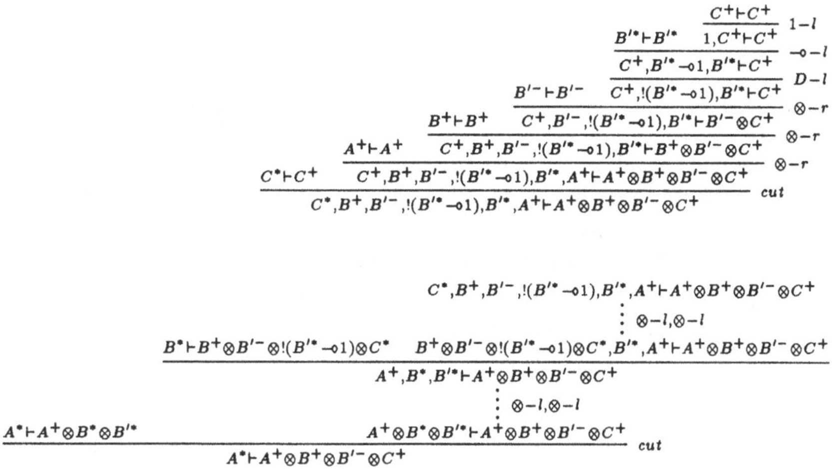

$$\dfrac{C^* \vdash C^+ \qquad \dfrac{A^+ \vdash A^+ \qquad \dfrac{B^+ \vdash B^+ \qquad \dfrac{B'^- \vdash B'^- \qquad \dfrac{\dfrac{B'^* \vdash B'^* \qquad \dfrac{C^+ \vdash C^+}{1, C^+ \vdash C^+}\ 1\text{-}l}{C^+, B'^* \multimap 1, B'^* \vdash C^+}\ \multimap\text{-}l}{C^+, !(B'^* \multimap 1), B'^* \vdash C^+}\ D\text{-}l}{C^+, B'^-, !(B'^* \multimap 1), B'^* \vdash B'^- \otimes C^+}\ \otimes\text{-}r}{C^+, B^+, B'^-, !(B'^* \multimap 1), B'^* \vdash B^+ \otimes B'^- \otimes C^+}\ \otimes\text{-}r}{C^+, B^+, B'^-, !(B'^* \multimap 1), B'^*, A^+ \vdash A^+ \otimes B^+ \otimes B'^- \otimes C^+}\ \otimes\text{-}r}{C^*, B^+, B'^-, !(B'^* \multimap 1), B'^*, A^+ \vdash A^+ \otimes B^+ \otimes B'^- \otimes C^+}\ cut$$

$$\dfrac{A^* \vdash A^+ \otimes B^* \otimes B'^* \qquad \begin{array}{c} \dfrac{B^* \vdash B^+ \otimes B'^- \otimes !(B'^* \multimap 1) \otimes C^* \qquad \begin{array}{c} C^*, B^+, B'^-, !(B'^* \multimap 1), B'^*, A^+ \vdash A^+ \otimes B^+ \otimes B'^- \otimes C^+ \\ \vdots\ \otimes\text{-}l, \otimes\text{-}l \\ B^+ \otimes B'^- \otimes !(B'^* \multimap 1) \otimes C^*, B'^*, A^+ \vdash A^+ \otimes B^+ \otimes B'^- \otimes C^+ \end{array}}{A^+, B^*, B'^* \vdash A^+ \otimes B^+ \otimes B'^- \otimes C^+} \\ \vdots\ \otimes\text{-}l, \otimes\text{-}l \\ A^+ \otimes B^* \otimes B'^* \vdash A^+ \otimes B^+ \otimes B'^- \otimes C^+ \end{array}}{A^* \vdash A^+ \otimes B^+ \otimes B'^- \otimes C^+}\ cut$$

Figure 2: Standard proof of the sequent $A^* \vdash A^+ \otimes B^+ \otimes B'^- \otimes C^+$

Lemma 7 If a simple sequent is provable, then it accepts a standard proof.

Proof see annex 6 lemma 27.

Example (2 continued) The figure 2 is a (standard) proof of the provable simple sequent $A^* \vdash A^+ \otimes B^+ \otimes B'^- \otimes C^+$.

Quasi-simple sequents occur in the (standard) proofs of simple sequents. A sequent is *quasi-simple* iff it is of the following form: $C_1, \ldots, C_q, M_1, \ldots, M_r \vdash X$ where C_i are non-strict literals, M_i are monomials and X is a signed monomial without incompatible literals. The notion of standard proof can obviously be extended to quasi-simple sequents.

3.2 Linear theories and nets

A proof contains many useless properties in its contexts. Girard has defined proof-structures [Gir87] and then particular proof structures which are proof-nets and which correspond exactly to proofs in (some fragment of) linear logic. Similarly, we define *pseudo-nets* and associate a pseudo-net to any (standard) proof of a simple sequent of $\mathcal{T}(\mathcal{N})$, which is said to be its net. The main criterion for a pseudo-net to be a net is that its vertices can be (partially) ordered. The notion of *pseudo-nets* and of *nets* in $\mathcal{T}(\mathcal{N})$ are closely connected with the notions of pseudo-plans and plans used in [Mas93] and we use the same terminology.

Definition 8 A *pseudo-net* (in $\mathcal{T}(\mathcal{N})$) is a finite graph composed of coloured vertices and of coloured oriented edges as follows.

- Each *vertex* is labelled by a triple (A, p_A, n_A) A being an atom of $\mathcal{T}(\mathcal{N})$, p_A and n_A being two integers. We say that A is the *name* of the vertex labelled by (A, p_A, n_A); in a pseudo-net there exists at most one vertex named A. Either the vertex (A, p_A, n_A) is white, n_A is null and the vertex is provided with the exits $xB_1, \ldots, xB_m, xC_1^-, \ldots, xC_p^-$: in this case A^* is the antecedent of the axiom $A^* \vdash A^+ \otimes B_1^* \otimes \ldots \otimes B_m^* \otimes C_1^- \otimes !(C_1^* \multimap 1) \otimes \ldots \otimes C_p^- \otimes !(C_p^* \multimap 1)$. Or the vertex (A, p_A, n_A) is black, n_A is strictly positive and the vertex has no exit: in this case A^- occurs in a consequent of an axiom of $\mathcal{T}(\mathcal{N})$.

- An *oriented* edge accepts an exit xO as its *origin* and a vertex named A as its *end*. An edge is given with the colour *white* if $O \equiv A$ is an atom or the colour *black* if $O \equiv A^-$. An edge such that its origin is an exit xA of the vertex $(B, p_B, 0)$ has the *weight* p_B.

- The *number-of-white-edges* w_A (resp. *number-of-black-edges* b_A) of white (resp. black) edges whose end is the vertex named A is the sum of weights of white (resp. black) edges whose end is the vertex named A. We still have $w_A \leq p_A$ and $b_A \leq n_A$.

- The *entries* of a pseudo-net are the vertices (A, p_A, n_A) such that $w_A < p_A$.

- A pseudo-net has no *exit*: all the exits of a vertex are connected with another vertex. The *orientation* of a pseudo-net is the transitive closure of the relation $\prec$ defined over the set of vertices by $(A, p_A, n_A) \prec (B, p_B, n_B)$ (or put simply $A \prec B$) iff there exists an exit xB of the vertex named A connected with the vertex named B.

Definition 9 To each subproof $\mathbb{D}$ of a standard proof of a simple sequent of $\mathcal{T}(\mathcal{N})$ we associate a pseudo-net $\mathbf{D}$ in the following way.

- If $\mathbb{D}$ is an identity axiom, then $\mathbf{D}$ is empty.

- If $\mathbb{D}$ is obtained from $\mathbb{E}$ by application of a $\otimes$-l rule, a 1-l rule or a D-l rule then $\mathbf{D}$ is identical to $\mathbf{E}$.

- If $\mathbb{D}$ is obtained from $\mathbb{D}_1$ and $\mathbb{D}_2$ by application of a $\otimes$-r rule remark that $\mathbf{D}_1$ and $\mathbf{D}_2$ only have black vertices or are empty, then $\mathbf{D}$ is the "union" of $\mathbf{D}_1$ and $\mathbf{D}_2$ defined in the following way: if a vertex (A, p_A, n_A) belongs to $\mathbf{D}_2$ and not to $\mathbf{D}_1$ then add this vertex to $\mathbf{D}_1$ else if the vertex (A, p_A, n_A) belongs to $\mathbf{D}_1$ and the vertex (A, p'_A, n'_A) belongs to $\mathbf{D}_2$ then replace in $\mathbf{D}_1$ the vertex (A, p_A, n_A) by the vertex $(A, p_A + p'_A, n_A + n'_A)$.

- If $\mathbb{D}$ is obtained from $\mathbb{E}$ by application of a $\multimap$-l rule on the formula $(A^* \multimap 1)$, then:
  - if a (black) vertex labelled (A, p_A, n_A) belongs to $\mathbf{E}$, then $\mathbf{D}$ is the pseudo-net obtained from $\mathbf{E}$ by replacing the label (A, p_A, n_A) by $(A, p_A + 1, n_A + 1)$;
  - if no vertex labelled (A, p_A, n_A) belongs to $\mathbf{E}$, then $\mathbf{D}$ is the pseudo-net obtained from $\mathbf{E}$ by adding the black vertex $(A, 1, 1)$.
- If $\mathbb{D}$ is obtained from $\mathbb{E}$ by application of a W-l rule on the formula $!(A^* \multimap 1)$ then:
  - if a (black) vertex labelled (A, p_A, n_A) belongs to $\mathbf{E}$, then $\mathbf{D}$ is the pseudo-net obtained from $\mathbf{E}$ by replacing the label (A, p_A, n_A) by $(A, p_A, n_A + 1)$;
  - if no vertex labelled (A, p_A, n_A) belongs to $\mathbf{E}$, then $\mathbf{D}$ is the pseudo-net obtained from $\mathbf{E}$ by adding the black vertex $(A, 0, 1)$.
- If $\mathbb{D}$ is obtained from $\mathbb{E}$ by application of a C-l rule on the formula $!(A^* \multimap 1)$ then $\mathbf{D}$ is the pseudo-net obtained from $\mathbf{E}$ by replacing the label (A, p_A, n_A) by $(A, p_A, n_A - 1)$.
- If $\mathbb{D}$ is obtained by application of a cut-rule between the sequent $A^* \vdash A^+ \otimes B_1^* \otimes \ldots \otimes B_m^* \otimes C_1^- \otimes !(C_1^* \multimap 1) \otimes \ldots \otimes C_p^- \otimes !(C_p^* \multimap 1)$ and the conclusion sequent of the proof $\mathbb{E}$, then:
  - if a (white) vertex labelled $(A, p_A, 0)$ belongs to $\mathbf{E}$, then $\mathbf{D}$ is the pseudo-net obtained from $\mathbf{E}$ by replacing the label $(A, p_A, 0)$ by $(A, p_A + 1, 0)$;
  - if no vertex labelled (A, p_A, n_A) belongs to $\mathbf{E}$, then $\mathbf{D}$ is the pseudo-net obtained from $\mathbf{E}$ by adding the white vertex $(A, 1, 0)$ and by linking each exit xB_j (resp. xC_j^-) to the vertex named B_j (resp. C_j) of $\mathbf{E}$.

Lemma 10 The above construction is correct: namely the graph that we have associated to a standard proof is a pseudo-net. Moreover, if $\mathbb{D}$ is the (standard) proof of the quasi-simple sequent $B_1^*, \ldots, B_p^*, !(C_1^* \multimap 1), \ldots, !(C_q^* \multimap 1) \vdash D_1^+ \otimes \ldots \otimes D_t^+ \otimes E_1^- \otimes \ldots \otimes E_s^-$, then the following conditions hold:

- Each D_i $(1 \leq i \leq t)$ is the name of a white vertex of $\mathbf{D}$; if A is the name of a white vertex of $\mathbf{D}$, then A^+ occurs exactly p_A times among $D_1^+, \ldots, D_t^+$.

- Each E_i ($1 \leq i \leq s$) is the name of a black vertex of **D**; for each black vertex named A, A^- occurs exactly b_A times among $E_1^-, \ldots, E_s^-$.
- Each B_i ($1 \leq i \leq p$) is an entry of **D**; all the entries of **D** are among $B_1, \ldots, B_p$; each entry A of **D** occurs exactly $p_A - w_A$ times among $B_1, \ldots, B_p$.
- Each C_i ($1 \leq i \leq q$) is the name of a black vertex (C_i, p_{C_i}, n_{C_i}) of **D** such that $b_{C_i} < n_{C_i}$; for each black vertex named A such that $b_A < n_A$, $!(A^* \multimap 1)$ occurs exactly $n_A - b_A$ times among $!(C_1^* \multimap 1), \ldots, !(C_q^* \multimap 1)$.

proof: by induction on a standard proof $\mathbb{D}$.

Definition 11 A *net* (in $\mathcal{T}(\mathcal{N})$) is a pseudo-net constructed from a (standard) proof of a simple sequent $A_1^*, \ldots, A_r^* \vdash D_1^+ \otimes \ldots \otimes D_t^+ \otimes E_1^- \otimes \ldots \otimes E_s^-$

Remark Since a net **D** is constructed from a proof of a simple sequent $A_1^*, \ldots, A_r^* \vdash D_1^+ \otimes \ldots \otimes D_t^+ \otimes E_1^- \otimes \ldots \otimes E_s^-$ then all the entries of **D** (namely the vertices A_i) are white and for each black vertex (B, p_B, n_B) of **D** we have $b_B = n_B$.

Lemma 12 A pseudo-net is a net if and only if all the entries are $A_1, \ldots, A_r$, are white, $p_{A_i} = 1$ ($1 \leq i \leq r$), for each black vertex (B, p_B, n_B) of **D** we have $b_B = n_B$ and if its orientation $\prec$ is an ordering.

Proof see annex 6 lemma 28.

Example 17 This is the net associated to a proof of the simple sequent $W^* \vdash W^+ \otimes C^+ \otimes M^+ \otimes P^+ \otimes A^+ \otimes L^-$.

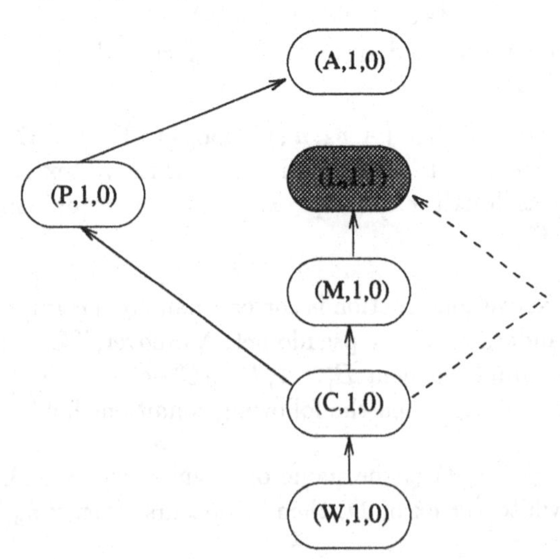

Example (2 continued) In the standard proof of the simple sequent $A^* \vdash A^+ \otimes B^+ \otimes B'^- \otimes C^+$ (figure 2), note that a $\multimap$-rule on the formula $(B'^* \multimap 1)$ is first applied; hence, to construct the net associated to the proof, a black vertex labelled $(B', 1, 1)$ is introduced; then, since a cut on the axiom $C^* \vdash C^+$ is performed, a white vertex labelled by $(C, 1, 0)$ is added; the cut on the axiom $B^* \vdash B^+ \otimes B'^- \otimes !(B'^* \multimap 1) \otimes C^*$ is translated into a white vertex $(B, 1, 0)$, the white link with the vertex named C and the black link with the vertex named B' are added. Finally the cut on the axiom $A^* \vdash A^+ \otimes B^* \otimes B'^*$ introduces the white vertex $(A, 1, 0)$ and the white links with vertices named B and B'.

Hence, the following net is associated to the proof of the figure 2.

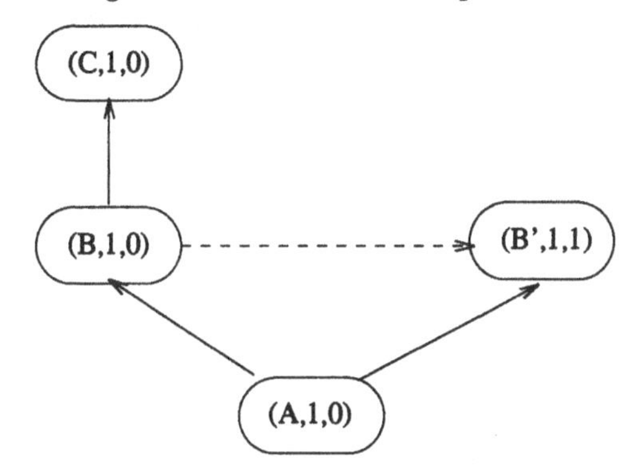

Note that this net can also been constructed following the informal reasoning of the section 3.1: first the axiom $A^* \vdash A^+ \otimes B^* \otimes B'^*$ is used introducing the white vertex $(A, 1, 0)$ and the two white exits B and B'; then the axiom $B^* \vdash B^+ \otimes B' - \otimes !(B'^* \multimap 1) \otimes C^*$ is used introducing the white vertex $(B, 1, 0)$, a white exit C and a black exit B'. Since one occurrence of B'^* is deleted using once $!(B'^* \multimap 1)$ the black vertex named B' is labelled by $(B', 1, 1)$. Finally the use of the axiom $C^* \vdash C^+$ is translated by the introduction of the white vertex $(C, 1, 0)$.

Example (3 continued) Since there does not exist a simple sequent provable in $\mathcal{T}(\mathcal{N}_3)$ there does not exist any net in $\mathcal{T}(\mathcal{N}_3)$.

Let us remark that there is a correspondence between proofnets of a taxonomical linear theory and the completely correct subgraphs of the corresponding taxonomical network. This correspondence is precisely described by the following definition and lemmas:

Definition 13 To each net $\mathbf{D}$ we associate a subgraph

$$\mathcal{R}(\mathbf{D}) = \langle R(\mathbf{D}), K_{\mathcal{R}(\mathbf{D})}, \rightarrow_{\mathcal{R}(\mathbf{D})}, \dashrightarrow_{\mathcal{R}(\mathbf{D})} \rangle$$

of $\mathcal{N}$ as follows:

- $K_{\mathcal{R}(\mathbf{D})} = \{A_1, \ldots, A_r\}$;
- each vertex (A, p_A, n_A) is translated into the node A;
- if there exists a white edge with origin a vertex named A and with end a vertex named B, then $(A, B) \in \rightarrow_{\mathcal{R}(\mathbf{D})}$;
- if there exists a black edge with origin a vertex named A and with end a vertex named B, then $(A, B) \in \dashrightarrow_{\mathcal{R}(\mathbf{D})}$.

Note that $\mathcal{R}(\mathbf{D})$ is a subgraph of $\mathcal{N}$.

Lemma 14 Let $\mathbf{D}$ be a net; then $\mathcal{R}(\mathbf{D}) = \langle R(\mathbf{D}), K_{\mathcal{R}(\mathbf{D})}, \rightarrow_{\mathcal{R}(\mathbf{D})}, \dashrightarrow_{\mathcal{R}(\mathbf{D})}\rangle$ is a completely correct subgraph of $\mathcal{N} = \langle N, K_{\mathcal{N}}, \rightarrow_{\mathcal{N}}, \dashrightarrow_{\mathcal{N}}\rangle$.

Proof see annex 6 lemma 29

Lemma 15 Let $\mathcal{M} = \langle M, K_{\mathcal{M}}, \rightarrow_{\mathcal{M}}, \dashrightarrow_{\mathcal{M}}\rangle$ be a completely correct subgraph of $\mathcal{N} = \langle N, K_{\mathcal{N}}, \rightarrow_{\mathcal{N}}, \dashrightarrow_{\mathcal{N}}\rangle$; then there exists a net $\mathbf{D}$ such that $\mathcal{R}(\mathbf{D}) \equiv \mathcal{M}$.

Proof see annex 6 lemma 30

Theorem 16 The nodes $B_1, \ldots, B_n$ of the TNE $\mathcal{N}$ whose kernel is $\{A_1, \ldots, A_r\}$ are compatible if and only if there exists a simple sequent $A_1^*, \ldots, A_r^* \vdash X$ such that $B_1^+, \ldots, B_n^+$ occur in X, provable in $\mathcal{T}(\mathcal{N})$.

Proof It is an immediate consequence of the two previous lemmas.

Note that nodes of a TNE are compatible if and only if they are the names of white vertices of a net.

4 Taxonomic networks in LU

The reader may consult [Gir92b] for an introduction on Unified Logic (**LU**) and its notations. Now we are interested in 'classical' reasoning involving our taxonomic network with exceptions. All atoms of $L_{\mathcal{N}}$ have to become classical, what is done by prefixing them with a symbol '!' and the language $L_{\mathcal{N}}$ is enriched into a langage L by adding new atoms.

Definition 17 The langage L is obtained by adding to $L_{\mathcal{N}}$ new propositional variables. If A is an atom of L, the formula $!A$ is said to be a *classical quasi-atom*. The *classical formulae* of L are the formulae built from the classical quasi-atoms of L by means of the classical connectives.

More precisely we are interested in all possible classical consequences of our TNE and of a set of classical formulae of L representing classical knowledge involving the classical quasi-atoms of the TNE. This will be done in theorems 21 and 23.

4.1 Integration of a linear theory into a (linear) deductive system

We add to the fragment of linear logic already considered the additive connective "with" (&). Let $\Sigma_{\mathcal{N}}$ be the following formula:

$$\Sigma_{\mathcal{N}} \equiv \otimes_{A \in N}!(A^{+} \multimap A \& 1) \& !(A^{-} \multimap 1).$$

Lemma 18 The sequent $\Sigma_{\mathcal{N}}, A_1^*, \ldots, A_r^* \vdash B_1 \otimes \ldots \otimes B_n$ is provable in $\mathcal{T}(\mathcal{N})$ if and only if $B_1, \ldots, B_n$ are compatible nodes of $\mathcal{N}$.

Proof see annex 6 lemma 31.

Remark Note that the sequent $\Sigma_{\mathcal{N}}, A_1^*, \ldots, A_r^* \vdash B$ is provable in $\mathcal{T}(\mathcal{N})$ iff there exists a correct path in $\mathcal{N}$ whose end is B.

4.2 Integration of a linear theory into LU

Let Γ be a proper axiom $A^* \vdash Y$ (Y being a monomial) of $\mathcal{T}(\mathcal{N})$; we denote by $\overline{\Gamma}$ the formula $A^* \multimap Y$. Let $\Gamma_{\mathcal{N}} = \overline{\Gamma_1}, \ldots, \overline{\Gamma_q}$ such that $\{\Gamma_1, \ldots, \Gamma_q\}$ is the set of axioms of $\mathcal{T}(\mathcal{N})$. All atoms and symbols $+$, $-$, $*$ are assumed to be neutral; hence all classical quasi-atoms are positive whereas the others are neutral. All the formulae of $\Gamma_{\mathcal{N}}$ are neutral.

Lemma 19 Let K^* be the formula $A_1^* \otimes \ldots \otimes A_r^*$. The following four conditions are equivalent:

- The sequent $\Sigma_{\mathcal{N}}, !K^* \vdash B_1 \otimes \ldots \otimes B_n$ is provable in $\mathcal{T}(\mathcal{N})$;
- $B_1, \ldots, B_n$ are compatible nodes of $\mathcal{N}$;
- The sequent $\Sigma_{\mathcal{N}}; K^*, \Gamma_{\mathcal{N}} \vdash ; !B_1 \wedge \ldots \wedge !Bn$ is provable in **LU**;
- The sequent $\Sigma_{\mathcal{N}}; K^*, \Gamma_{\mathcal{N}} \vdash ; B_1 \otimes \ldots \otimes B_n$ is provable in **LU**.

Proof see annex 6 lemma 32

Definition 20 A classical formula of the form $!C_1 \wedge \ldots \wedge !C_p \Rightarrow !D$ (resp. $!C_1 \wedge \ldots \wedge !C_p \Rightarrow !D_1 \vee \ldots \vee !D_q$) each C_i $(1 \leq i \leq p)$ and D (resp. each D_j $(1 \leq j \leq q)$) being an atom, is said to be a *Horn* (resp. a *clausal*) formula.

Now a set Δ of Horn formulae of L is considered and all classical consequences (under clausal form) of Δ and of our network will be deduced.

Theorem 21 Let $\mathcal{N} = \langle N, K_{\mathcal{N}}, \rightarrow_{\mathcal{N}}, \dashrightarrow_{\mathcal{N}} \rangle$ be a TNE, Δ be a set of Horn formulae of L and T be a clausal formula of L. The sequent $\Sigma_{\mathcal{N}}; K^*, \Gamma_{\mathcal{N}}, \Delta \vdash T;$ is provable in **LU** if and only if either T is a classical consequence of Δ or there exist compatible nodes $B_1, \ldots, B_n$ of $\mathcal{N}$ such that T is a classical consequence of Δ and of $!B_1 \wedge \ldots \wedge !B_n$.

Proof see annex 6 theorem 33

Remark The fact that the formulae of Δ are Horn formulae is essential as it is shown by the following example: the TNE of example 2 is considered and we set $\Delta \equiv \{!B \wedge !E \Rightarrow !D, !B' \wedge !F \Rightarrow !D, !E \vee !F\}$. The sequent $\Sigma_{\mathcal{N}_2}; A^*, \Gamma_{\mathcal{N}_2}, \Delta \vdash !D;$ can be proved though $!D$ is neither a classical consequence of Δ and $!A \wedge !B \wedge !C$ nor of Δ and $!A \wedge !B' \wedge !C'$ (note that compatible nodes of $\mathcal{N}_2$ are on the one hand A, B and C and on the other hand A, B', C'). But if the TNE $\mathcal{N} = \langle N, K_{\mathcal{N}}, \rightarrow_{\mathcal{N}}, \dashrightarrow_{\mathcal{N}} \rangle$ is such that the relation $\mathcal{R} \equiv \rightarrow_{\mathcal{N}} \cup \dashrightarrow_{\mathcal{N}}$ is acyclic then the theorem can be extended to the case where the formulae of Δ are arbitrary classical formulae (under conjunctive normal form): if the TNE $\mathcal{N}$ is such that $\mathcal{R}$ is acyclic, then, as it is proved in the following lemma, $\mathcal{N}$ admits one and only one completely correct subgraph, and this fact is essential to obtain the result.

Lemma 22 Let $\mathcal{N} = \langle N, K_{\mathcal{N}}, \rightarrow_{\mathcal{N}}, \dashrightarrow_{\mathcal{N}} \rangle$ be a TNE such that the relation $\mathcal{R} \equiv \rightarrow_{\mathcal{N}} \cup \dashrightarrow_{\mathcal{N}}$ is acyclic and such that there do not exist $a \in K_{\mathcal{N}}, b \in K_{\mathcal{N}}$ and $(a, b) \in \dashrightarrow_{\mathcal{N}}$; then $\mathcal{N}$ admits one and only one completely correct subgraph. Such a TNE is said to be *normal.*

Proof see annex 6 lemma 34

Theorem 23 Let Δ be a set of classical formulae of L and T be a classical formula of L (all classical formulae of Δ and T being under conjunctive normal form). Assume that $\mathcal{N} = \langle N, K_{\mathcal{N}}, \rightarrow_{\mathcal{N}}, \dashrightarrow_{\mathcal{N}} \rangle$ is a normal TNE. Then the sequent $\Sigma_{\mathcal{N}}; K^*, \Gamma_{\mathcal{N}}, \Delta \vdash T;$ is provable in **LU** if and only if either T is a classical consequence of Δ or there exist compatible nodes $B_1, \ldots, B_n$ of $\mathcal{N}$ such that T is a classical consequence of Δ and $!B_1 \wedge \ldots \wedge !B_n$.

Proof see annex 6 theorem 35

5 Conclusion

In previous sections, we presented an axiomatization for taxonomical networks with default and exception links. We have already improved our approach to deal with strict links or with counterexceptions [FV94]. One immediate extension appears to be disjunction. Since this feature is essential to knowledge representation, its formalization in Linear Logic could be a real challenge. The extension of the proofnet/ subnetwork relation to the disjunctive case could be studied.

Acknowledgments

We are undebted to Jean-Yves Girard for the fruitful discussions we had at the Linear Logic Workshop, in particular for calling our attention the formula $\Sigma_{\mathcal{N}}$.

References

[FV94] C. Fouqueré and J. Vauzeilles. Linear logic and exceptions. *Journal of Logic and Computation*, 4(6), 1994.

[Gir87] J.-Y. Girard. Linear logic. *Theoretical Computer Science*, 50:1–102, 1987.

[Gir92a] J.-Y. Girard. Logic and exceptions: A few remarks. *Journal of Logic and Computation*, 2:111–118, 1992.

[Gir92b] J.-Y. Girard. On the unity of logic. *Annals of Pure and Applied Logic*, 59:201–217, 1992.

[Mas93] M. Masseron. Generating plans in linear logic II: A geometry of conjunctive actions. *Theoretical computer Science*, 113:371–375, 1993.

[Tro92] A.S. Troelstra. *Lectures on Linear Logic.* Number 29 in Lecture Notes. CSLI, 1992.

6 Annex

6.1 Simple sequents

Lemma 24 If a simple sequent is provable then it admits a proof such that for all identity axioms $B \vdash B$ occurring in the proof, B is a quasi-atomic formula.

Proof Let Π be a proof of the simple sequent (with only proper cuts) in which all the identity axioms $A \vdash A$ are restricted to the case where A is atomic. Then, by permutation of some rules, the proof Π can be transformed into a proof Π' of the same sequent such that all $\multimap$-l rules on formulae $*\multimap A$ (resp. $+\multimap A$, $-\multimap A$) are applied to identity axioms and all $\multimap$-r rules on such formulae are applied exactly after $\multimap$-l rules. Now note that these rules are applied between identity axioms $* \vdash *$ (resp. $+ \vdash +$, $- \vdash -$) and $A \vdash A$.

Therefore, the identity axioms $B \vdash B$ can be restricted to the case where B is a quasi-atomic formula.

Notation • Let M be a monomial, we note $[M]$ the sequence of literals while replacing the connector $\otimes$ by the comma sign ','.

- Let Γ be a proper axiom $A^* \vdash M$; M will sometimes be denoted by $CONS(\Gamma)$.
- Let $M \equiv A_1 \otimes \ldots \otimes A_n$ be a formula such that each A_i is a strict literal; we set $A_i \in M$ to say that A_i occurs in M.

Lemma 25 Suppose there exists a proof Π (with only proper cuts) of the quasi-simple sequent $C_1, \ldots, C_q, M_1, \ldots, M_r \vdash X$ then a proof Π' of the sequent $C_1, \ldots, C_q, [M_1], \ldots, [M_r] \vdash X$ can be constructed.

Proof it is a well-known permutability property of $\otimes$-l rules.

Lemma 26 Suppose Π is a proof (with only proper cuts) of a quasi-simple sequent $C_1, \ldots, C_q, A_1, \ldots, A_r \vdash X$, where A_i are strict literals, and C_j are non-strict literals. Then we can exhibit a proof $\mathbb{D}'$ of the same sequent whose last rule is a cut on an axiom of $\mathcal{T}(\mathcal{N})$ (if any cut is used in the proof).

Proof by induction on Π:

- If only one rule is used and if it is a cut or if the last rule is a cut then the result holds.
- Let $\mathcal{R}$ be the last rule of Π:
  - if $\mathcal{R}$ is a $\multimap$-l rule: (with $C_1 : A \multimap 1$)

$$\frac{A_1, A_2, \ldots, A_q \vdash A \qquad 1, C_2, \ldots, C_q, A_{q+1}, \ldots, A_r \vdash X}{C_1, C_2, \ldots, C_q, A_1, A_2, \ldots A_r \vdash X}$$

    * if any cut is used in the proof of $1, C_2, \ldots, C_q, A_{q+1}, \ldots, A_r \vdash X$ then by induction hypothesis, we have a proof whose last rule is the following (with $\Gamma : A_{q+1} \vdash CONS(\Gamma)$)

$$\frac{A_{q+1} \vdash CONS(\Gamma) \qquad CONS(\Gamma), 1, C_2, \ldots, C_q, A_3, \ldots, A_r \vdash X}{1, C_2, \ldots, C_q, A_2, A_3, \ldots A_r \vdash X}$$

    we can transpose these two rules:

$$\frac{A_{q+1} \vdash CONS(\Gamma) \qquad \dfrac{A_1, A_2, \ldots, A_q \vdash A \qquad 1, CONS(\Gamma), C_2, \ldots, C_q, A_3, \ldots, A_r \vdash X}{CONS(\Gamma), C_1, \ldots, C_q, A_1, A_3, \ldots, A_r \vdash X}}{C_1, \ldots, C_q, A_1, A_2, A_3, \ldots, A_r \vdash X}$$

    thus the result holds;
    * else if any cut is used in the proof of $A_1, A_2, \ldots, A_q \vdash A$ we use induction hypothesis and proceed similarly.

- If $\mathcal{R}$ is a weakening rule, rules are transposed.
- If $\mathcal{R}$ is a !-l, rules are transposed.
- If $\mathcal{R}$ is a $\otimes$-r, one at least of the premises is of the previous form. Thus, using hypothesis induction, a $\otimes$-r has to be done before a cut.

Lemma 27 If a quasi-simple sequent is provable, then it accepts a standard proof.

Proof it is a consequence of the two above lemmas. Remark that if the sequent $A \vdash X$ is simple, then at least one cut is used in the proof.

6.2 Linear theories and nets

Lemma 28 A pseudo-net is a net if and only if all the entries are $A_1, \ldots, A_r$, are white, $p_{A_i} = 1$ $(1 \leq i \leq r)$, for each black vertex (B, p_B, n_B) of $\mathbf{D}$ we have $b_B = n_B$ and if its orientation $\prec$ is an ordering.

Proof
- We prove by induction on a standard proof $\mathbb{D}$ of a simple sequent that the construction of the net $\mathbf{D}$ gives an orientation which is an ordering. The other conditions are satisfied.
- Conversely, let $\mathbf{D}$ be a net; we construct a standard proof $\mathbb{D}$ of a simple sequent such that the net associated to $\mathbb{D}$ is $\mathbf{D}$:
  - let X be the following signed monomial:

    for each black vertex (A, p_A, n_A) A^- occurs n_A times in X;

    for each white vertex $(A, p_A, 0)$ A^+ occurs p_A times in X;
  - we construct a proof of $[X] \vdash X$;
  - for each black vertex (A, p_A, n_A), apply 1-l, $\multimap$-l, D-l, W-l, C-l rules to introduce at the left of the sequent p_A times the formulae A^* and n_A times the formula $!(A^* \multimap 1)$;
  - we set $S = \{A/(A, p_A, n_A)$ is a black vertex of $\mathbf{D}\}$;
  - let $(A, p_A, 0)$ be a white vertex of $\mathbf{D}$ which is maximal for the ordering $\prec$, or such that all its successors for the ordering $\prec$ are in S; if A is the name of a white vertex, then apply p_A times a cut between $\Gamma(A)$ and the above proof (after having applied the necessary $\otimes$-l rules); add A to S; we inductively apply this step until all A such that (A, p_A, n_A) is a vertex of $\mathbf{D}$ are in S.
  - we obtain a proof $\mathbb{D}$ of the sequent $A_1^*, \ldots, A_r^* \vdash X$ such that its net is $\mathbf{D}$.

Lemma 29 Let $\mathbf{D}$ be a net; then $\mathcal{R}(\mathbf{D}) = \langle R(\mathbf{D}), K_{\mathcal{R}(\mathbf{D})}, \rightarrow_{\mathcal{R}(\mathbf{D})}, \dashrightarrow_{\mathcal{R}(\mathbf{D})} \rangle$ is a completely correct subgraph of $\mathcal{N} = \langle N, K_{\mathcal{N}}, \rightarrow_{\mathcal{N}}, \dashrightarrow_{\mathcal{N}} \rangle$.

Proof • first remark that, by definition of a quasi-net, all paths are correct;

- $\mathcal{R}(\mathbf{D})$ is a $\mathcal{N}$-complete subgraph of $\mathcal{N}$:
  - $K_{\mathcal{R}(\mathbf{D})} = K_{\mathcal{N}} = \{A_1, \ldots, A_r\}$;
  - suppose that $A \in R(\mathbf{D})$; by the properties of a net we conclude that there exists a correct path $a_1, \ldots, a_n$ in $\mathcal{R}(\mathbf{D})$ such that $a_1 \equiv A_j$ $(1 \leq j \leq r)$ and $a_n \equiv A$;
  - if $A \in R(\mathbf{D})$ then A is the translated of (A, p_A, n_A); if (A, p_A, n_A) is white then by definition of the axiom $\Gamma(A)$ and of the exits of (A, p_A, n_A), we conclude that $\{B; (A, B) \in \rightarrow_{\mathcal{R}(\mathbf{D})}\} = \{B; (A, B) \in \rightarrow_{\mathcal{N}}\}$ and that $\{B; (A, B) \in \dashrightarrow_{\mathcal{R}(\mathbf{D})}\} = \{B; (A, B) \in \dashrightarrow_{\mathcal{N}}\}$; if (A, p_A, n_A) is black, then (A, p_A, n_A) has no exit and $\{B; (A, B) \in \rightarrow_{\mathcal{R}(\mathbf{D})}\} = \{B; (A, B) \in \dashrightarrow_{\mathcal{R}(\mathbf{D})}\} = \emptyset$;
- the above conditions prove that $\mathcal{R}(\mathbf{D})$ is a completely correct subgraph of $\mathcal{N}$.

Lemma 30 Let $\mathcal{M} = \langle M, K_{\mathcal{M}}, \rightarrow_{\mathcal{M}}, \dashrightarrow_{\mathcal{M}} \rangle$ be a completely correct subgraph of $\mathcal{N} = \langle N, K_{\mathcal{N}}, \rightarrow_{\mathcal{N}}, \dashrightarrow_{\mathcal{N}} \rangle$; then there exists a net $\mathbf{D}$ such that $\mathcal{R}(\mathbf{D}) \equiv \mathcal{M}$.

Proof We translate $\mathcal{M}$ into a net $\mathbf{D}$ as follows:

- the elements $A_1, \ldots, A_r$ of the kernel are translated into the white vertices $(A_1, 1, 0), \ldots, (A_r, 1, 0)$;
- let B be a node of $\mathcal{M}$ and suppose that there does not exist a node D of $\mathcal{M}$ such that $(D, B) \in \dashrightarrow_{\mathcal{M}}$; let $C_1, \ldots, C_q$ be all the nodes of $\mathcal{M}$ such that $(C_i, B) \in \rightarrow_{\mathcal{M}}$ $(1 \leq i \leq q)$; since $\mathcal{M}$ is a completely correct subgraph of $\mathcal{N}$, we can suppose that each C_i has been translated into a white vertex $(C_i, p_{C_i}, 0)$; then, we translate B into the white vertex $(B, p_B, 0)$ with $p_B = \sum_{1 \leq i \leq q} p_{C_i}$; each edge $\rightarrow_{\mathcal{M}}$ between C_i and B is translated into a white edge between the vertices $(C_i, p_{C_i}, 0)$ and $(B, p_B, 0)$;
- suppose that we have translated all the nodes B of $\mathcal{M}$ such that there does not exist a node D of $\mathcal{M}$ with $(D, B) \in \dashrightarrow_{\mathcal{M}}$; let E be a node of $\mathcal{M}$ and suppose that there exist nodes D of $\mathcal{M}$ such that $(D, E) \in \dashrightarrow_{\mathcal{M}}$: let $C_1, \ldots, C_q$ be all the nodes of $\mathcal{M}$ such that $(C_i, E) \in \rightarrow_{\mathcal{M}}$ $(1 \leq i \leq q)$ and let $D_1, \ldots, D_s$ be all the nodes of $\mathcal{M}$ such that $(D_i, E) \in \dashrightarrow_{\mathcal{M}}$

$(1 \leq i \leq s)$; since $\mathcal{M}$ is a completely correct subgraph of $\mathcal{N}$, all the nodes $C_1, \ldots, C_q, D_1, \ldots, D_s$ have been translated into white vertices $(C_1, p_{C_1}, 0), \ldots, (C_q, p_{C_q}, 0), (D_1, p_{D_1}, 0), \ldots, (D_s, p_{D_s}, 0)$; then the node E is translated into the black vertex (E, p_E, n_E) with $p_E = \sum_{1 \leq i \leq q} p_{C_i}$ and $n_E = \sum_{1 \leq i \leq s} p_{D_i}$; each edge $\rightarrow_{\mathcal{M}}$ (resp. $\dashrightarrow_{\mathcal{M}}$) between C_i (resp. D_i) and E is translated into a white (resp. black) edge between the vertices $(C_i, p_{C_i}, 0)$ (resp. $(D_i, p_{D_i}, 0)$) and (E, p_E, n_E).

6.3 Integration of a linear theory in LU

Lemma 31 The sequent $\Sigma_{\mathcal{N}}, A_1^*, \ldots, A_r^* \vdash B_1 \otimes \ldots \otimes B_n$ is provable in $\mathcal{T}(\mathcal{N})$ if and only if $B_1, \ldots, B_n$ are compatible nodes of $\mathcal{N}$.

Proof • Suppose that Π is a proof (with only proper cuts) in $\mathcal{T}(\mathcal{N})$ of the sequent $\Sigma_{\mathcal{N}}, A_1^*, \ldots, A_r^* \vdash B_1 \otimes \ldots \otimes B_n$. We can suppose that no $\multimap$-rule whose main formula is a quasi-atom is used in Π (extension of lemma 24). A proof Π' of a simple sequent $A_1^*, \ldots, A_r^* \vdash X$ such that $B_1^+, \ldots, B_n^+$ occur in X will be constructed, by induction on Π. Moreover Π' has the following property: each subproof Π_1 of the proof Π, of a sequent $\Sigma_1, \Gamma_1 \vdash M$ (resp. $\Sigma_1, \Gamma_1 \vdash C$) where M is a product of atoms (resp. C is a quasi-atomic formula), Σ_1 is a sequence of subformulae of $\Sigma_{\mathcal{N}}$ and Γ_1 is the sequence of others formulae, is replaced in the proof Π' by a subproof Π_1' of the sequent $\Sigma_1', \Gamma_1 \vdash M'$ (resp. $\Sigma_1', \Gamma_1 \vdash C$) where Σ_1' is a sequence of quasi-atomic formulae and M' is a signed monomial such that if A occurs in M then A^+ occurs in M'. The proof Π' is constructed inductively as follows:

- Each axiom $A \vdash A$ such A is an atom is replaced by the axiom $A^+ \vdash A^+$.
- If the last rule of the subproof of Π is a rule such that the main formula is not a subformula of $\Sigma_{\mathcal{N}}$ then the same rule is applied to construct Π': for instance, if the last rule is a $\otimes$-r rule applied to two subproofs Π_1 and Π_2 of the sequents $\Sigma_1, \Gamma_1 \vdash M_1$ and $\Sigma_2, \Gamma_2 \vdash M_2$ then apply a $\otimes$-r rule to the two subproofs Π_1' and Π_2' of the sequents $\Sigma_1', \Gamma_1 \vdash M_1'$ and $\Sigma_2', \Gamma_2 \vdash M_2'$.
- if the last rule of the subproof of Π is a $\multimap$-l rule whose main formula is a subformula of $\Sigma_{\mathcal{N}}$ then
  * if the main formula is $A^+ \multimap A \& 1$ then let Π_1 and Π_2 be the subproofs of Π of the sequents $\Sigma_1, \Gamma_1 \vdash A^+$ and $A \& 1, \Sigma_2, \Gamma_2 \vdash M_2$ which are the premises of the $\multimap$-l rule, and Π_1' and Π_2' be the proofs of the sequents $\Sigma_1', \Gamma_1 \vdash A^+$ and $A^+, \Sigma_2', \Gamma_2 \vdash M_2'$

or $\Sigma'_2, \Gamma_2 \vdash M'_2$ that we have already constructed. In the first case a cut between Π'_1 and Π'_2 is applied to obtain a proof of $\Sigma'_1, \Sigma'_2, \Gamma_1, \Gamma_2 \vdash M'_2$. In the second case a $\otimes$-r rule is applied to Π'_1 and Π'_2 to obtain a proof of $\Sigma'_1, \Sigma'_2, \Gamma_1, \Gamma_2 \vdash A^+ \otimes M'_2$;

* if the main formula is $A^- \multimap 1$ then let Π_1 and Π_2 be the subproofs of Π of the sequents $\Sigma_1, \Gamma_1 \vdash A^-$ and $1, \Sigma_2, \Gamma_2 \vdash M_2$ which are the premises of the $\multimap$-l rule, and Π'_1 and Π'_2 be the proofs of the sequents $\Sigma'_1, \Gamma_1 \vdash A^-$ and $\Sigma'_2, \Gamma_2 \vdash M'_2$ that we have already constructed. A $\otimes$-r rule is applied to Π'_1 and Π'_2 to obtain a proof of $\Sigma'_1, \Sigma'_2, \Gamma_1, \Gamma_2 \vdash A^- \otimes M'_2$;
* if the rule is a 1-l rule, a &-l rule, a !-l rule, a W-l rule, a C-l rule, a D-l rule then do nothing.

Hence a proof Π' of a sequent $A^*_1, \ldots, A^*_r \vdash X$ such that X is a signed monomial is obtained. Now it can be proved that X does not contain two incompatible literals: if a literal A^+ occurs in X then it becomes

- either from the translation of an identity axiom $A \vdash A$ of Π; in this case, since no atom A occurs in any proper axiom neither in the conclusion of Π, at least one $\multimap$-l rule whose main formula is $A^+ \multimap A\&1$ is used in Π;
- or from the translation of a $\multimap$-l rule by a $\otimes$-r rule (see above);

in the previous two cases, since $\multimap$-l rule whose main formula is $A^+ \multimap A\&1$ is used in Π, no $\multimap$-l rule whose main formula is $A^- \multimap 1$ can be used in Π (whereas the form of the formula $\Sigma_{\mathcal{N}}$); hence A^- cannot occur in X.

Since Π' is a proof of a simple sequent $A^*_1, \ldots, A^*_r \vdash X$ such that $B^+_1, \ldots, B^+_n$ occur in X, it can be concluded, by theorem 3.1, that $B_1, \ldots, B_n$ are compatible nodes of $\mathcal{N}$.

- Suppose that $B_1, \ldots, B_n$ are compatible nodes of $\mathcal{N}$; then by theorem 3.1 there exists a proof Π of a simple sequent $A^*_1, \ldots, A^*_r \vdash X$ such that $B^+_1, \ldots, B^+_n$ occur in X. Since X is a signed monomial without incompatible literals, a proof Π' of the sequent $\Sigma_{\mathcal{N}}, X \vdash B_1 \otimes \ldots \otimes B_n$ can be exhibited. Now, using a cut between Π and Π' the result holds.

If $\Gamma; \Delta \vdash \Lambda; \Pi$ is a sequent of **LU** then Γ (resp. Π) is said to be the linear antecedent (resp. consequent) of the sequent.

Lemma 32 Let K^* be the formula $A^*_1 \otimes \ldots \otimes A^*_r$. The following four conditions are equivalent:

i) The sequent $\Sigma_{\mathcal{N}}, !K^* \vdash B_1 \otimes \ldots \otimes B_n$ is provable in $\mathcal{T}(\mathcal{N})$;

ii) $B_1, \ldots, B_n$ are compatible nodes of $\mathcal{N}$;

iii) The sequent $\Sigma_{\mathcal{N}}; K^*, \Gamma_{\mathcal{N}} \vdash; !B_1 \wedge \ldots \wedge !B_n$ is provable in **LU**;

iv) The sequent $\Sigma_{\mathcal{N}}; K^*, \Gamma_{\mathcal{N}} \vdash; B_1 \otimes \ldots \otimes B_n$ is provable in **LU**.

Proof i) $\Rightarrow$ ii): First note that, since star quasi-formulae occur positively and negatively in the axioms of $\mathcal{T}(\mathcal{N})$, $!K^*$ cannot come uniquely by a W-l rule; now it can be easily deduced from the proof of $\Sigma_{\mathcal{N}}, !K^* \vdash B_1 \otimes \ldots \otimes B_n$, a proof of $\Sigma_{\mathcal{N}}, K^*, \ldots, K^* \vdash B_1 \otimes \ldots \otimes B_n$ (K^* occurring p times) for some integer p. Then, as in lemma 32, a proof of a sequent $K^*, \ldots, K^* \vdash Y$ can be constructed, Y being a signed monomial without incompatible literals and such that $B_1^+, \ldots, B_n^+$ occur in Y. By an easy extension of results of the section 3, it can be deduced that $B_1, \ldots, B_n$ are the names of white vertices of a net and then, by lemma 28, that $B_1, \ldots, B_n$ are compatible nodes of $\mathcal{N}$.

ii) $\Rightarrow$ iii): If $B_1, \ldots, B_n$ are compatible nodes of $\mathcal{N}$ then by theorem3.1 there exists a proof Π (in $\mathcal{T}(\mathcal{N})$) of a simple sequent $A_1^*, \ldots, A_r^* \vdash X$ such that $B_1^+, \ldots, B_n^+$ occur in X. Let $\Gamma'_1, \ldots, \Gamma'_q$ be the sequence (possibly with repetitions) of axioms of $\mathcal{T}(\mathcal{N})$ used in the proof Π. Then $A_1^*, \ldots, A_r^*, \overline{\Gamma'_1}, \ldots, \overline{\Gamma'_q} \vdash X$ is provable in linear logic (i.e. without proper axioms). Hence, if $X \equiv B_1^+ \otimes \ldots \otimes B_n^+ \otimes D_1^+ \otimes \ldots \otimes D_t^+ \otimes E_1^- \otimes \ldots \otimes E_s^-$ then the sequent

$$\begin{array}{c} !(B_1^+ \multimap B_1 \& 1), \ldots, !(B_n^+ \multimap B_n \& 1), \\ !(D_1^+ \multimap D_1 \& 1), \ldots, !(D_t^+ \multimap D_t \& 1), \\ !(E_1^- \multimap 1), \ldots, !(E_s^- \multimap 1), \\ K^*, \\ \overline{\Gamma'_1}, \ldots, \overline{\Gamma'_q} \end{array} \quad \vdash \quad B_i$$

is provable (for each i such that $1 \leq i \leq n$); then the sequent

$$\begin{array}{c} !(B_1^+ \multimap B_1 \& 1), \ldots, !(B_n^+ \multimap B_n \& 1), \\ !(D_1^+ \multimap D_1 \& 1), \ldots, !(D_t^+ \multimap D_t \& 1), \\ !(E_1^- \multimap 1), \ldots, !(E_s^- \multimap 1), \\ K^*, \\ \overline{\Gamma'_1}, \ldots, \overline{\Gamma'_q} \end{array} \quad ; \vdash \quad ; B_i$$

is provable in **LU**, so the sequent $\Sigma_{\mathcal{N}} \; ; \; K^*, \Gamma_{\mathcal{N}} \vdash \; ; \; !B_1 \wedge \ldots \wedge !B_n$ is provable in **LU**.

iii) $\Rightarrow$ iv): Since the sequent $!B_1 \wedge \ldots \wedge !B_n \; ; \vdash \; ; \; B_1 \otimes \ldots \otimes B_n$ is a provable sequent of **LU**, the result is obvious.

iv) $\Rightarrow$ i): If the sequent $\Sigma_{\mathcal{N}} \; ; \; K^*, \Gamma_{\mathcal{N}} \vdash \; ; \; B_1 \otimes \ldots \otimes B_n$ is provable in **LU** then the sequent $\Sigma_{\mathcal{N}}, !K^*, !\Gamma_{\mathcal{N}} \; ; \vdash \; ; \; B_1 \otimes \ldots \otimes B_n$ is also provable in **LU**. But all the formulae of the sequent are linear formulae and so the $\Sigma_{\mathcal{N}}, !K^*, !\Gamma_{\mathcal{N}} \vdash B_1 \otimes \ldots \otimes B_n$ is provable in linear logic and the sequent $\Sigma_{\mathcal{N}}, !K^* \vdash B_1 \otimes \ldots \otimes B_n$ is provable in $\mathcal{T}(\mathcal{N})$.

Theorem 33 Let $\mathcal{N} = \langle N, K_{\mathcal{N}}, \rightarrow_{\mathcal{N}}, \dashrightarrow_{\mathcal{N}} \rangle$ be a TNE, Δ be a set of Horn formulae of L and T be a clausal formula of L. The sequent $\Sigma_{\mathcal{N}}\ ;\ K^*, \Gamma_{\mathcal{N}}, \Delta \vdash T\ ;$ is provable in **LU** if and only if either the sequent $;\ \Delta \vdash T\ ;$ is provable in **LU** or there exist compatible nodes $B_1, \ldots, B_n$ of $\mathcal{N}$ such that T is a classical consequence of Δ and of $!B_1 \wedge \ldots \wedge !B_n$.

Proof

- Suppose that there exist compatible nodes $B_1, \ldots, B_n$ of $\mathcal{N}$ such that T is a classical consequence of Δ and of $!B_1 \wedge \ldots \wedge !B_n$. On the one hand the sequent $;\ !B_1 \wedge \ldots \wedge !B_n, \Delta \vdash T\ ;$ is provable in **LU**. On the other hand, by lemma 32 the sequent $\Sigma_{\mathcal{N}}\ ;\ K^*, \Gamma_{\mathcal{N}} \vdash\ ;\ !B_1 \wedge \ldots \wedge !B_n$ is provable in **LU**. Then, using structural rules and a cut, the result holds. If the sequent $;\ \Delta \vdash T\ ;$ is provable in **LU** then a proof of $\Sigma_{\mathcal{N}}\ ;\ K^*, \Gamma_{\mathcal{N}}, \Delta \vdash T\ ;$ can be constructed.

- First, note that $!B_1 \wedge \ldots \wedge !B_n$ is an interplolant between the linear and the classical part of the proved sequent. Suppose that the sequent $\Sigma_{\mathcal{N}}\ ;\ K^*, \Gamma_{\mathcal{N}}, \Delta \vdash T\ ;$ is provable in **LU**. Let T be the clausal formula $!C_1 \wedge \ldots \wedge !C_p \Rightarrow !D_1 \vee \ldots \vee !D_q$; then the sequent $\Sigma_{\mathcal{N}}\ ;\ K^*, \Gamma_{\mathcal{N}}, \Delta \vdash T\ ;$ is provable if and only if the sequent $\Sigma_{\mathcal{N}}\ ;\ K^*, \Gamma_{\mathcal{N}}, \Delta, !C_1, \ldots, !C_p \vdash !D_1 \vee \ldots \vee !D_q\ ;$ is provable.

  Let Π be a cut-free proof of this sequent.

  - Lemma 24 can be extended to this case: we can assume that all identity axioms $B\ ;\ \vdash\ ;\ B$ are restricted to the cases where B is an atom or a quasi-atom (no $\multimap$-rule whose main formula is a quasi-atom is used in the proof). Moreover if a D-l rule is applied to an atom A then A occurs in the linear consequent of this sequent and necessarily a S-r rule will be applied later on this right occurrence of A; note that the right and the left occurrences of A come from an identity axiom $A\ ;\ \vdash\ ;\ A$ which can be replaced by the identity axiom $!A\ ;\ \vdash\ ;\ !A$. Now it can be assumed that in the proof Π no D-l rule is applied to any atom.
  - Note that each formula occurring in the proof is either an atom or a linear formula (for instance non classical quasi-atoms) or a classical formula. An atom A will be considered as a linear formula while $!A$ (A being an atom) will be considered as a classical formula. If an atom A occurs in the right part of a sequent then a S-r right rule will be after applied in the proof to this occurrence of A. Each sequent

occurring in the proof Π is of the form $\Gamma", \Delta"\ ;\ \Gamma', \Delta' \vdash D\ ;\ E$ such that:

* $\Gamma", \Gamma'$ (resp. $\Delta", \Delta'$) are sequences (possibly empty) of linear (resp. classical) formulae; we can assume that all formulae of Δ' are negative and all formulae of $\Delta"$ are positive;
* D is a sequence (possibly empty) of formulae $!D_1 \vee \ldots \vee !D_q$;
* E is an atom or a quasi-atom or a classical formula or is empty;

note that D and E cannot be empty at the same time. For each subproof Π' of the sequent $\Gamma", \Delta"\ ;\ \Gamma', \Delta' \vdash D\ ;\ E$,

* if E is an atom or a non-classical quasi-atomic formula then it is constructed a proof $L(\Pi')$ of the sequent $\Gamma"\ ;\ \Gamma' \vdash\ ;\ E$;
* if E is a classical formula or if E is empty then there are constructed two proofs $L(\Pi')$ and $C(\Pi')$ respectively of the sequent
$$\Gamma"\ ;\ \Gamma' \vdash\ ;\ !B_1 \wedge \ldots \wedge !B_s$$
and of the sequent
$$!B_1 \wedge \ldots \wedge !B_s, \Delta"\ ;\ \Delta' \vdash D\ ;\ E$$
(each B_i $(1 \leq i \leq s)$ being an atom of L or the constant 1).

- These constructions are handled inductively on the proof Π as follows:
  * if Π' is an axiom $A\ ;\ \vdash\ ;\ A$ and A is a non classical quasi-atom or an atom then $L(\Pi') \equiv \Pi'$;
  * if Π' is an axiom $A\ ;\ \vdash\ ;\ A$ and A is a classical quasi-atom then $L(\Pi')$ is a proof of $;\ \vdash\ ;\ !1$ and $C(\Pi')$ is a proof of $!1, A\ ;\ \vdash\ ;\ A$;
  * if the last rule of Π' is a $\multimap$-l rule applied to the two subproofs Π_1 and Π_2 then note that the linear consequent of the conclusion of Π_1 is a non-classical quasi-atom. Then apply the $\multimap$-l rule to the same formulae of $L(\Pi_1)$ and $L(\Pi_2)$ to obtain $L(\Pi')$; if the consequent E of the conclusion of Π' is a classical formula and if Π_1 (resp. Π_2) proves the sequent $\Gamma"_1, \Delta"_1\ ;\ \Gamma', \Delta' \vdash D\ ;\ E_1$ (resp. $\Gamma"_2, \Delta"_2\ ;\ \Gamma', \Delta' \vdash D\ ;\ E$) then apply to $C(\Pi_2)$ left weakening rules followed by left permeability rules to formulae of $\Delta"_1$;
  * if the last rule of Π' is a $\otimes$-l or a &-l or a 1-l or a D-l rule applied to the subproof Π_1 then apply the same rule to $L(\Pi_1)$ and set $C(\Pi') \equiv C(\Pi_1)$;
  * if the last rule of Π' is a S-r rule applied to the subproof Π_1 and whose main formula is $!A$ then note that $\Gamma"$ and $\Delta"$ are empty; apply the S-r rule to $L(\Pi_1)$ to get $L(\Pi')$ and take for $C(\Pi')$ a proof of the sequent $!A\ ;\ \Delta' \vdash D\ ;\ !A$;

* if the last rule of Π' is a $\Rightarrow$-l rule applied to the two subproofs Π_1 and Π_2 then E is empty: suppose that Π_1 (resp. Π_2) proves the sequent $\Gamma", \Delta_1^" \,;\, \Gamma', \Delta' \vdash D \,;\, E_1$ (resp. $!B \,;\, \Gamma', \Delta' \vdash D \,;$) and that the main formula of the last rule of Π' is $E_1 \Rightarrow !B$. Then $L(\Pi_1)$ (resp. $L(\Pi_2)$) proves the sequent $\Gamma" \,;\, \Gamma' \vdash \,;\, !B_1 \wedge \ldots \wedge !B_s$ (resp. $\,;\, \Gamma' \vdash \,;\, !B'_1 \wedge \ldots \wedge !B'_t$) and $C(\Pi_1)$ (resp. $C(\Pi_2)$) proves the sequent $!B_1 \wedge \ldots \wedge !B_s, \Delta_1^" \,;\, \Delta' \vdash D \,;\, E_1$ (resp. $!B'_1 \wedge \ldots \wedge !B'_t, !B \,;\, \Delta' \vdash D \,;$). Then the proof $L(\Pi')$ of the sequent $\Gamma" \,;\, \Gamma' \vdash \,;\, !B_1 \wedge \ldots \wedge !B_s \wedge !B'_1 \wedge \ldots \wedge !B'_t$ and also the proof $C(\Pi')$ of $!B_1 \wedge \ldots \wedge !B_s \wedge !B'_1 \wedge \ldots \wedge !B'_t, E_1 \Rightarrow !B, \Delta_1^" \,;\, \Delta' \vdash D \,;$ can easily be constructed;
* the cases where the last rule of Π' is a $\wedge$-l or a $\wedge$-r or a $\vee$-r or a structural rule are left to the reader.

Note that no $\vee$-l rule is used in the proof Π.

- We have obtained formulae $B_1, \ldots, B_t$ (each B_i $(1 \leq i \leq t)$ being an atom of L or the constant 1) and two proofs $L(\Pi)$ and $C(\Pi)$ respectively of the sequent $\Sigma_{\mathcal{N}} \,;\, K^*, \Gamma_{\mathcal{N}} \vdash \,;\, !B_1 \wedge \ldots \wedge !B_t$ and of the sequent $!B_1 \wedge \ldots \wedge !B_t \,;\, \Delta, !C_1, \ldots, !C_p \vdash !D_1 \vee \ldots \vee !D_q \,;$ If all the B_i $(1 \leq i \leq t)$ are the constant 1 then note that the sequent $\,;\, \Delta, !C_1, \ldots, !C_p \vdash !D_1 \vee \ldots \vee !D_q \,;$ is provable and hence the sequent $\,;\, \Delta \vdash T \,;$ is provable too; else assume that, for instance, $B_1, \ldots, B_n$ $(n \leq t)$ are different from 1; in this latter case the sequents $\Sigma_{\mathcal{N}} \,;\, K^*, \Gamma_{\mathcal{N}} \vdash \,;\, !B_1 \wedge \ldots \wedge !B_n$ and $!B_1 \wedge \ldots \wedge !B_n \,;\, \Delta \vdash T \,;$ are provable. Then, by lemma 32, $B_1, \ldots, B_n$ are compatible nodes of $\mathcal{N}$, and T is a classical consequence of Δ and of $!B_1 \wedge \ldots \wedge !B_n$.

Lemma 34 Let $\mathcal{N} = \langle N, K_{\mathcal{N}}, \rightarrow_{\mathcal{N}}, \dashrightarrow_{\mathcal{N}} \rangle$ be a normal TNE; then $\mathcal{N}$ admits one and only one completely correct subgraph.

Proof It is constructed by induction on $\mathcal{R}$ a set E of nodes of $\mathcal{N}$ and a subgraph $\mathcal{M} = \langle M, K_{\mathcal{M}}, \rightarrow_{\mathcal{M}}, \dashrightarrow_{\mathcal{M}} \rangle$ of $\mathcal{N}$ satisfying the following properties:

- if $a \in E$ then for each b such that $(b, a) \in \mathcal{R}$ we have $b \in E$;
- if $a \in E$ and if all paths whose end is a are correct (in $\mathcal{N}$) then $a \in M$; moreover if all paths whose end is a are default paths (in $\mathcal{N}$) then for each b such that $(a, b) \in \mathcal{R}$ we have $b \in M$ and if $(a, b) \in \rightarrow_{\mathcal{N}}$ (resp. $\dashrightarrow_{\mathcal{N}}$) then $(a, b) \in \rightarrow_{\mathcal{M}}$ (resp. $\dashrightarrow_{\mathcal{M}}$).

Now E and $\mathcal{M}$ are constructed:

- First take $E = M = K_{\mathcal{N}}$; if $a \in K_{\mathcal{N}}$ and $(a, b) \in \mathcal{R}$ we set $M := M \in \{b\}$ and if $(a, b) \in \rightarrow_{\mathcal{N}}$ (resp. $\dashrightarrow_{\mathcal{N}}$) then we set $(a, b) \in \rightarrow_{\mathcal{M}}$ (resp. $\dashrightarrow_{\mathcal{M}}$);

- let $a \in N$ such that if $(b,a) \in \mathcal{R}$ we have $b \in E$; then $E := E \cup \{a\}$; if all paths whose end is a are default paths (in $\mathcal{N}$) then note that we have $a \in M$, and in this case $M := M \cup \{b; (a,b) \in \mathcal{R}\}$ and if $(a,b) \in \rightarrow_{\mathcal{N}}$ (resp. $\dashrightarrow_{\mathcal{N}}$) then we set $(a,b) \in \rightarrow_{\mathcal{M}}$ (resp. $\dashrightarrow_{\mathcal{M}}$). This process is done inductively. Finally $E = N$ and it is easy to verify that $\mathcal{M}$ is the only completely correct subgraph of $\mathcal{N}$.

Theorem 35 Let Δ be a set of classical formulae of L and T be a classical formula of L (all classical formulae of Δ and T being under conjunctive normal form). Assume that $\mathcal{N} = \langle N, K_{\mathcal{N}}, \rightarrow_{\mathcal{N}}, \dashrightarrow_{\mathcal{N}}\rangle$ is a normal TNE. Then the sequent $\Sigma_{\mathcal{N}}$; $K^*, \Gamma_{\mathcal{N}}, \Delta \vdash T$; is provable in **LU** if and only if either the sequent ; $\Delta \vdash T$; is provable in **LU** or there exist compatible nodes $B_1, \ldots, B_n$ of $\mathcal{N}$ such that T is a classical consequence of Δ and of $!B_1 \wedge \ldots \wedge !B_n$.

Proof It can be assumed, without any loss of generality, that all formulae of Δ are clausal formulae.

- First assume that T is a clausal formula. The proof is essentially handled as in theorem 2.1 but the case where the last rule of Π' is a $\vee$-l rule (when Π' is a subproof of Π) must be added in the construction of $L(\Pi)$ and of $C(\Pi)$.
  - As a matter of fact, if Π' is a subproof of Π of the sequent

    $$\Gamma", \Delta" \ ; \ \Gamma', \Delta' \vdash D \ ; \ E$$

    such that k $\vee$-l rules are used in Π' then it will be constructed from a sequence of proofs $L_1(\Pi'), \ldots, L_{k+1}(\Pi')$ of the sequent $\Gamma"$; $\Gamma' \vdash$; E if E is an atom or a non-classical quasi-atom, and of $\Gamma"$; $\Gamma' \vdash$; $!B_{11} \wedge \ldots \wedge !B_{1s_1}, \ldots, \Gamma"$; $\Gamma' \vdash$; $!B_{k+11} \wedge \ldots \wedge !B_{k+1s_{k+1}}$ if E is a classical formula or if E is empty. All cases are handled as in theorem 2.1. except if the last rule is a $\vee$-l rule:
    * if the last rule of Π' is a $\vee$-l rule applied to the two subproofs Π_1 and Π_2 and E is an atom or a non-classical quasi-atom: suppose that Π_1 (resp. Π_2) proves the sequent $!B, \Gamma", \Delta_1"$; $\Gamma', \Delta' \vdash D$; E (resp. $!C, \Gamma", \Delta_1"$; $\Gamma', \Delta' \vdash D$; E) and that the main formula of the last rule of Π' is $!B \vee !C$. Then for each i $(1 \leq i \leq t+1)$ we set $L_i(\Pi) = L_i(\Pi_1)$ and for each j $(1 \leq j \leq m+1)$ we set $L_{t+1+j}(\Pi) = L_j(\Pi_2)$ (if it is assumed that there is in Π_1 (resp. in Π_2) t (resp. m) $\vee$-l rules (note that $k = t + m + 1$);
    * if the last rule of Π' is a $\vee$-l rule applied to the two subproofs Π_1 and Π_2 and E is a classical formula or is empty: suppose that Π_1 (resp. Π_2) proves the sequent $!B, \Gamma", \Delta_1"$; $\Gamma', \Delta' \vdash D$; E

(resp. $!C, \Gamma", \Delta_1" \ ; \ \Gamma', \Delta' \vdash D \ ; \ E$) and that the main formula of the last rule of Π' is $!B \vee !C$. Then, by induction hypothesis, $L_1(\Pi_1), \ldots, L_{t+1}(\Pi_1)$ prove respectively the sequents

$$\Gamma" \ ; \ \Gamma' \vdash \ ; \ !B_{11} \wedge \ldots \wedge !B_{1s_1}$$

$$\ldots$$

$$\Gamma" \ ; \ \Gamma' \vdash \ ; \ !B_{t+11} \wedge \ldots \wedge !B_{t+1s_{t+1}}$$

and $L_1(\Pi_2), \ldots, L_{m+1}(\Pi_2)$ prove respectively the sequents

$$\Gamma" \ ; \ \Gamma' \vdash \ ; \ !C_{11} \wedge \ldots \wedge !C_{1v_1}$$

$$\ldots$$

$$\Gamma" \ ; \ \Gamma' \vdash \ ; \ !C_{m+11} \wedge \ldots \wedge !C_{m+1v_{m+1}}$$

(if it is assumed that there is in Π_1 (resp. in Π_2) t (resp. m) $\vee$-l rules; note that $k = t + m + 1$); then for each i ($1 \leq i \leq t+1$) we set $L_i(\Pi) = L_i(\Pi_1)$ and for each j ($1 \leq j \leq m+1$) we set $L_{t+1+j}(\Pi) = L_j(\Pi_2)$.

- If r $\vee$-l rules are used in Π we have constructed $r+1$ proofs $L_1(\Pi), \ldots, L_{r+1}(\Pi)$ of the sequents

$$\Sigma_{\mathcal{N}} \ ; \ K^*, \Gamma_{\mathcal{N}} \vdash \ ; \ !B_{11} \wedge \ldots \wedge !B_{1s_1}$$

$$\ldots$$

$$\Sigma_{\mathcal{N}} \ ; \ K^*, \Gamma_{\mathcal{N}} \vdash \ ; \ !B_{r+11} \wedge \ldots \wedge !B_{r+1s_{r+1}}.$$

  We can assume that all the B_{ij} are different from 1 (otherwise see the proof of theorem 4.1). By lemma 32, for each i ($1 \leq i \leq r+1$) $B_{i1}, \ldots, B_{is_i}$ are compatible nodes of $\mathcal{N}$. Hence, by lemma 2.1, all the B_{ij} ($1 \leq i \leq r+1$ and $1 \leq j \leq s_i$) are compatible. Let $B_1, \ldots, B_n$ be an enumeration of all the nodes B_{ij}.

- Now, as in theorem 33, a proof $C(\Pi)$ of the sequent $!B_1 \wedge \ldots \wedge !B_n \ ; \ \Delta, \vdash T$; can be constructed (by induction on the proof Π).

- Assume that T is under conjunctive normal form $T_1 \wedge \ldots \wedge T_k$; the previous result proves that there exist compatible nodes $B_1^i, \ldots, B_{n_i}^i$ such that $!B_1^i \wedge \ldots \wedge !B_{n_i}^i \ ; \ \Delta, \vdash T_i$; are provable sequents (for each i such that $1 \leq i \leq k$); since all B_j^i ($1 \leq i \leq k$ and $1 \leq j \leq n_i$) are compatible nodes, let $B_1, \ldots, B_n$ be an enumeration of all the nodes B_j^i, then a proof of the sequent $!B_1 \wedge \ldots \wedge !B_n \ ; \ \Delta, \vdash T$; can easily be constructed. And by lemma 32, the sequent $\Sigma_{\mathcal{N}} \ ; \ K^*, \Gamma_N \vdash \ ; \ !B_1 \wedge \ldots \wedge !B_n$ is provable.

On the Fine Structure of the Exponential Rule

Simone Martini   Andrea Masini

Abstract

We present natural deduction systems for fragments of intuitionistic linear logic obtained by dropping weakening and contractions also on !-prefixed formulas. The systems are based on a two-dimensional generalization of the notion of sequent, which accounts for a clean formulation of the introduction/elimination rules of the modality. Moreover, the different subsystems are obtained in a modular way, by simple conditions on the elimination rule for !. For the proposed systems we introduce a notion of reduction and we prove a normalization theorem.

1 Introduction

Proof theory of modalities is a delicate subject. The shape of the rules governing the different modalities in the overpopulated world of modal logics is often an example of what a good rule should *not* be. In the context of sequent calculus, if we want cut elimination, we are often forced to accept rules which are neither left nor right rules, and which completely destroy the deep symmetries the calculus is based upon. In the context of natural deduction the situation is even worse, since we have to admit deduction trees whose subtrees are not deductions, or, in the best case, elimination rules containing in their premise(s) the eliminated connective. On top of this, any such rule do not characterize (in a universal way, as category theoreticians would say) the modality it "defines": two different modality with the same rules bear no relation among each other (cf. Section 4.1).

S. Martini: Dipartimento di Matematica e Informatica, Università di Udine, Via Zanon, 6, I-33100 Udine – Italy; `martini@dimi.uniud.it`.

A. Masini: Dipartimento di Informatica, Università di Pisa, Corso Italia, 40, I-56125 Pisa – Italy; `masini@di.unipi.it`.

Work partially supported by HCM Project CHRX-CT93-0046 Lambda Calcul Typé and CNR-GNSAGA (Martini), and ESPRIT BRA 8130 LOMAPS (Masini).

As long as modalities are concerned, linear logic is no exception to this situation. Its (exponential) rules are not as bad as that of other modal logics, since they correspond to the modalities of S4, one the most civilized modal logics; still, as the attempt to define a natural deduction system shows (see, e.g., [BBdPH93] for a good discussion of a calculus for the multiplicatives and the exponentials), the rules are asymmetric and do not characterize the modalities. The situation becomes worse if one is interested in *subsystems* of linear logic where the power of the modalities is, in some way, limited (for instance dropping the property $!A \multimap A$ and/or the property $!A \multimap !!A$).

In [Mas92, Mas93] a generalization of the Gentzen format for sequents has been proposed, with the aim of giving a better proof theory for the minimal normal modal logic. The technique, however, is general and applies to a variety of situations where a notion of modality is present. It is the aim of this paper to extend it to some *completely linear* fragments of intuitionistic multiplicative linear logic, in the sense that we forbid weakening and contraction even on exponential formulas. The resulting systems are weak, but their study seems to provide insight on the *fine structure* of the exponentials, that is how the several properties they enjoy in the full logic interact.

Cutting down the power of the exponentials, indeed, appears as a promising subject of investigation. Exponentials are the culprits both for undecidability of linear logic and for the hyperexponential complexity of its cut-elimination, and it would be an important result to isolate natural fragments where only one of these "features" appears. An early proposal in this thread is bounded linear logic [GSS92]; an approach similar to our own work is presented in Section 5 of [DJS93] (see also [Sch94, Chap. 7]), where exponentials are indexed over a set equipped with a binary relation.

The main step taken in the natural deduction approach of [Mas93] is a two-dimensional generalization of the notion of sequent. Let us denote formulas with lowercase Greek letters $\alpha, \beta, \sigma, \ldots$, and sequences of formulas with uppercase Greek letters Γ, while ε will be the empty sequence. An *intuitionistic 2-sequent* is an expression of the form

$$\begin{array}{ccc} \Gamma_1 & & \varepsilon \\ \Gamma_2 & & \varepsilon \\ \vdots & \vdash & \vdots \\ \Gamma_k & & \sigma \end{array} \tag{1}$$

whose intended meaning is the formula

$$\bigwedge \Gamma_1 \supset !(\bigwedge \Gamma_2 \supset \ldots !(\bigwedge \Gamma_k \supset \sigma) \ldots).$$

Note that the formula σ, the *conclusion* of the deduction, lies at a *level*, k, which is greater than, or equal to, the level of any assumption it depends on (any of the Γ_j's may be empty, of course). The propositional rules act over these two dimensional structures in the expected way, just "respecting the levels". Formulas may change their level only by means of modal rules:

$$\frac{\begin{array}{ccc}\Gamma_1 & & \varepsilon \\ \vdots & \vdash & \vdots \\ \Gamma_k & & \varepsilon \\ \varepsilon & & \sigma\end{array}}{\begin{array}{ccc}\Gamma_1 & & \varepsilon \\ \vdots & \vdash & \vdots \\ \Gamma_k & & !\sigma\end{array}}\,!\mathcal{I} \qquad \frac{\begin{array}{ccc}\Gamma_1 & & \varepsilon \\ \vdots & \vdash & \vdots \\ \Gamma_k & & !\sigma\end{array}}{\begin{array}{ccc}\Gamma_1 & & \varepsilon \\ \vdots & \vdash & \vdots \\ \Gamma_k & & \varepsilon \\ \varepsilon & & \sigma\end{array}}\,!\mathcal{E}$$

Thus, the only way to introduce a modality on a formula occurrence σ at level k is that σ be the only formula present at that level. As a result of the rule, the introduced formula is lifted one level up. Vice versa, the elimination rule pushes a formula down one level (but there is no restriction on its premise). The levels thus represent in the calculus a notion of modal dependence: the conclusion σ at level k modally depends on the assumptions at the same level.

Before going into the details of the systems, we adopt a more compact representation for 2-sequents. Instead of writing two-dimensional judgements, we will denote each formula σ at level k with σ^k and write the judgment (1) as $\Gamma \vdash \sigma^k$, where Γ will be seen as a multiset $\{\tau_1^{i_1}, \ldots, \tau_n^{i_n}\}$. The reader should always bear in mind, however, that the indexes on formulas are only a metatheoretical notation for two dimensional structures.

2 Completely linear calculi

The standard sequent rules for ! can be divided in two classes: the *logical* ones (promotion and dereliction), in a sense defining the meaning of the connective; and the *structural* ones (weakening and contraction), allowing to recover in the context of linear logic the full intuitionistic and classical logic. An important subsystem of linear logic, thus, is the one where we retain the logical !-rules, but we drop the structural ones, giving rise to a "fully linear exponential". The best way to characterize this system is looking at the properties ! has to satisfy in any model. It is well known (e.g. [Tro91]) that a set of axioms for this purpose is:

K $!(\alpha \multimap \beta) \multimap (!\alpha \multimap !\beta)$
T $!\alpha \multimap \alpha$
4 $!\alpha \multimap !!\alpha$

From the point of view of the categorical interpretation of linear logic, **K** says that ! is a functor, while **T** and **4** specify, respectively, the two natural transformations, ε and δ, of a comonad.

The level machinery allows a clear and modular study of the several fragments we obtain selecting only some of those axioms. In particular, functoriality of ! will provide the basic introduction/elimination rules, while the natural transformations ε and δ will be obtained by suitable modifications of just the elimination rules. For the systems obtained in this way we will prove normalization.

2.1 Basic definitions

Formulas are built out of *atoms* (ranged over by p) and the constant $\mathbf{1}$; compound formulas are obtained with the connectives: ! (unary), $\otimes$ and $\multimap$ (binary). Any formula of the calculus will be marked with a *level index*, varying in $\mathbb{N}^+$; an indexed formula σ of level i will be written σ^i. A *set of assumptions* is a set $\Gamma = \{\sigma_1^{i_1}, \ldots, \sigma_n^{i_n}\}$; for a set of assumptions Γ, define $\#\Gamma = \max\{k_j \mid \sigma_j^{k_j} \in \Gamma\}$; $\#\Gamma = 0$ when Γ is empty.

Definition 2.1 [Calculus L^ℓ] *Deductions* are inductively defined by the following rules.

$$\sigma^k$$

$$\frac{}{\mathbf{1}^k}\,\mathbf{1}\mathcal{I} \qquad \frac{\begin{array}{cc}\Gamma & \Delta\\ \vdots & \vdots\\ \sigma^k & \mathbf{1}^j\end{array}}{\sigma^k}\,\mathbf{1}\mathcal{E}\ {}_{j\leq k}$$

$$\frac{\begin{array}{cc}\Gamma & \Delta\\ \vdots & \vdots\\ \sigma^k & \tau^k\end{array}}{\sigma\otimes\tau^k}\,\otimes\mathcal{I} \qquad \frac{\begin{array}{cc}\Gamma & [\sigma^j]\,[\tau^j]\,\Delta\\ \vdots & \vdots\\ \sigma\otimes\tau^j & \rho^k\end{array}}{\rho^k}\,\otimes\mathcal{E}\ {}_{j\leq k}$$

$$\frac{\begin{array}{c}\Gamma\,[\sigma^k]\\ \vdots\\ \tau^k\end{array}}{\sigma\multimap\tau^k}\,\multimap\mathcal{I} \qquad \frac{\begin{array}{cc}\Gamma & \Delta\\ \vdots & \vdots\\ \sigma\multimap\tau^k & \sigma^k\end{array}}{\tau^k}\,\multimap\mathcal{E}$$

where in $\multimap \mathcal{I}$ and $\otimes\mathcal{E}$ each discharging is compulsory and involves exactly one formula occurrence.

If S is one of the systems introduced below, we write $\Gamma \vdash_S \sigma^i$ when in S there is a deduction of σ^i from the assumptions of the multiset Γ.

2.2 Minimal linear exponential

The first fragment we have in mind is the one where only **K** is required; the system $L^\ell K$ is obtained from L^ℓ by adding the exponential rules:

$$\frac{\begin{array}{c}\Gamma\\ \vdots\\ \sigma^j\end{array}}{!\sigma^{j-1}}\ !\mathcal{I}\ {\scriptstyle j>\#\Gamma} \qquad\qquad \frac{\begin{array}{c}\Gamma\\ \vdots\\ !\sigma^j\end{array}}{\sigma^{j+1}}\ !\mathcal{E}$$

We thus have only functoriality of !; axiom **K** is obtained as follows.

$$\dfrac{\dfrac{\dfrac{\dfrac{\dfrac{[!\alpha^1]}{\alpha^2}\,!\mathcal{E} \qquad \dfrac{[!(\alpha\multimap\beta)^1]}{\alpha\multimap\beta^2}\,!\mathcal{E}}{\beta^2}\multimap\mathcal{E}}{!\beta^1}\,!\mathcal{I}}{!\alpha\multimap !\beta^1}\multimap\mathcal{I}}{!(\alpha\multimap\beta)\multimap(!\alpha\multimap !\beta)^1}\multimap\mathcal{I}$$

Observe that the deduction proceeds exactly as the first order derivation of the formula $\forall x(\alpha \supset \beta) \supset (\forall x\alpha \supset \forall x\beta)$.

2.3 T-Linear exponential

Axiom **T**, that is the comonad natural transformation $\varepsilon : ! \rightarrow \mathrm{Id}$, allows the "elimination" of a ! without incrementing the level index of a formula: $\varepsilon_{\alpha^j} : !\alpha^j \rightarrow \alpha^j$. The system $L^\ell KT$ is thus obtained extending the $!\mathcal{E}$ rule of $L^\ell K$ to take care of this possibility:

$$\frac{\begin{array}{c}\Gamma\\ \vdots\\ !\sigma^j\end{array}}{\sigma^k}\ !\mathcal{E}\ {\scriptstyle k\in\{j,j+1\}}$$

Axiom **T** can now be proved in a simple way:

$$\dfrac{\dfrac{[!\alpha^1]}{\alpha^1}\,!\mathcal{E}}{!\alpha \multimap \alpha^1} \multimap \mathcal{I}$$

2.4 4-Linear exponential

Axiom **4**, finally, specifies the natural transformation $\delta : ! \rightarrow !!$. Following the pattern of the previous case, we are thus looking for a rule allowing the derivation

$$\begin{array}{c} !\alpha^j \\ \vdots \\ !!\alpha^j \end{array}$$

At this point we could apply twice the minimal rule $!\mathcal{E}$ of $L^\ell K$, obtaining a deduction of α^{j+2} from the assumption $!\alpha^j$. We can pack this deduction in a single rule, and, since we can clearly apply this pattern many times, we may generalize the rule to obtain α^{j+k} from the assumption $!\alpha^j$. The system $L^\ell K4$ is thus obtained from $L^\ell K$ by extending its $!\mathcal{E}$ rule as:

$$\begin{array}{c} \Gamma \\ \vdots \\ \dfrac{!\sigma^j}{\sigma^k}\,!\mathcal{E} \quad {\scriptstyle k>j} \end{array}$$

The crucial fact is that this rule is in fact sufficient to obtain a proof of axiom **4**:

$$\dfrac{\dfrac{\dfrac{\dfrac{[!\alpha^1]}{\alpha^3}\,!\mathcal{E}}{!\alpha^2}\,!\mathcal{I}}{!!\alpha^1}\,!\mathcal{I}}{!\alpha \multimap !!\alpha^1} \multimap \mathcal{I}$$

2.5 Full linear exponential

The full power of the linear exponential (axioms **K**, **T**, and **4**) is recovered in system $L^\ell KT4$, whose rule $!\mathcal{E}$ encompasses the rules of the same name of the previous systems:

$$\dfrac{\begin{matrix}\Gamma\\ \vdots\\ !\sigma^j\end{matrix}}{\sigma^k}\ !\mathcal{E}\quad {\scriptstyle k\geq j}$$

REMARK. The four systems we have considered, $L^\ell K$, $L^\ell KT$, $L^\ell K4$, and $L^\ell KT4$, are distinct, since it can be shown that by relaxing the linearity constraints they respectively yield natural deduction systems for the (positive fragments of the) modal logics **K**, **KT**, **K4**, **S4** (after interpreting ! as necessity).

3 Normalization

It remains to show that the proposed calculi can be given a computational interpretation. For each calculus we define, as usual, a correct notion of reduction on deductions, proving the existence of a normalization strategy.

The redexes are defined as expected. In particular, the propositional contractions are standard, since they do not interfere with the level structure. More care has to be taken with the !-rules; in general, a redex is given by the following proof-figure, where the possible values of j depend on the specific calculus.

$$\dfrac{\dfrac{\begin{matrix}\mathcal{D}\\ \alpha^k\end{matrix}}{!\alpha^{k-1}}\ !\mathcal{I}}{\alpha^{k-1+j}}\ !\mathcal{E}$$

The level index of the conclusion may be different from the level index of the premise of $!\mathcal{I}$. The same situation arises in the first order calculus, where a redex is:

$$\dfrac{\dfrac{\begin{matrix}\mathcal{D}\\ \alpha\end{matrix}}{\forall x\alpha}\ !\mathcal{I}}{[t/x]\alpha}\ !\mathcal{E}$$

and its reduction requires the substitution of t for x in $\mathcal{D}$. The exact definition of this substitution, though conceptually non deep, is non trivial (see, e.g., [TvD88]). The following definition is for the level indexes what the usual substitution is for terms and variables.

Definition 3.1 [Level substitution] Let $\mathcal{D}$ be a deduction of conclusion α^v; let $i \geq 2$ and $n \in \{-1\} \cup \mathbb{N}$ if $v \geq 2$; let $i \geq 1$ and $n \in \mathbb{N}$ if $v \geq 1$. We inductively define the deduction $[n]_i\mathcal{D}$ (read: increment by n any level greater than or equal to i).

If $v < i$:

$$[n]_i \left\{ \begin{array}{c} \mathcal{D} \\ \alpha^v \end{array} \right. = \begin{array}{c} \mathcal{D} \\ \alpha^v \end{array}$$

If $v \geq i$:

- $[n]_i\alpha^v = \alpha^{v+n}$
- Let $\dfrac{\begin{array}{cc} \mathcal{D}' & \mathcal{D}'' \\ \alpha^k & \beta^j \end{array}}{\gamma^v}$ be a generic rule (possibly unary). Then:

$$[n]_i \left\{ \dfrac{\begin{array}{cc} \mathcal{D}' & \mathcal{D}'' \\ \alpha^k & \beta^j \end{array}}{\gamma^v} \right. = \dfrac{[n]_i \left\{ \begin{array}{c} \mathcal{D}' \\ \alpha^k \end{array} \right. \quad [n]_i \left\{ \begin{array}{c} \mathcal{D}'' \\ \beta^j \end{array} \right.}{\gamma^{v+n}}$$

Observe that the level renaming, in general, affects also the hypotheses of a deduction; the constraint on the application of $!\mathcal{I}$ will guarantee that this does not happen during the actual reduction of a redex. One easily proves, moreover, that $[0]_k\mathcal{D} = \mathcal{D}$. We write $[n]_i\Gamma$ for the application of the level substitution to each formula of Γ.

Before defining the redexes and their contractions, we show that the calculi introduced in the previous section are closed under level substitution.

Lemma 3.2 *1. Let $\mathcal{D}$ be a deduction of $\Gamma \vdash_{L^\ell KT} \alpha^j$, with $j \geq 2$. Then, for any $2 \leq i \leq j$, $[-1]_i\mathcal{D}$ is a deduction of $[-1]_i\Gamma \vdash_{L^\ell KT} \alpha^{j-1}$.*

2. Let $\mathcal{D}$ be a deduction of $\Gamma \vdash_{L^\ell K4} \alpha^j$, with $j \geq 1$. Then, for any $1 \leq i \leq j$ and any $k \geq 1$, $[k]_i\mathcal{D}$ is a deduction of $[k]_i\Gamma \vdash_{L^\ell K4} \alpha^{j+k}$.

3. Both 1 and 2 hold for deductions in $L^\ell KT4$.

PROOF. The statements are proved by induction on $\mathcal{D}$. We show the most interesting case of (1); the others are similar and left to the reader. Let $\mathcal{D}$ be

$$\dfrac{\begin{array}{c} \Gamma \\ \vdots \\ !\alpha^r \end{array}}{\alpha^j}$$

where $j = r$ or $j = r + 1$. Then,

$$[-1]_i\mathcal{D} = \dfrac{[-1]_i \left\{ \begin{array}{c} \Gamma \\ \vdots \\ !\alpha^r \end{array} \right.}{\alpha^{j-1}}$$

We have two subcases, depending on the value of i:

1. $i \leq r$. By induction hypothesis (note that $i \leq r$ and $2 \leq i$ imply $r \geq 2$), we have:

$$[-1]_i\mathcal{D} = \dfrac{\begin{array}{c} [-1]_i\Gamma \\ \vdots \\ !\alpha^{r-1} \end{array}}{\alpha^{j-1}}$$

2. $r < i \leq j$. Then it is the case that $j = r + 1$; note, moreover, that the indicated level substitution is the identity and that $\alpha^{j-1} = \alpha^r$. The thesis follows by an application of the proper elimination rule of $L^\ell KT$.■

Definition 3.3 [Reduction] We define the redexes and their contractions; the boxed formula is the *principal formula* of the redex.

- ***β-contractions***:

$$\dfrac{\begin{array}{c} [\alpha^j]\ [\beta^j] \\ \mathcal{D} \\ \gamma^v \end{array} \qquad \dfrac{\begin{array}{cc} \mathcal{E}' & \mathcal{E}'' \\ \alpha^j & \beta^j \end{array}}{\boxed{\alpha \otimes \beta^j}}\otimes\mathcal{I}}{\gamma^v}\otimes\mathcal{E} \quad \triangleright \quad \begin{array}{c} \mathcal{E}' \quad \mathcal{E}'' \\ \alpha^j \quad \beta^j \\ \mathcal{D} \\ \gamma^v \end{array}$$

$$\dfrac{\begin{array}{c} \mathcal{D}' \\ \alpha^v \end{array} \qquad \dfrac{\begin{array}{c} [\alpha^v] \\ \mathcal{D}'' \\ \beta^v \end{array}}{\boxed{\alpha \multimap \beta^v}}\multimap\mathcal{I}}{\beta^v}\multimap\mathcal{E} \quad \triangleright \quad \begin{array}{c} \mathcal{D}' \\ \alpha^v \\ \mathcal{D}'' \\ \beta^v \end{array}$$

$$\dfrac{\dfrac{\begin{array}{c} \mathcal{D} \\ \alpha^k \end{array}}{\boxed{!\alpha^{k-1}}}!\mathcal{I}}{\alpha^{k-1+j}}!\mathcal{E} \quad \triangleright \quad \begin{array}{c} [j-1]_k\mathcal{D} \\ \alpha^{k-1+j} \end{array}, \text{ where: } \left\{ \begin{array}{ll} j = 1 & \text{in } L^\ell K \\ 0 \leq j \leq 1 & \text{in } L^\ell KT \\ 1 \leq j & \text{in } L^\ell K4 \\ 0 \leq j & \text{in } L^\ell KT4 \end{array} \right.$$

- *Commutative contractions*: With $\frac{\alpha^j \quad \vdots}{\beta^{j+v}}r$ we denote a generic rule in $\{\otimes\mathcal{E}, \multimap \mathcal{E}, !\mathcal{E}\}$, where α^j is the principal premise of the rule r, the vertical dots $\vdots$ represent some (possible) non principal premise, and β^{j+v} is the conclusion.

$$\dfrac{\dfrac{\begin{matrix}\mathcal{D}'\\ \alpha^j\end{matrix} \quad \begin{matrix}\mathcal{D}''\\ \mathbf{1}^k\end{matrix}}{\boxed{\alpha^j}}\,1\mathcal{E} \qquad \vdots}{\beta^{j+v}}\,r \quad \rhd \quad \dfrac{\dfrac{\begin{matrix}\mathcal{D}'\\ \alpha^j\end{matrix} \quad \vdots}{\beta^{j+v}}\,r \qquad \begin{matrix}\mathcal{D}''\\ \mathbf{1}^k\end{matrix}}{\beta^{j+v}}\,1\mathcal{E}$$

$$\dfrac{\dfrac{\begin{matrix}\mathcal{D}'\\ \alpha\otimes\beta^k\end{matrix} \quad \begin{matrix}[\alpha^k]\;[\beta^k]\\ \mathcal{D}''\\ \gamma^j\end{matrix}}{\boxed{\gamma^j}}\,\otimes\mathcal{E} \qquad \vdots}{\delta^{j+v}}\,r \quad \rhd \quad \dfrac{\begin{matrix}\mathcal{D}'\\ \alpha\otimes\beta^k\end{matrix} \qquad \dfrac{\begin{matrix}[\alpha^k]\;[\beta^k]\\ \mathcal{D}''\\ \gamma^j\end{matrix} \quad \vdots}{\delta^{j+v}}\,r}{\delta^{j+v}}\,\otimes\mathcal{E}$$

The *degree* of a redex is the structural complexity of its principal formula.

We need to show that the different systems are stable under normalization.

Proposition 3.4 *Let S be a system in $\{L^\ell K, L^\ell KT, L^\ell K4, L^\ell KT4\}$. The contractions of Definition 3.3 are correct for S.*

PROOF. Correctness of propositional and commutative contractions is obvious; we thus take into account only !-contraction. If the LHS of the !-contraction is a deduction in S establishing $\Gamma \vdash_S \alpha^{k-1+j}$, then, by Lemma 3.2, the contractum is still a deduction in S, establishing $[j-1]_k\Gamma \vdash_S \alpha^{k-1+j}$. But the side condition on the application of $!\mathcal{I}$ ensures that k is greater than the level of any formula in Γ, namely $[j-1]_k\Gamma = \Gamma$. ∎

The compatible closure of $\rhd$ (or *one-step reduction*) is denoted with $\rightarrow$; its transitive closure is $\xrightarrow{+}$, while $\twoheadrightarrow$ (the *reduction relation*) is the transitive and reflexive closure.

Theorem 3.5 *Let $\mathcal{D}$ be a deduction of $\Gamma \vdash_{L^\ell KT4}$ and let $\mathcal{D} \twoheadrightarrow \mathcal{D}'$. Then also $\mathcal{D}'$ is a deduction of $\Gamma \vdash_{L^\ell KT4}$.*

PROOF. By induction on the length of the reduction, using Proposition 3.4. ∎

Theorem 3.6 (Normalization) *The $\twoheadrightarrow$ relation is normalizing.*

PROOF. (Sketch: For a detailed discussion of the standard intuitionistic case, which is technically the same, see [TvD88, 530–537].) Define the *degree*, $d(\mathcal{D})$, of a deduction $\mathcal{D}$ as the maximum of the degrees of the redexes in $\mathcal{D}$. Set now $\mu(\mathcal{D}) = (d(\mathcal{D}), m)$, where m is the sum of the lengths of the maximal redexes (i.e. redexes of degree $d(\mathcal{D})$) in $\mathcal{D}$. By induction on $\mu(\mathcal{D})$ (under the lexicographic ordering) we prove that, selected a redex of maximal degree in $\mathcal{D}$, we can reduce $\mu(\mathcal{D})$. ■

4 Discussion

4.1 On the universality of !

It is well known that usual presentations of modal logics (both in sequent calculi and in natural deduction systems) do no allow to prove the equivalence of two modalities enjoying exactly the same rules. Standard presentations of linear logic have the same defect; for example it is possible to add to linear logic another exponential, say ¡, with exactly the same rules of !, but it is impossible to prove that $\vdash !\alpha \multimap\!\circ\; ¡\alpha$. In fact, the introduction of a ! is constrained not by structural properties of proofs (or in the case of sequent formulations, by structural properties of sequents), but by the syntactical shape of the premises.

At a first level, our formulation has not this kind of problems, as the following proposition shows.

Proposition 4.1 *Let S be one of the linear logic systems previously defined. Let us duplicate the exponential rules of S with a new exponential ¡, and call $S¡$ the resulting system. Then*

$$\vdash_{S¡} !\alpha \multimap\!\circ\, ¡\alpha^j$$

for each level j.

PROOF. Let us prove that $¡\alpha \multimap !\alpha^j$.

$$\dfrac{\dfrac{\dfrac{[¡\alpha^j]}{\alpha^{j+1}}\,¡\mathcal{E}}{!\alpha^j}\,!\mathcal{I}}{¡\alpha \multimap !\alpha^j}\multimap \mathcal{I}$$

■

This happens because ! and ¡ share the same level structure; if we added another level structure, different from the first, and we defined ¡ in terms of the new levels, then the anomaly would be there again.

4.2 Linear lambda-calculi

The choice of a natural deduction presentation allows a simple computational interpretation of the proposed fragments of linear logic. It is straightforward to decorate the deductions with λ-terms. A suitable syntax for the modal rules and their contraction is the following:

$$\dfrac{\dfrac{\begin{array}{c}\mathcal{D}\\ M : \alpha^k\end{array}}{\mathsf{gen}(M)^{k-1} : !\alpha^{k-1}}\,!\mathcal{I}}{\mathsf{ungen}(\mathsf{gen}(M)^{k-1})^{k-1+j} : \alpha^{k-1+j}}\,!\mathcal{E} \quad \rhd \quad \begin{array}{c}[j-1]_k\mathcal{D}\\ [j-1]_k M : \alpha^{k-1+j}\end{array}$$

As an example, the following are the terms representing the given proofs for the three exponential axioms **K**, **T** and **4**.

$$\lambda y^1.\lambda x^1.\mathsf{gen}((\mathsf{ungen}(y^1)^2\mathsf{ungen}(x^1)^2)^2)^1 : \Box(\alpha \supset \beta) \supset (\Box\alpha \supset \Box\beta)^1$$

$$(\lambda x^1.\mathsf{ungen}(x^1)^1)^1 : \Box\alpha \supset \alpha^1$$

$$(\lambda x^1.\mathsf{gen}(\mathsf{gen}(\mathsf{ungen}(x^1)^3)^2)^1)^1 : \Box\alpha \supset \Box\Box\alpha^1$$

4.3 Relations with proof-nets

An anonymous referee suggested an interesting interpretation of our work in terms of proof-nets, as proposed in [Reg92]. For the purpose of this discussion, we may assume such nets as composed of links of two kinds, ? and !; each time a link ! is added to a net, also a box is given, whose ports different from the ! link are all ? links. It should be clear that the creation of a !-link corresponds to !-introduction, while the creation of ?-links corresponds to !-elimination.

On these nets we may investigate the *lift* of a ?-link, that is the number of boxes the link may cross; setting suitable constraints on the allowed lifts results in constraining !-elimination. The lift of a ?-link exactly corresponds, in our approach, to the difference between the level of the conclusion and the level of the premise in the !-elimination rule. Thus, if we only allow lifts of 1, we obtain $L^\ell K$; lifts in $\{0,1\}$ yield $L^\ell KT$, lifts in $\{p \mid p \geq 1\}$ yield $L^\ell K4$, while unrestricted lifts gives full linear exponentials. It is not too difficult to check that these four sets of allowed lifts are all stable under proof-net reduction, a result matching our Lemma 3.2. As a corollary, one obtains immediately strong normalization and confluence for the proposed systems.

More systems enjoy the same properties. For $L \subseteq \mathbb{N}$ (the set of *allowed lifts*), let $L^\ell L$ be the system with lifts only in L (thus $L^\ell KT$ is $L^\ell\{0,1\}$, etc.); it can be shown that $L^\ell L$ is closed under reduction iff $L = \{0\}$ or $\{i+j-1 \mid i,j \in L \wedge j \neq 0\} \subseteq L$. In terms of levels, this results in the following generalization of Lemma 3.2.

Proposition 4.2 *Let $\mathcal{D}$ be a deduction of $\Gamma \vdash_{L^\ell L} \alpha^j$.*

1. *If $0 \in L$, then for any $2 \leq i \leq j$, $[-1]_i\mathcal{D}$ is a deduction of $[-1]_i\Gamma \vdash_{L^\ell L} \alpha^{j-1}$.*

2. *For any $p \in L$, $p \neq 0$, and any $1 \leq i \leq j$, $[p-1]_i\mathcal{D}$ is a deduction of $[p-1]_i\Gamma \vdash_{L^\ell L} \alpha^{j+p-1}$.*

We thus have infinitely many fragments (e.g. L may be the set of odd numbers), all closed under reduction and enjoying strong normalization and confluence, although none of them seems to bear a real interest (the "characteristic formulas" of $L^\ell L$ are $!A \multimap !^pA$, for $p \in L$).

Acknowledgments: Section 4.3 is essentially due to an anonymous referee, which we sincerely thank.

References

[BBdPH93] N. Benton, G. Bierman, V. de Paiva, and J.M.E. Hyland. Linear λ-calculus and categorical models revisited. In E. Börger et al., editors, *Computer Science Logic 1992*, volume 702 of *Lectures Notes in Computer Science*, pages 61–84. Springer-Verlag, 1993. San Miniato, September.

[DJS93] V. Danos, J.-B. Joinet, and H. Schellinx. The structure of exponentials: uncovering the dynamics of linear logic proofs. In G. Gottlob et al., editors, *Computational Logic and Proof Theory*, volume 713 of *Lecture Notes in Computer Science*, pages 159–171. Springer-Verlag, 1993. Proc. of Third Kurt Gödel Colloquium, Brno, August 1993.

[GSS92] J.-Y. Girard, A. Scedrov, and P.J. Scott. Bounded linear logic: A modular approach to polynomial time computability. *Theoretical Computer Science*, 97:1–66, 1992.

[Mas92] Andrea Masini. 2-sequent calculus: A proof theory of modalities. *Annals of Pure and Applied Logic*, 58:229–246, 1992.

[Mas93] Andrea Masini. 2-sequent calculus: Intuitionism and natural deduction. *Journal of Logic and Computation*, 3:533–562, 1993.

[Reg92] Laurent Regnier. *Lambda-Calcul et Réseaux.* Thèse de doctorat, Université Paris 7, 1992.

[Sch94] H. Schellinx. *The Noble Art of Linear Decorating.* ILLC dissertation series 1994-1, Institute for Logic, Language and Computation, Universiteit Amsterdam, 1994.

[Tro91] Anne S. Troelstra. *Lectures on Linear Logic*, volume 29 of *CSLI Lecture Notes.* Chicago University Press, Chicago, Illinois, 1991.

[TvD88] Anne S. Troelstra and Dirk van Dalen. *Constructivism in Mathematics*, volume II. North-Holland, 1988.

LKQ and **LKT**: Sequent calculi for second order logic based upon dual linear decompositions of classical implication

by

Vincent Danos[*], Jean-Baptiste Joinet[†] & Harold Schellinx[‡]
Équipe de Logique Mathématique, Université Paris VII
Faculteit Wiskunde en Informatica, Universiteit van Amsterdam

As is well known, one can recover intuitionistic logic within linear logic. Indeed, linear logic found its origin in a semantical decomposition of intuitionistic type constructors corresponding, in the sense of the 'Curry-Howard-de Bruijn isomorphism' (Howard(1980)), to the intuitionistic connectives (see Girard et al.(1988) for details). (A such decomposition in fact already appears in the simple set-theoretical models for the untyped lambda calculus.) Conversely this decomposition gives rise to Girard's embedding of intuitionistic into linear logic, which, for (the $\rightarrow, \forall_2$-fragment of) second order propositional intuitionistic logic, a.k.a.[1] system $\mathcal{F}$, is inductively defined as follows.

For atomic p let $p^\star := p$; then put

$$\begin{aligned} (A \rightarrow B)^\star &:= \; !A^\star \multimap B^\star \\ (\forall \alpha A)^\star &:= \forall \alpha A^\star. \end{aligned}$$

This mapping of intuitionistic to linear formulas is an example of what we call a(n) *(inductive) modal translation*: the translation of a given formula is obtainable by inductively prefixing the formula's subformulas by *modalities*, i.e. (possibly empty) strings of exponentials '!', '?'.[2]

[*]danos@logique.jussieu.fr; CNRS URA 753

[†]joinet@logique.jussieu.fr; Université Paris I

[‡]harold@fwi.uva.nl; supported by an HCM Research Training Fellowship of the European Economic Community.

[1]I.e. *'also known as'*

[2]Note that a modal translation $(\cdot)^\checkmark$ will be *compatible with substitution* (i.e. for all A, B the formulas $(A[B/\alpha])^\checkmark$ and $A^\checkmark[B^\checkmark/\alpha]$ are identical) if and only if it is the identity on atoms.

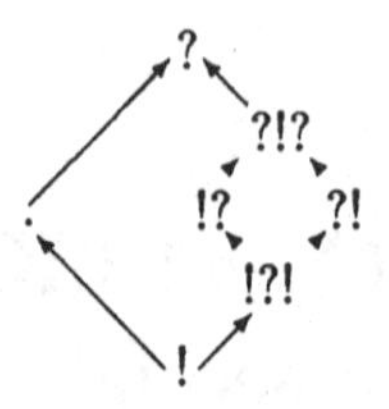

Figure 1: The lattice of linear modalities

Observe that for all such modalities μ both $!A \Rightarrow \mu A$ and $\mu A \Rightarrow ?A$ are derivable, whatever A: starting from an axiom $A \Rightarrow A$ and an application of L!, we can without restriction apply R? and R! to obtain μA in the first, and by the unrestricted possibility of using L! and L? starting from an axiom followed by an application of R? in the second case. Otherwise said: in the partial ordering on modalities induced by linear derivability ($\mu \preceq \nu$ iff $\vdash \mu A \Rightarrow \nu A$ is derivable in linear logic, for any A) we have that "!" is minimal, "?" is maximal.

If we consider the equivalence relation induced by this ordering, we find *seven* equivalence classes: calling $\cdot, ?, !, !?, ?!, !?!$ and $?!?$ (where "$\cdot$" stands for the void modality) *basic* modalities, one easily shows (e.g. using the *idempotency* of these basic modalities) that for any modality μ there is a *unique* basic μ_0 such that $CLL \vdash \mu A \Longleftrightarrow \mu_0 A$ for all A. So,

> *modulo provable linear equivalence, there are precisely seven modalities in linear logic.*

(Consequently, in *intuitionistic* linear logic, there is modulo linear equivalence just *one* non-trivial modality: "!".) Basic modalities are related as in figure 1, where an arrow from μ to ν indicates that $\mu \prec \nu$ (see Joinet(1993)).

As an easy corollary we then find that *all* modalities are idempotent: $CLL \vdash \mu A \Longleftrightarrow \mu\mu A$ for all μ, A.

Girard's embedding is both sound- and faithful, i.e. if $\Gamma \Rightarrow A$ is derivable in system $\mathcal{F}$, then so is $!\Gamma^* \Rightarrow A^*$ in classical linear logic; conversely, if CLL derives $!\Gamma^* \Rightarrow A^*$, then $\mathcal{F}$ will prove $\Gamma \Rightarrow A$ (see Schellinx(1994)).

Moreover, if in a linear derivation π of $!\Gamma^* \Rightarrow A^*$ we replace the linear connectives by their non-linear analogues, and simply forget about the exponentials, delete resulting repetitions of sequents, then what we find will in general be an intuitionistic derivation of $\Gamma \Rightarrow A$. We make the proviso "in general", for example because of the devious behaviour of the linear constant

0 (cf. Schellinx(1991)), which is linearly equivalent to $\forall\alpha.\alpha$. Anyway, it is a pretty trivial remark that this *collapsing* of a linear derivation π results in a derivation in *classical logic*, which we will refer to as π's *skeleton*. We express this as follows:

> *the* skeleton $sk(\pi)$ *of a linear derivation* π *is a derivation in (intuitionistic or) classical logic.*

Despite its obviousness, we consider this observation to be essential, not in the least because we will show that it has a converse:

CLAIM. *Any derivation in intuitionistic or classical sequent calculus occurs as the skeleton of a derivation in linear logic.*

This provides an interesting argument supporting linear logic's claim to being a *refinement* of intuitionistic and classical sequent calculus. Note e.g. that it enables us to obtain eliminability of cut from classical and intuitionistic derivations as a corollary to the linear cut elimination theorem.

DEFINITION. A *decoration* of a (classical, intuitionistic) derivation π is a linear derivation $\partial(\pi)$ such that $sk(\partial(\pi)) = \pi$; by a decoration-strategy for a given (sequent)calculus we mean a uniform procedure (algorithm) that outputs a decoration for any given derivation in the calculus. □

How to prove our claim? Well, one simply shows, given any intuitionistic or classical derivation π, how to produce a decoration $\partial(\pi)$. One possibility is to start from π and 'linearize' it by tracing the effects of occurrences of structural rules in π. We looked at this option in detail for intuitionistic implicational logic in Danos et al.(1993a) (see also Joinet(1993)).

A second possibility is to try to transform a given derivation π into a linear derivation $\partial(\pi)$, by inductively applying a modal translation to the sequents occurring in π.

DEFINITION. Let a modal translation $(\cdot)^{\checkmark}$ and modalities μ, ν be given. We say that the triple $\langle(\cdot)^{\checkmark}, \mu, \nu\rangle$ defines an *inductive decoration strategy* for a sequent calculus $\mathcal{S}$ if

1/ for all $\mathcal{S}$-axioms $\Gamma \Rightarrow \Delta$ it holds that $\mu\Gamma^{\checkmark} \Rightarrow \nu\Delta^{\checkmark}$ is a CLL-axiom or obtainable from such solely by means of zero or more applications of exponential contextual and/or dereliction rules;

2/ for all $\mathcal{S}$-rules with conclusion $\Gamma \Rightarrow \Delta$ and premiss(es) $\Gamma_i \Rightarrow \Delta_i$ we can derive $\mu\Gamma^{\checkmark} \Rightarrow \nu\Delta^{\checkmark}$ in linear logic from $\mu\Gamma_i^{\checkmark} \Rightarrow \nu\Delta_i^{\checkmark}$ by an application of the

Identity axiom and cut rule:

$$\text{Ax} \quad A \Rightarrow A \qquad\qquad \text{cut}\ \frac{\Gamma \Rightarrow \Delta, A \qquad A, \Gamma' \Rightarrow \Delta'}{\Gamma, \Gamma' \Rightarrow \Delta, \Delta'}$$

Logical rules:

$$\text{L}\multimap \frac{\Gamma \Rightarrow \Delta, A \qquad B, \Gamma' \Rightarrow \Delta'}{\Gamma, \Gamma', A \multimap B \Rightarrow \Delta, \Delta'} \qquad \text{R}\multimap \frac{\Gamma, A \Rightarrow B, \Delta}{\Gamma \Rightarrow A \multimap B, \Delta}$$

Rules for the second order quantifier (β not free in Γ, Δ):

$$\text{L}\forall_2 \frac{\Gamma, A[X/\alpha] \Rightarrow \Delta}{\Gamma, \forall\alpha\, A \Rightarrow \Delta} \qquad \text{R}\forall_2 \frac{\Gamma \Rightarrow \Delta, A[\beta/\alpha]}{\Gamma \Rightarrow \Delta, \forall\alpha\, A}$$

Exponential structural rules:

$$\text{W!} \frac{\Gamma \Rightarrow \Delta}{\Gamma, !A \Rightarrow \Delta} \qquad \text{W?} \frac{\Gamma \Rightarrow \Delta}{\Gamma \Rightarrow ?A, \Delta} \qquad \text{C!} \frac{\Gamma, !A, !A \Rightarrow \Delta}{\Gamma, !A \Rightarrow \Delta} \qquad \text{C?} \frac{\Gamma \Rightarrow ?A, ?A, \Delta}{\Gamma \Rightarrow ?A, \Delta}$$

Exponential contextual rules:

$$\text{L?} \frac{!\Gamma, A \Rightarrow ?\Delta}{!\Gamma, ?A \Rightarrow ?\Delta} \qquad \text{R!} \frac{!\Gamma \Rightarrow A, ?\Delta}{!\Gamma \Rightarrow !A, ?\Delta}$$

Exponential dereliction rules:

$$\text{R?} \frac{\Gamma \Rightarrow A, \Delta}{\Gamma \Rightarrow ?A, \Delta} \qquad \text{L!} \frac{\Gamma, A \Rightarrow \Delta}{\Gamma, !A \Rightarrow \Delta}$$

Table 1: Linear logic, the $\{!, ?, \multimap, \forall_2\}$-fragment

Identity axiom and cut rule:

$$\text{Ax}\quad A \Rightarrow A \qquad\qquad \text{cut}\ \frac{\Gamma \Rightarrow \Delta, A \qquad A, \Gamma' \Rightarrow \Delta'}{\Gamma, \Gamma' \Rightarrow \Delta, \Delta'}$$

Logical rules:

$$\text{L}\!\rightarrow \frac{\Gamma \Rightarrow \Delta, A \qquad B, \Gamma' \Rightarrow \Delta'}{\Gamma, \Gamma', A \rightarrow B \Rightarrow \Delta, \Delta'} \qquad \text{R}\!\rightarrow \frac{\Gamma, A \Rightarrow B, \Delta}{\Gamma \Rightarrow A \rightarrow B, \Delta}$$

Rules for the second order quantifier (β not free in Γ, Δ):

$$\text{L}\forall_2\ \frac{\Gamma, A[X/\alpha] \Rightarrow \Delta}{\Gamma, \forall\alpha\, A \Rightarrow \Delta} \qquad \text{R}\forall_2\ \frac{\Gamma \Rightarrow \Delta, A[\beta/\alpha]}{\Gamma \Rightarrow \Delta, \forall\alpha\, A}$$

Structural rules:

$$\text{WL}\ \frac{\Gamma \Rightarrow \Delta}{\Gamma, A \Rightarrow \Delta} \qquad \text{WR}\ \frac{\Gamma \Rightarrow \Delta}{\Gamma \Rightarrow A, \Delta} \qquad \text{CL}\ \frac{\Gamma, A, A \Rightarrow \Delta}{\Gamma, A \Rightarrow \Delta} \qquad \text{CR}\ \frac{\Gamma \Rightarrow A, A, \Delta}{\Gamma \Rightarrow A, \Delta}$$

Table 2: CL, the $\{\rightarrow, \forall_2\}$-fragment

corresponding CLL-rule preceded and/or followed by zero or more applications of exponential contextual and/or dereliction rules. □

Obviously, by definition, if $\langle (\cdot)^{\checkmark}, \mu, \nu \rangle$ is an inductive decoration strategy for a calculus $\mathcal{S}$, then, given an $\mathcal{S}$-derivation π of a sequent $\Gamma \Rightarrow \Delta$, we can apply the translation $(\cdot)^{\checkmark}$ inductively to π and derive $\mu\Gamma^{\checkmark} \Rightarrow \nu\Delta^{\checkmark}$ by means of a linear derivation $\pi^{\checkmark}$ which is a decoration of the original one.

In the present paper we will be mainly interested in the $\{\rightarrow, \forall_2\}$-fragments of second order propositional classical and intuitionistic logic. To be precise, we consider the corresponding fragments of sequent calculi for these logics. For brevity's sake we will simply refer to the fragments as CL, respectively IL. The calculus for CL is described in table 2. The calculus for IL is that for CL, subjected to the usual intuitionistic restriction of the succedents to singletons. The corresponding fragment of CLL is given as table 1. (Note that extending these fragments with rules for a *first order* universal quantifier is completely straightforward, and all results stated in what follows hold for these extensions. In proofs and definitions the case of the first order quantifier is completely analogous to that of the second order one.)

As we observed in Danos et al.(1993a), Girard's translation does not extend to sequent-calculus derivations: $\langle (\cdot)^*, !, \cdot \rangle$ does not define an inductive decoration strategy for IL. Using this translation, the inductive transformation of proofs will introduce cuts at several points. To be precise (in the case of system

$\mathcal{F}$), each time we encounter an application of L$\rightarrow$, we apply a cut with the canonical derivation of $!(!A^* \multimap B^*) \Rightarrow !A^* \multimap !B^*$; and each time we encounter an application of L$\forall$, we cut with the canonical derivation of $!\forall\alpha A^* \Rightarrow \forall\alpha!A$. We call these the *'correction cuts'*.

However, note that the inductive application of Girard's translation to IL-derivations in fact indicates a modified (and, with respect to the number of shrieks introduced, less economical) translation $(\cdot)^{\circledast}$ that *does* define an inductive decoration strategy $\langle(\cdot)^{\circledast}, !, \cdot\rangle$ (and, of course, also is a sound- and faithful embedding of system $\mathcal{F}$ into CLL, which can be shown e.g. using the fact that $!A^* \Longleftrightarrow !A^{\circledast}$ is linearly provable). It is inductively given as follows. For atomic p let $p^{\circledast} := p$; then put

$$\begin{array}{rcl}(A \rightarrow B)^{\circledast} & := & !A^{\circledast} \multimap !B^{\circledast} \\ (\forall\alpha A)^{\circledast} & := & \forall\alpha!A^{\circledast}.\end{array}$$

Consequently we established the first, intuitionistic, half of our claim.

But, as a matter of fact, *no* strategy can lead to decorations of IL-derivations that are always *subGirardian*, i.e. do not shriek (sub)formulas that are not banged in the $(\cdot)^*$-translation. This is shown by the following example, where each decoration of the skeleton will necessarily contain this, *minimal*, decoration.

$$\dfrac{C \Rightarrow C \qquad \dfrac{A \Rightarrow A}{!B, A \Rightarrow A}}{C, C \multimap !B, A \Rightarrow A}$$

The exclamation mark appearing in front of B is forced by the use of the structural rule of weakening. Deleting it results in a non-linear derivation.

The 'root of all evil' apparently is that intuitionistic sequent calculus allows applications of e.g. the rules L$\rightarrow$, L$\forall_2$ in case the active formula in the (right) premiss has been subjected to structural manipulation.

The *correctness* of the $(\cdot)^*$-translation shows that we can do without that property: the collection of derivations that do *not* use it is complete for intuitionistic logic. Indeed, this is an immediate corollary to the subformula-property and the fact that the skeleton of a cut free intuitionistic linear derivation of $!\Gamma^* \Rightarrow A^*$ is an IL-derivation.

This suggests a formulation of intuitionistic sequent calculus in which the use of these rules on such, non-linear, formulas is forbidden, and for which as a consequence Girard's translation should be a decoration-strategy. Such

Identity axiom:

$$A; \Rightarrow A$$

Logical rules:

$$\text{L}\!\rightarrow \frac{;\Gamma \Rightarrow A \qquad B;\Gamma' \Rightarrow C}{A \rightarrow B;\Gamma,\Gamma' \Rightarrow C} \qquad \text{R}\!\rightarrow \frac{\Pi;\Gamma, A \Rightarrow B}{\Pi;\Gamma \Rightarrow A \rightarrow B}$$

Rules for the second order quantifier (β not free in Γ, Π):

$$\text{L}\forall_2 \frac{A[X/\alpha];\Gamma \Rightarrow B}{\forall\alpha\, A;\Gamma \Rightarrow B} \qquad \text{R}\forall_2 \frac{\Pi;\Gamma \Rightarrow A[\beta/\alpha]}{\Pi;\Gamma \Rightarrow \forall\alpha\, A}$$

Structural rules:

$$\text{WL} \frac{\Pi;\Gamma \Rightarrow A}{\Pi;\Gamma, B \Rightarrow A} \qquad \text{CL} \frac{\Pi;\Gamma, B, B \Rightarrow A}{\Pi;\Gamma, B \Rightarrow A} \qquad \text{D} \frac{B;\Gamma \Rightarrow A}{;B,\Gamma \Rightarrow A}$$

Table 3: ILU, the cut-free fragment.

a formulation can be found by a rather straightforward abstraction of the structure of linear derivations of sequents of the form $!\Gamma^* \Rightarrow A^*$ (table 3). [3]

In a sequent $\Pi;\Gamma \Rightarrow A$ the symbol Π denotes a multiset containing *at most* one (the *head*-)formula whose occurrence in a sequent is distinguished by means of the ";". In the linear interpretation it corresponds to a formula that is not (yet) shrieked. The structural rule D is the equivalent of L!, the linear dereliction rule.

Included we find the *neutral* fragment of intuitionistic implicational logic as it appears in Girard's system of Unified Logic (LU, Girard(1993)). For this reason we refer to the above calculus as ILU.

$\Pi;\Gamma \Rightarrow A$ is derivable in ILU if and only if $\Pi^*, !\Gamma^* \Rightarrow A^*$ is derivable in the $\{!, \multimap, \forall_2^*\}$-fragment of linear logic, where $\forall_2^*$ indicates abstraction limited to formulas of the form X^* (observe that if a sequent $\Pi^*, !\Gamma^* \Rightarrow !\Sigma^*, \Delta^*$ is derivable in this fragment, then $|\,\Pi \cup \Sigma\,| \leq 1$ and $|\,\Sigma \cup \Delta\,| = 1$). Moreover, by construction, Girard's translation $(\cdot)^*$ determines an inductive decoration strategy (in the sense of the above definition adapted to ILU-sequents in the obvious[4] way) for ILU-derivations π.

[3]Note that the instances of rules that we will get rid of have no direct equivalent in the natural deduction formulation of intuitionistic logic. Therefore this modified sequent calculus will be closer to natural deduction and the simply typed λ-calculus than the standard formulation. (The reader will find that the 'natural' way to interpret a natural deduction derivation in sequent calculus is as an ILU-derivation!)

[4]One merely replaces 'S' by 'ILU', '$\Gamma \Rightarrow \Delta$' by '$\Pi;\Gamma \Rightarrow \Delta$', and '$\mu\Gamma^{\checkmark} \Rightarrow \nu\Delta^{\checkmark}$' by $\Pi^{\checkmark};\mu\Gamma^{\checkmark} \Rightarrow \nu\Delta^{\checkmark}$'.

Thus we found a first example of what we might launch as a slogan but in fact is a

FACT. *Linear logic suggests restrictions on derivations in its underlying calculi, restrictions leading to subsets of the collection of these proofs that nevertheless are* complete.

It is not difficult to show that the collection of $(\cdot)^*$-decorated ILU-sequents is closed under cut (see Danos et al.(1993a)), from which it follows that ILU is closed under the rules

$$head - cut \quad \frac{\Pi; \Gamma_1 \Rightarrow A \qquad A; \Gamma_2 \Rightarrow B}{\Pi; \Gamma_1, \Gamma_2 \Rightarrow B}$$

$$mid - cut \quad \frac{; \Gamma_1 \Rightarrow A \quad \Pi; A, \Gamma_2 \Rightarrow C}{\Pi; \Gamma_1, \Gamma_2 \Rightarrow C}$$

where the cut elimination procedure for ILU is the obvious analogue (the 'reflection') of the linear procedure, whence, using the terminology introduced in Danos et al.(1993b), the $(\cdot)^*$-decorations of ILU-derivations are *strong* decorations. As, moreover, ILU is complete for provability in intuitionistic logic, in fact what we obtained is a proof system for intuitionistic logic which is a proper fragment of CLL.

PROPOSITION. *If π is a derivation in $\{!, \multimap, \forall_2^*\}$ of $\Pi^*, !\Gamma^* \Rightarrow !\Sigma^*, \Delta^*$ in which all cutformulas are of the form A^* or $!A^*$, and all identity axioms of the form $A^* \Rightarrow A^*$, then $sk(\pi)$ is an ILU-derivation of $\Pi; \Gamma \Rightarrow \Sigma \cup \Delta$.* □

Thus ILU inherits the computational properties of CLL.

How about classical logic? Let us try to define a modal translation $(\cdot)^{\checkmark}$ of *classical* logic that, like the $(\cdot)^{\circledast}$-translation for intuitionistic logic, can be extended to an inductive decoration strategy for CL. In order to do so, we have to interpret sequents $\Gamma \Rightarrow \Delta$ as $\mu\Gamma^{\checkmark} \Rightarrow \nu\Delta^{\checkmark}$, where μ, ν are modalities. Then observe that, in order to satisfy condition 2 in the definition of decoration strategy,

1. in case of the structural rules we need that $\mu \equiv !\mu'$ and $\nu \equiv ?\nu'$, for modalities μ', ν';
2. in case of an application of cut, we have to be able to 'unify' the decorations $\mu A^{\checkmark}$ and $\nu A^{\checkmark}$ of the cut formula by some series of applications of dereliction- and/or promotion-rules. Clearly this can be done if and only if either μ is a suffix of ν or ν is a suffix of μ.

We will call a pair of modalities (μ, ν) satisfying these two conditions *adequate*.

PROPOSITION. *Let (μ, ν) be a pair of modalities. There exists a modal translation $(\cdot)^{\checkmark}$ such that $\langle(\cdot)^{\checkmark}, \mu, \nu\rangle$ is an inductive decoration strategy for CL if and only if (μ, ν) is adequate.*

PROOF: That adequacy is a necessary condition has already been shown. It is also sufficient: given an adequate pair (μ, ν) define $p^{\circledcirc} := p$ for p atomic; then take

$$\begin{aligned}(A \to B)^{\circledcirc} &:= max(\mu,\nu)A^{\circledcirc} \multimap max(\mu,\nu)B^{\circledcirc}\\ (\forall \alpha A)^{\circledcirc} &:= \forall\alpha\, max(\mu,\nu)A^{\circledcirc},\end{aligned}$$

where $max(\mu, \nu)$ denotes the longest of the two modalities. It is not difficult to verify that $\langle(\cdot)^{\circledcirc}, \mu, \nu\rangle$ is an inductive decoration strategy for CL. □

The proposition proves the second, classical, half of our claim, as obviously there exist adequate pairs of modalities.

The modal translations corresponding to the *two* simplest possible adequate pairs, namely (!, ?!) and (!?, ?), will be called the **q**-, respectively the **t**-translation. In fact, **q** and **t** are, in a way, the *unique* inductive decoration strategies for CL: in an inductive decoration strategy $\langle(\cdot)^{\circledcirc}, \mu, \nu\rangle$ as in the proof of the proposition, by adequacy, either (1) $\mu \equiv\, !\alpha?\beta$ and $\nu \equiv\, ?\beta$, or (2) $\mu \equiv\, !\beta$ and $\nu \equiv\, ?\alpha!\beta$, for modalities α, β; using the terminology and techniques introduced in Danos et al.(1993b) one then shows that in $\pi^{\circledcirc}$ exponentials in the classes induced by α and β are always superfluous, and can be stripped, hence resulting in either π^{t} (case 1) or π^{q} (case 2). (Details are in Schellinx(1994).)

Girard's embedding $(\cdot)^*$ can be seen as an optimization of the decorating embedding $(\cdot)^{\circledast}$. It appears that similar optimizations are possible in the classical case, for the **q**- as well as for the **t**-translation. To see this, note that the following are linearly derivable, for any A, B:

$$\begin{aligned}!(!A \multimap ?!B) &\Rightarrow ?!A \multimap ?!B\\ !?(!?A \multimap ?B) &\Rightarrow !?A \multimap !?B\\ !?\forall\alpha ?A &\Rightarrow \forall\alpha !?A.\end{aligned}$$

These suggest the following translations:

- the Q-translation, which maps atoms to atoms, then

$$\begin{aligned}(A \to B)^{Q} &:= !A^{Q} \multimap ?!B^{Q}\\ (\forall\alpha A)^{Q} &:= \forall\alpha ?!A^{Q};\end{aligned}$$

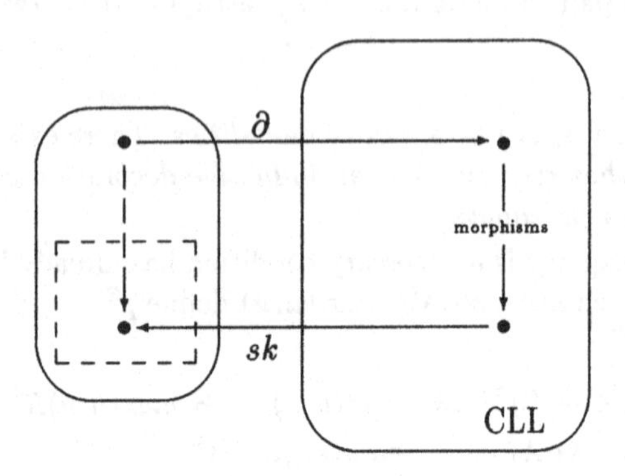

Figure 2: Transformation of proofs by means of constrictive morphisms

- the T-translation, which maps atoms to atoms, then

$$\begin{array}{rcl}(A \rightarrow B)^T & := & !?A^T \multimap ?B^T \\ (\forall \alpha A)^T & := & \forall \alpha ?A^T.\end{array}$$

Using the canonical derivations of the sequents above to 'inject' correction-cuts at the appropriate places[5] in q- respectively t-decorated CL-derivations, we find that both Q and T are sound- and faithful embeddings of CL into linear logic:

$$CLL \vdash !\Gamma^Q \Rightarrow ?!\Delta^Q \quad \text{iff} \quad CL \vdash \Gamma \Rightarrow \Delta \quad \text{iff} \quad CLL \vdash !?\Gamma^T \Rightarrow ?\Delta^T.$$

And as for Girard's embedding in the case of intuitionistic logic, the existence of these translations suggests restrictions on CL-derivations, defining a subcollection that is complete for provability.

As a matter of fact, starting from a q- or t-decorated CL-derivation π and introducing correction cuts at the appropriate places, *elimination* of precisely these cuts will result in a CL-derivation π' satisfying these restrictions. To take an example, the T-translation tells us that the succedent in an implication need never be subject to structural manipulations to the left of the entailment sign. One *constructs* a CL-derivation satisfying this restriction for a given occurrence of an implication by eliminating the correction cut with the canonical derivation of $!?(!?A \multimap ?B) \Rightarrow !?A \multimap !?B$, and taking the skeleton. We therefore speak of *'constrictive morphisms'*.

[5]Being instances of L→ for q, instances of L→, L$\forall_2$ for t.

Identity axiom:

$$A \Rightarrow\, ; A$$

Logical rules:

$$\text{L}\!\rightarrow \frac{\Gamma \Rightarrow \Delta; A \qquad B, \Gamma' \Rightarrow \Delta';}{\Gamma, \Gamma', A \rightarrow B \Rightarrow \Delta, \Delta';} \qquad \text{R}\!\rightarrow \frac{\Gamma, A \Rightarrow \Delta, B;}{\Gamma \Rightarrow \Delta; A \rightarrow B}$$

Rules for the second order quantifier (β not free in Γ, Δ):

$$\text{L}\forall_2 \frac{\Gamma, A[X/\alpha] \Rightarrow \Delta;}{\Gamma, \forall\alpha\, A \Rightarrow \Delta;} \qquad \text{R}\forall_2 \frac{\Gamma \Rightarrow \Delta, A[\beta/\alpha];}{\Gamma \Rightarrow \Delta; \forall\alpha\, A}$$

Structural rules:

$$\text{D} \frac{\Gamma \Rightarrow \Delta; A}{\Gamma \Rightarrow \Delta, A;}$$

$$\text{LW} \frac{\Gamma \Rightarrow \Delta; \Pi}{\Gamma, A \Rightarrow \Delta; \Pi} \qquad \text{RW} \frac{\Gamma \Rightarrow \Delta; \Pi}{\Gamma \Rightarrow A, \Delta; \Pi} \qquad \text{LC} \frac{\Gamma, A, A \Rightarrow \Delta; \Pi}{\Gamma, A \Rightarrow \Delta; \Pi} \qquad \text{RC} \frac{\Gamma \Rightarrow A, A, \Delta; \Pi}{\Gamma \Rightarrow A, \Delta; \Pi}$$

Cut rules:

$$\text{tail} \frac{\Gamma \Rightarrow \Delta; A \qquad A, \Gamma' \Rightarrow \Delta'; \Pi}{\Gamma, \Gamma' \Rightarrow \Delta, \Delta'; \Pi} \qquad \text{mid} \frac{\Gamma \Rightarrow \Delta, A; \Pi \qquad A, \Gamma' \Rightarrow \Delta';}{\Gamma, \Gamma' \Rightarrow \Delta, \Delta'; \Pi}$$

Table 4: The calculus LKQ

The general pattern of this proof transformation is schematized in figure 2. In case the sequent calculus of departure is IL, we start by applying inductively the $(\cdot)^*$-translation, add correction cuts whenever necessary, and then eliminate these. The skeleton of the reduct is an ILU-derivation (see Schellinx(1994) for more details).

We call the calculi corresponding to the Q- and the T-translation respectively LKQ and LKT. One obtains LKQ-, respectively LKT-derivations of a given sequent by eliminating the appropriate correction cuts applied to the **q**-, respectively the **t**-decoration of a CL-derivation of the sequent. The formulation of the rules in the calculi is found by abstraction of the structure of linear derivations of the form $!\Gamma^Q \Rightarrow ?!\Delta^Q$ respectively $!?\Gamma^T \Rightarrow ?\Delta^T$.

The calculus LKQ (table 4) has sequents $\Gamma \Rightarrow \Delta; \Pi$, where, as in ILU, the symbol Π denotes a multi-set containing *at most* one, the *'queue'* (whence Q) or *tail*-formula whose occurrence in the succedent of a sequent is distinguished by means of the ";". In the linear interpretation it corresponds to a formula that has not (yet) been questioned. Again we find a dereliction rule D, this time the equivalent of the linear R?-rule.

Identity axiom:

$$A; \Rightarrow A$$

Logical rules:

$$\text{L}\!\to \frac{;\Gamma \Rightarrow \Delta, A \qquad B;\Gamma' \Rightarrow \Delta'}{A \to B;\Gamma,\Gamma' \Rightarrow \Delta,\Delta'} \qquad \text{R}\!\to \frac{\Pi;\Gamma, A \Rightarrow \Delta, B}{\Pi;\Gamma \Rightarrow \Delta, A \to B}$$

Rules for the second order quantifier (β not free in Π, Γ, Δ):

$$\text{L}\forall_2 \frac{A[X/\alpha];\Gamma \Rightarrow \Delta}{\forall\alpha\, A;\Gamma \Rightarrow \Delta} \qquad \text{R}\forall_2 \frac{\Pi;\Gamma \Rightarrow \Delta, A[\beta/\alpha]}{\Pi;\Gamma \Rightarrow \Delta, \forall\alpha\, A}$$

Structural rules:

$$\text{D} \frac{A;\Gamma \Rightarrow \Delta}{;A,\Gamma \Rightarrow \Delta}$$

$$\text{LW} \frac{\Pi;\Gamma \Rightarrow \Delta}{\Pi;\Gamma, A \Rightarrow \Delta} \qquad \text{RW} \frac{\Pi;\Gamma \Rightarrow \Delta}{\Pi;\Gamma \Rightarrow A, \Delta} \qquad \text{LC} \frac{\Pi;\Gamma, A, A \Rightarrow \Delta}{\Pi;\Gamma, A \Rightarrow \Delta} \qquad \text{RC} \frac{\Pi;\Gamma \Rightarrow A, A, \Delta}{\Pi;\Gamma \Rightarrow A, \Delta}$$

Cut rules:

$$\text{head} \frac{\Pi;\Gamma \Rightarrow \Delta, A \qquad A;\Gamma' \Rightarrow \Delta'}{\Pi;\Gamma,\Gamma' \Rightarrow \Delta,\Delta'} \qquad \text{mid} \frac{;\Gamma \Rightarrow \Delta, A \qquad \Pi; A,\Gamma' \Rightarrow \Delta'}{\Pi;\Gamma,\Gamma' \Rightarrow \Delta,\Delta'}$$

Table 5: The calculus LKT

Derivability of $\Gamma \Rightarrow \Delta; \Pi$ in LKQ corresponds precisely to linear derivability of $!\Gamma^Q \Rightarrow ?!\Delta^Q; \Pi^Q$, and $\langle (\cdot)^Q, !, ?! \rangle$ is an inductive decoration strategy for LKQ.

The calculus LKT (table 5) appears as the classical equivalent of ILU (as this intuitionistic calculus is obtained from LKT by the usual intuitionistic restriction of the succedents to singletons; moreover, the $(\cdot)^*$-translation is obtained by deleting all occurrences of '?' in the T-translation). Here we find sequents $\Pi; \Gamma \Rightarrow \Delta$, with Π containing at most one distinguished formula, the '*tête*' (whence T) or *head*-formula. As in ILU, it corresponds to a formula that has not yet been subjected to non-linear manipulations on the left. Included we find the *negative* fragment of classical implicational logic as it appears in LU (Girard(1993)).

Derivability of $\Pi; \Gamma \Rightarrow \Delta$ in LKT corresponds precisely to linear derivability of $\Pi^T; !?\Gamma^T \Rightarrow ?\Delta^T$ in the $\{!, ?, \multimap, \forall_2^T\}$-fragment of linear logic, where $\forall_2^T$ indicates abstraction restricted to formulas of the form X^T (observe that if a sequent $\Gamma_1^T, ?\Gamma_2^T, !?\Gamma_3^T \Rightarrow !?\Delta_3^T, ?\Delta_2^T, \Delta_1^T$ is derivable in this fragment, then $|\, \Gamma_1 \cup \Gamma_2 \cup \Delta_3 \,| \leq 1$), and $\langle (\cdot)^T, !?, ? \rangle$ is an inductive decoration strategy for LKT.

The calculus LKT, like ILU, like LKQ, inherits the computational properties of linear logic: the T-decorations of LKT-derivations are *strong* decorations. Hence we found a proofsystem for *classical* logic as a proper fragment of CLL:

PROPOSITION. *If* π *is a derivation in* $\{!, ?, \multimap \forall_2^T\}$ *of a sequent*

$$\Gamma_1^T, ?\Gamma_2^T, !?\Gamma_3^T \Rightarrow !?\Delta_3^T, ?\Delta_2^T, \Delta_1^T$$

in which all cutformulas are of the form $A^T, ?A^T$ *or* $!?A^T$*, and all axioms are of the form* $A^T \Rightarrow A^T$*, then* $sk(\pi)$ *is an LKT-derivation of*

$$\Gamma_1 \cup \Gamma_2; \Gamma_3 \Rightarrow \Delta_3, \Delta_2, \Delta_1. \qquad \square$$

The distinction Q/T reflects the dichotomy *positive/negative* introduced in Girard(1991), even though identifying LKQ as a proper fragment of LU seems not to be possible. Maybe somewhat surprising is that we obtained two constructive calculi complete for classical logic each of which *stays* at a different 'side of the mirror'.

One final remark: the calculus LKT is closely related to the system of classical natural deduction and its term calculus $\lambda\mu$ as introduced and studied by Parigot(1992), a relation that we might state as

$$\frac{ILU}{\lambda_2} = \frac{LKT}{\lambda\mu},$$

an observation that is explored further in Danos et al.(1994), in which it is shown how a natural generalization of the techniques and results of the present paper can be used in order to classify and understand some of the recent, seemingly unrelated, solutions to the problem of devising classical second order calculi equipped with a strong (and, as much as possible, confluent) normalization, and a denotational semantics.

References

DANOS, V., JOINET, J.-B., AND SCHELLINX, H. (1993a). On the linear decoration of intuitionistic derivations. Prépublication 41, Équipe de Logique Mathématique, Université Paris VII.

DANOS, V., JOINET, J.-B., AND SCHELLINX, H. (1993b). The structure of exponentials: uncovering the dynamics of linear logic proofs. In Gottlob, G., Leitsch, A., and Mundici, D., editors, *Computational Logic and Proof Theory,* pages 159–171. Springer Verlag. Lecture Notes in Computer Science 713, Proceedings of the Third Kurt Gödel Colloquium, Brno, Czech Republic, August 1993.

DANOS, V., JOINET, J.-B., AND SCHELLINX, H. (1994). A new deconstructive logic: linear logic. Prépublication 52, Equipe de Logique Mathématique, Université Paris VII.

GIRARD, J.-Y. (1991). A new constructive logic: classical logic. *Mathemathical Structures in Computer Science*, 1(3):255–296.

GIRARD, J.-Y. (1993). On the unity of logic. *Annals of Pure and Applied Logic*, 59:201–217.

GIRARD, J.-Y., LAFONT, Y., AND TAYLOR, P. (1988). *Proofs and Types.* Cambridge Tracts in Theoretical Computer Science 7. Cambridge University Press.

HOWARD, W. A. (1980). The formulae-as-types notion of construction. In Seldin, J. P. and Hindley, J. R., editors, *To H.B. Curry: Essays on Combinatory Logic, Lambda Calculus and Formalism*, pages 479 – 490. Academic Press.

JOINET, J.-B. (1993). *Etude de la normalisation du calcul des séquents classique à travers la logique linéaire.* PhD thesis, Université Paris VII.

PARIGOT, M. (1992). $\lambda\mu$-Calculus: an algorithmic interpretation of classical natural deduction. In Voronkov, A., editor, *Logic Programming and Automated Reasoning*, pages 190–201. Springer Verlag. Lecture Notes in Artificial Intelligence 624, Proceedings of the LPAR, St. Petersburg, July 1992.

SCHELLINX, H. (1991). Some syntactical observations on linear logic. *Journal of Logic and Computation*, 1(4):537–559.

SCHELLINX, H. (1994). *The Noble Art of Linear Decorating.* ILLC Dissertation Series, 1994-1. Institute for Language, Logic and Computation, University of Amsterdam.

From Proof-Nets to Interaction Nets

Yves Lafont

Laboratoire de Mathématiques Discrètes
UPR 9016 du CNRS, 163 avenue de Luminy, case 930
F 13288 MARSEILLE CEDEX 9
lafont@lmd.univ-mrs.fr

1 Introduction

If we consider the interpretation of proofs as programs, say in intuitionistic logic, the question of equality between proofs becomes crucial: The syntax introduces meaningless distinctions whereas the (denotational) semantics makes excessive identifications. This question does not have a simple answer in general, but it leads to the notion of proof-net, which is one of the main novelties of linear logic. This has been already explained in [Gir87] and [GLT89].

The notion of interaction net introduced in [Laf90] comes from an attempt to implement the reduction of these proof-nets. It happens to be a simple model of parallel computation, and so it can be presented independently of linear logic, as in [Laf94]. However, we think that it is also useful to relate the exact origin of interaction nets, especially for readers with some knowledge in linear logic. We take this opportunity to give a survey of the theory of proof-nets, including a new proof of the sequentialization theorem.

2 Multiplicatives

First we consider the kernel of linear logic, with only two connectives: $\otimes$ (*times* or *tensor product*) and its dual $\wp$ (*par* or *tensor sum*). The first one can be seen as a conjunction and the second one as a disjunction. Each atom has a positive form p and a negative one $p^\perp$ (the *linear negation* of p). This linear negation is extended to all formulae according to the following laws:

$$p^{\perp\perp} = p, \quad (A \otimes B)^\perp = B^\perp \wp A^\perp, \quad (A \wp B)^\perp = B^\perp \otimes A^\perp.$$

Note that our negation reverses the order of subformulae, although it would only matter in the case of non-commutative linear logic. In the present case, a sequent $\vdash A_1, \ldots, A_n$ is a multiset, which means that the order of formulae is irrelevant. With this convention, there are only four deduction rules (*axiom*, *cut*, *times* and *par*):

$$\frac{}{\vdash A, A^\perp} \qquad \frac{\vdash A,\Gamma \quad \vdash A^\perp,\Delta}{\vdash \Gamma,\Delta} \qquad \frac{\vdash A,\Gamma \quad \vdash B,\Delta}{\vdash A\otimes B,\Gamma,\Delta} \qquad \frac{\vdash A,B,\Gamma}{\vdash A \wp B,\Gamma}$$

This sequent calculus satisfies the cut-elimination property, which means that any proof of a sequent $\vdash \Gamma$ reduces to a cut-free one. For instance, the sequent $\vdash p^\perp, p\otimes(q\otimes r), r^\perp \wp q^\perp$ has two *cut-free* proofs:

$$\frac{\dfrac{\dfrac{}{\vdash p^\perp,p} \quad \dfrac{\dfrac{}{\vdash q,q^\perp}\quad\dfrac{}{\vdash r,r^\perp}}{\vdash q\otimes r, r^\perp, q^\perp}}{\vdash p^\perp, p\otimes(q\otimes r), r^\perp, q^\perp}}{\vdash p^\perp, p\otimes(q\otimes r), r^\perp \wp q^\perp} \qquad \text{and} \qquad \frac{\dfrac{}{\vdash p^\perp,p}\quad \dfrac{\dfrac{\dfrac{}{\vdash q,q^\perp}\quad\dfrac{}{\vdash r,r^\perp}}{\vdash q\otimes r,r^\perp,q^\perp}}{\vdash q\otimes r, r^\perp\wp q^\perp}}{\vdash p^\perp,p\otimes(q\otimes r),r^\perp\wp q^\perp}$$

By cut-elimination, any proof of this sequent reduces to one of those two proofs. There are three basic reduction rules for cut-elimination:

$$\frac{\dfrac{}{\vdash A^\perp,A}\quad \begin{array}{c}\vdots\\ \vdash A,\Gamma\end{array}}{\vdash A,\Gamma} \longrightarrow \begin{array}{c}\vdots\\ \vdash A,\Gamma\end{array} \qquad \frac{\begin{array}{c}\vdots\\ \vdash A,\Gamma\end{array}\quad \dfrac{}{\vdash A^\perp,A}}{\vdash A,\Gamma} \longrightarrow \begin{array}{c}\vdots\\ \vdash A,\Gamma\end{array}$$

$$\frac{\dfrac{\begin{array}{c}\vdots\\ \vdash A,\Gamma\end{array}\quad \begin{array}{c}\vdots\\ \vdash B,\Delta\end{array}}{\vdash A\otimes B,\Gamma,\Delta} \quad \dfrac{\begin{array}{c}\vdots\\ \vdash B^\perp,A^\perp,\Lambda\end{array}}{\vdash B^\perp\wp A^\perp,\Lambda}}{\vdash \Gamma,\Delta,\Lambda} \longrightarrow \frac{\begin{array}{c}\vdots\\ \vdash A,\Gamma\end{array}\quad \dfrac{\begin{array}{c}\vdots\\ \vdash B,\Delta\end{array}\quad \begin{array}{c}\vdots\\ \vdash B^\perp,A^\perp,\Lambda\end{array}}{\vdash A^\perp,\Delta,\Lambda}}{\vdash \Gamma,\Delta,\Lambda}$$

The other reduction rules allow to move cuts upwards, as in the following case:

$$\frac{\dfrac{\begin{array}{c}\vdots\\ \vdash A,\Gamma\end{array}\quad \begin{array}{c}\vdots\\ \vdash B,C,\Delta\end{array}}{\vdash A\otimes B,C,\Gamma,\Delta}\quad \begin{array}{c}\vdots\\ \vdash C^\perp,\Lambda\end{array}}{\vdash A\otimes B,\Gamma,\Delta,\Lambda} \longrightarrow \frac{\begin{array}{c}\vdots\\ \vdash A,\Gamma\end{array}\quad \dfrac{\begin{array}{c}\vdots\\ \vdash B,C,\Delta\end{array}\quad \begin{array}{c}\vdots\\ \vdash C^\perp,\Lambda\end{array}}{\vdash B,\Delta,\Lambda}}{\vdash A\otimes B,\Gamma,\Delta,\Lambda}$$

This reduction terminates but is not confluent. In other words, the normal form of a proof is not unique in general. This comes from the fact that, in sequent calculus, the order of application of rules may be irrelevant. *Proof-nets* allow one to abstract from this irrelevant order.

Beside cut-elimination, there is another transformation for eliminating non-atomic axioms. It is an expansion rather than a reduction:

$$\frac{}{\vdash A \otimes B, B^{\perp} \wp A^{\perp}} \quad \longrightarrow \quad \frac{\dfrac{\dfrac{}{\vdash A, A^{\perp}} \quad \dfrac{}{\vdash B, B^{\perp}}}{\vdash A \otimes B, B^{\perp}, A^{\perp}}}{\vdash A \otimes B, B^{\perp} \wp A^{\perp}}$$

This means that, if a sequent is provable, it has a cut-free proof with only atomic axioms.

3 Proof-nets

Since our notion differs slightly from the original one in [Gir87], we shall give precise definitions. First, we introduce a *times cell* and a *par cell*:

Both have three *ports*: a *principal* one below and two *auxiliary* ones above. The two auxiliary ports are distinguished: There is a left one and a right one.

A *net* is just a finite graph built with times and par cells. In addition to the ports associated with those cells, there is an extra set of *free ports* for subsequent connections, and each port must be connected to another one by a *wire*. For instance, here is a net with 3 free ports:

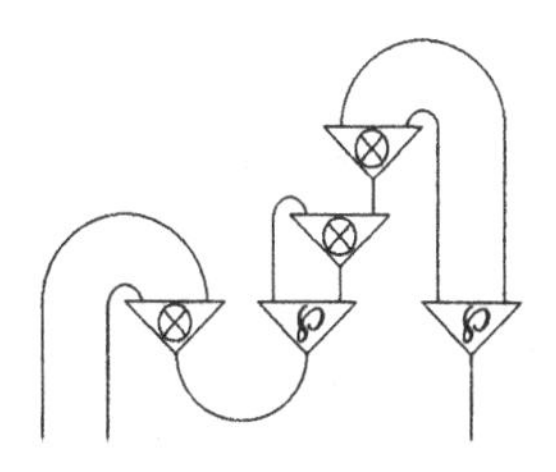

We say that a net is *well-formed* if it is of one of the following forms:

- a single wire,
- two well-formed nets connected by a single wire,
- two well-formed nets connected by a times cell through its auxiliary ports,
- a well-formed net connected to itself by a par cell through its auxiliary ports.

This is of course an inductive definition. Intuitively, a well-formed net is a kind of module obtained by plugging smaller modules according to certain rules. The free ports are the interface of this module. The four cases can be pictured as follows:

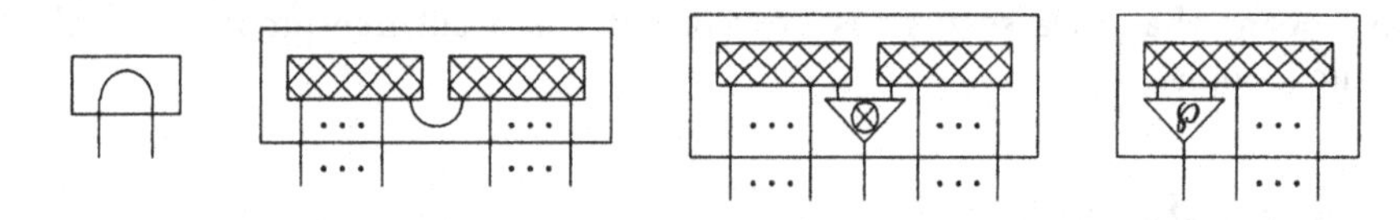

For instance, the above net is well-formed (see figure 1). In fact, the four cases correspond to the four rules of sequent calculus. Since we do not impose the planarity of nets, wires are allowed to cross each other, and the exchange rule is implicit:

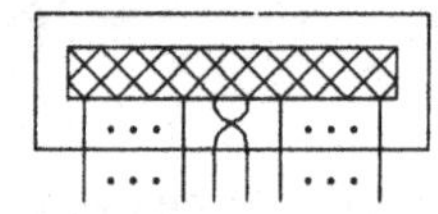

A *typed net* is a net where each wire is labelled with a formula. If a wire is explicitly labelled with A for a given orientation, then it is implicitly labelled with $A^\perp$ for the opposite orientation:

Furthermore, for each cell, the connecting wires must be typed as follows:

$$A \quad B \qquad\qquad A \quad B$$
$$A \otimes B \qquad\qquad A \wp B$$

For instance, our favorite net can be typed (see figure 2): It defines a *proof-net*, *i.e.* a well-formed typed net. Clearly, a proof Π of a sequent $\vdash A_1, \ldots, A_n$

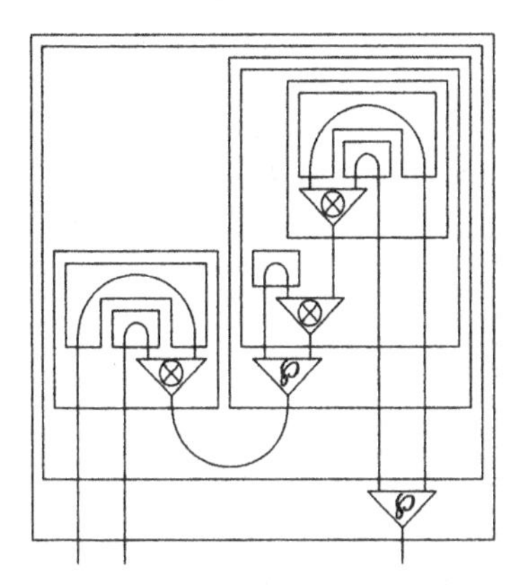

Figure 1: checking that a net is well-formed

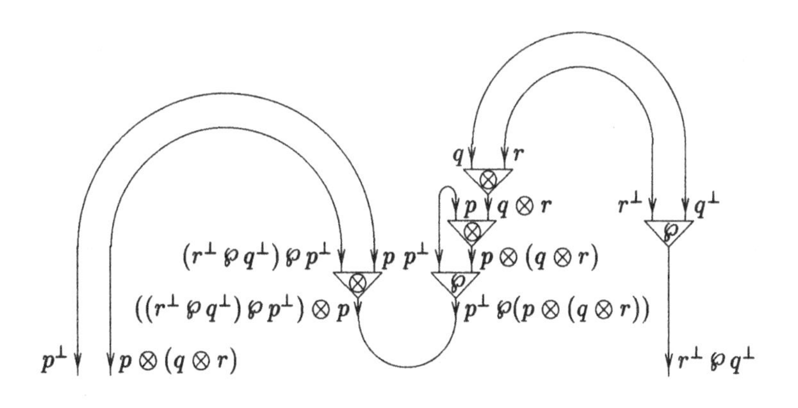

Figure 2: typing a net

defines a proof-net $\langle \Pi \rangle$ with n free ports typed as follows:

$$A_1 \cdots A_n$$

Conversely, any proof-net π is obtained in this way, although not uniquely. For instance, the two cut-free proofs of the sequent $\vdash p^\perp, p \otimes (q \otimes r), r^\perp \wp q^\perp$ define the same net:

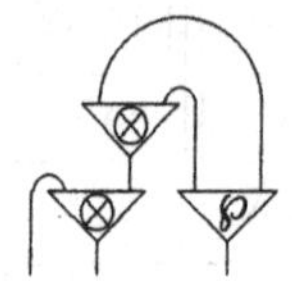

This means that the proof-net is more intrinsic than the proof. In fact, it is possible to show that two cut-free proofs define the same net if and only if they are equivalent modulo some permutations of rules of the following kind:

$$\dfrac{\dfrac{\begin{matrix}\vdots \\ \vdash A, \Gamma\end{matrix} \quad \begin{matrix}\vdots \\ \vdash B, C, D, \Delta\end{matrix}}{\vdash A \otimes B, C, D, \Gamma, \Delta}}{\vdash A \otimes B, C \wp D, \Gamma, \Delta} \quad \sim \quad \dfrac{\begin{matrix}\vdots \\ \vdash A, \Gamma\end{matrix} \quad \dfrac{\begin{matrix}\vdots \\ \vdash B, C, D, \Delta\end{matrix}}{\vdash B, C \wp D, \Gamma, \Delta}}{\vdash A \otimes B, C \wp D, \Gamma, \Delta}$$

However, it is much better to think in terms of proof-nets than in terms of equivalence classes of proofs.

4 Reducing

With nets, the cut-elimination becomes surprisingly simple. There is only one reduction rule:

$$x_1\ x_2\ y_1\ y_2 \longrightarrow x_1\ x_2\ y_1\ y_2$$

This means that, if the principal port of a times cell is connected to the principal port of a par cell, the two cells disappear and the remaining wires are joined together as indicated by the right member of the rule. Here is an example of reduction:

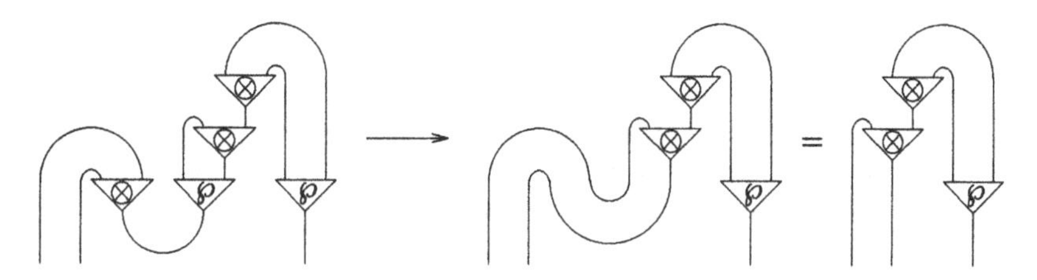

Of course, the reduction terminates since it decreases the number of cells, but it is also confluent in a strong sense:

Lemma 1 *If a net π reduces in one step to π' and to π'', with $\pi' \neq \pi''$, then π' and π'' reduce in one step to a common net.*

The point is that, in an instance of the left member of the rule, the two cells are connected through their principal ports, so that two such instances are necessarily disjoint. As a corollary, the number of steps for reducing a net to its normal form is independent of the reduction path. Note that our notion of net avoids the critical pairs of [Gir87] because we have no explicit cell for axiom and cut.

To be honest, there is a little problem with our definition. Indeed, if we apply the reduction rule to an arbitrary net, we may have to connect a wire with itself as in the following example:

To fix this problem, we must allow *closed wires* in the definition of nets. Fortunately, we shall see that those closed wires cannot appear if we start from a well-formed net.

Clearly, if a proof Π reduces to Π', then the net $\langle\Pi\rangle$ reduces to $\langle\Pi'\rangle$. In particular, if Π' is cut-free, then $\langle\Pi'\rangle$ is the normal form of $\langle\Pi\rangle$, which is uniquely determined by Π. Moreover, this normal form depends really on the proof, not only on the sequent. For instance, there are two distinct normal proof-nets for the sequent $\vdash p \otimes p, p^\perp, p^\perp$:

Conversely, it is possible to show that if $\langle\Pi\rangle$ reduces to a net π', then there is a proof Π' such that Π reduces to Π' and $\pi' = \langle\Pi'\rangle$. In particular, proof-nets are closed under reduction.

Now, we shall look more closely at the structure of irreducible nets. If $A[p_1, \ldots, p_n]$ is a formula where the atoms $p_1, \ldots, p_n$ occur positively, exactly once and in order, it defines a net $\langle A\rangle$ (the *tree* of A) with $n+1$ free ports. For instance, if A is $p_1 \otimes (p_2 \wp p_3)$, the tree $\langle A\rangle$ is the following net:

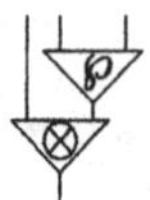

Note that, in general, $\langle A\rangle$ is not a proof-net, but it is clearly typable.

Lemma 2 *If π is an irreducible net, then at least one of the following statements holds:*

- *π contains a clash, i.e. two cells connected through their principal ports for which no rule applies:*

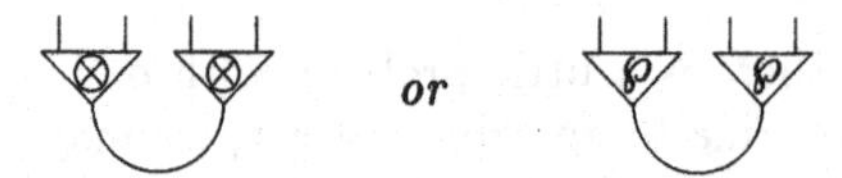

- *π contains a vicious circle, i.e. a sequence of cells $a_1, a_2, \ldots, a_n, a_{n+1} = a_1$ such that the principal port of a_i is connected to an auxiliary port of a_{i+1}:*

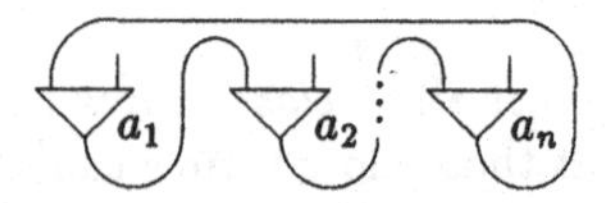

- *π is in reduced form, i.e. it can be uniquely decomposed as follows:*

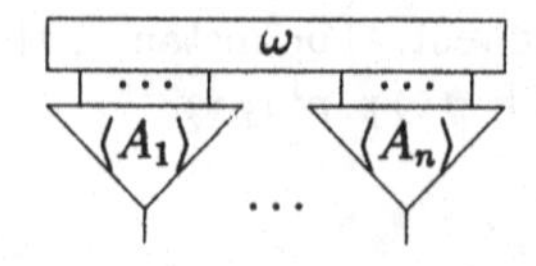

where $A_1, \ldots, A_n$ is a sequence of formulae and ω is a net without cell and without closed wire, i.e. a partition of the remaining ports into pairs.

Here, a closed wire is considered as a vicious circle of length $n = 0$. This lemma is proved by induction on the number of cells. The point is that, if π contains some cell but no clash and no vicious circle, then some free port must be connected to the principal port of a cell.

It is clear that a typed net contains no clash, and it is easy to see that a well-formed net contains no vicious circle.[1] Therefore, an irreducible proof-net is necessarily in reduced form. Conversely, if π is in reduced form, then π is typable although not necessarily well-formed: The types are obtained by unifying the atoms in $A_1, \ldots, A_n$ according to ω.

5 Criterion

Clearly, a net built only with times cells is well-formed if and only if it defines a connected acyclic graph, also called a tree. In general, it is still true that a well-formed net must be connected, but the acyclicity condition is neither necessary nor sufficient. Consider indeed the following nets:

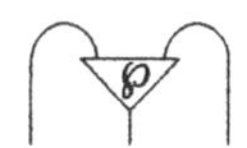

The first one is well-formed but not acyclic and the second one is acyclic but not well-formed. This means that an alternative condition is needed. A *switching* of a net is a graph obtained by replacing each par cell with one of the following two configurations:

For the two examples above, we get the following graphs:

In the first case, both are connected and acyclic, whereas in the second case, none is connected. All four switchings of our favorite proof-net are connected

[1] In fact, it also happens that a typed net contains no vicious circle of length $n > 0$, but it may contain closed wires.

and acyclic. Here is one of them:

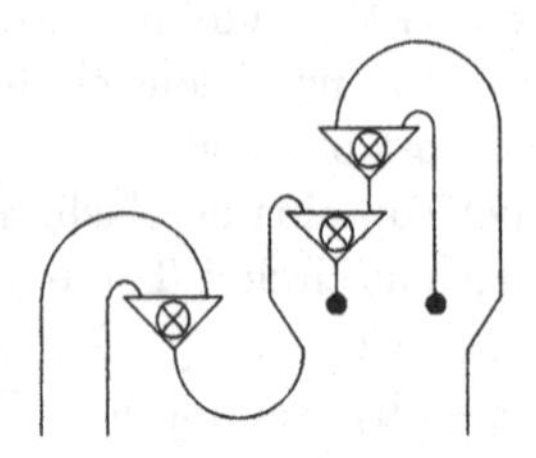

Theorem 1 *A net is well-formed if and only if every switching defines a connected acyclic graph.*

This is the *Danos-Regnier criterion* introduced in [DR89]. In the case of a typed net, we get the sequentialization theorem of [Gir87].

The proof of theorem 1 will be the object of the next section, but we can already make some comments. First, notice that if a net satisfies the criterion, then it contains no vicious circle: This is a direct consequence of the acyclicity condition. Moreover, if a net contains p par cells, it has 2^p switchings. Hence, the criterion itself does not provide an efficient algorithm for checking if a net is well-formed, but it is interesting from a theoretical viewpoint, because of its geometrical nature. For example, it is obviously preserved by reduction:

Lemma 3 *If π reduces to π' and π satisfies the criterion, so does π'.*

Indeed, let λ be a subnet of π consisting of a times cell and a par cell connected through their principal ports. The criterion ensures that any switching of the complement of λ consists of three connected components, two of them being connected to the times cell, and the third one to the par cell. If we apply the reduction to λ, we get a connected acyclic switching:

The lemma follows easily. By theorem 1, this means that well-formed nets are closed under reduction, and similarly for proof-nets, since typing is obviously preserved by reduction. Note that the converse of the lemma does not hold: π does not necessarily satisfy the criterion if π' does. Here is a simple counterexample:

6 Parsing

Typability of nets can be checked with a very simple unification algorithm. To check if a net is well-formed, there is of course a naive exponential algorithm which tries all possibilities, but Danos and Regnier noticed that this can be done in quadratic time (see also [Gal91]). In order to show this, we introduce the *parsing box*:

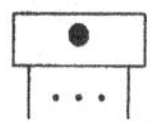

This is a special kind of cell with a variable number of non-distinguished ports. Clearly, a net is well-formed if and only if it reduces to a parsing box by the rules of figure 3: This is just a reformulation of the definition. Of course, this reduction has nothing to do with cut-elimination! Unfortunately, it does not terminate:

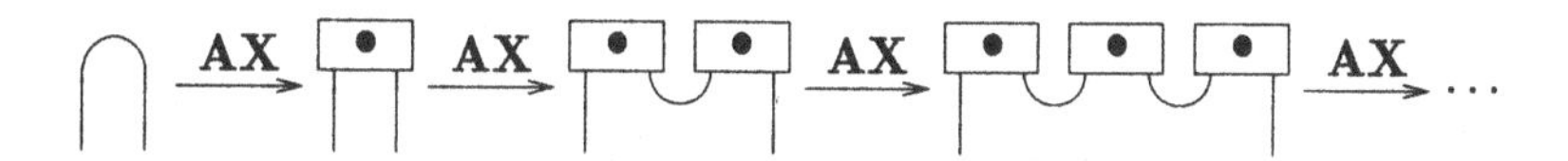

To avoid this problem, we exclude the degenerate case of a net without cell:

Lemma 4 *A non-degenerate net is well-formed if and only if it reduces to a parsing box by the rules of figure 4.*

Proof: First notice that *times* is derivable from $TIMES$ and AX. Similarly, *par* is derivable from PAR and AX. This shows that a net which reduces to a parsing box by the rules of figure 4 is well-formed. The converse is proved by case analysis, using in particular the fact that $TIMES$ is derivable from *times* and CUT. **Q.e.d.**

Starting from a net with n cells, the parsing terminates in at most $2n - 1$ steps. Furthermore, it is confluent (see figure 5), and so it gives an efficient algorithm testing if a net is well-formed. In case of success, and if the net is typable, a proof of sequent calculus can be extracted from the reduction.

We can use this algorithm to prove theorem 1. First, we notice that the Danos-Regnier criterion makes sense in the case of nets with parsing boxes, and it is invariant by the parsing rules:

Lemma 5 *If π reduces to π' by one of the rules of figure 4, then π satisfies the criterion if and only if π' does.*

Proof: For any of the four rules, every switching of the left member is a tree, and similarly for the right member. Now, replacing a subtree by a tree in a tree yields a tree. **Q.e.d.**

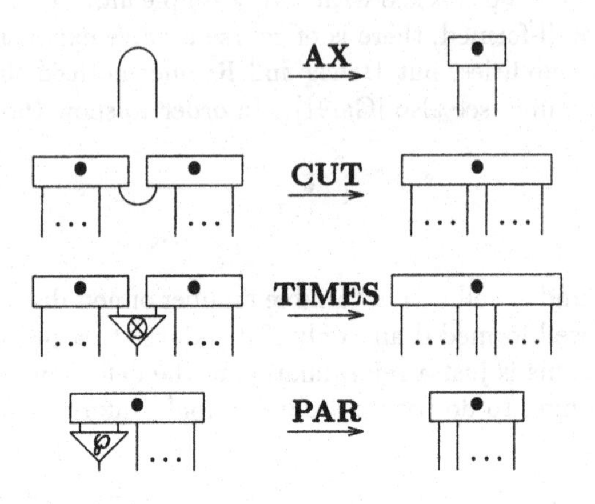

Figure 3: grammar for well-formed nets

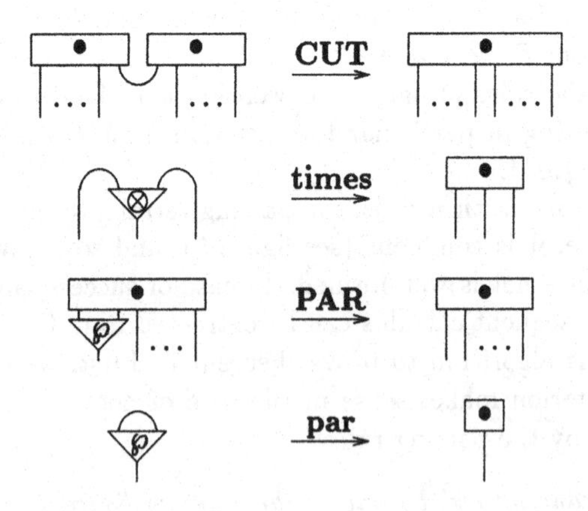

Figure 4: grammar for non-degenerate well-formed nets

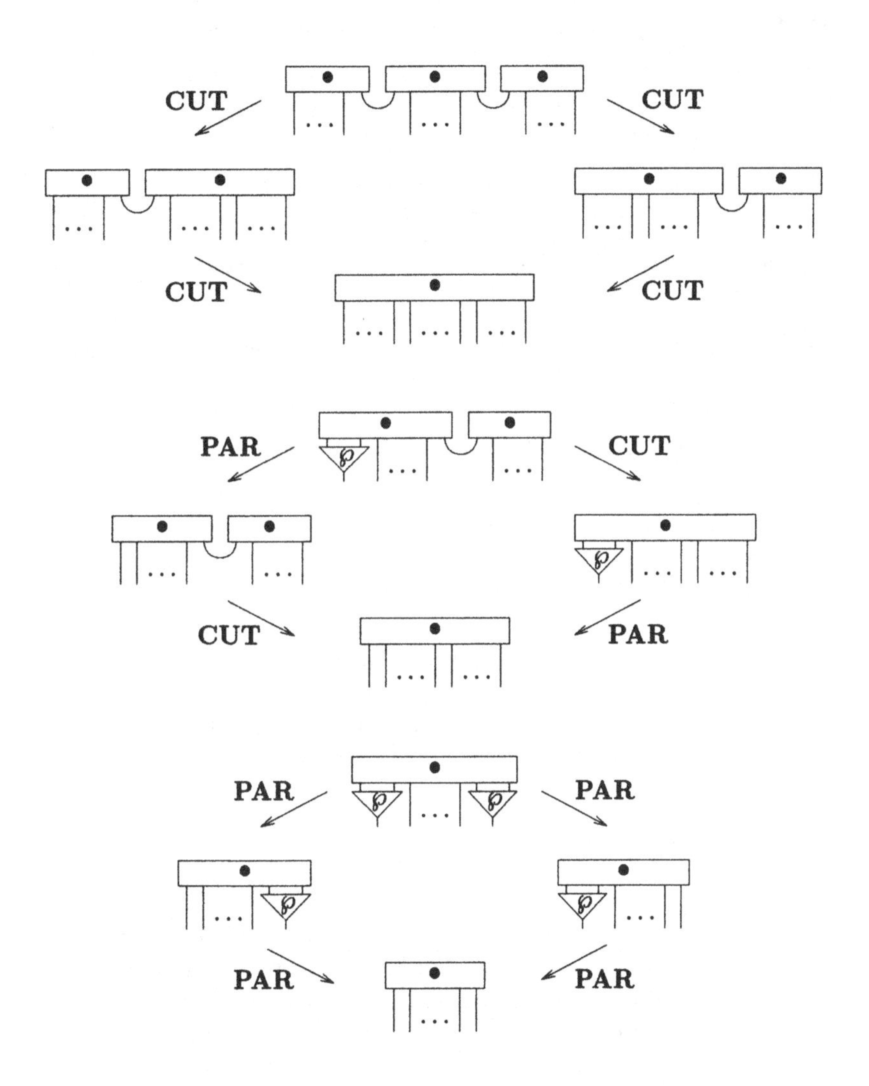

Figure 5: confluence of parsing

The theorem follows from lemmas 4, 5 and the following argument, which was first pointed out in [Dan90]:

Lemma 6 *If a non-degenerate net π satisfies the Danos-Regnier criterion and is irreducible by the rules of figure 4, then π is a parsing box.*

Proof: Since π is irreducible by *times*, it contains no times cell. If a and b are cells in π, we say that a is *dominated by* b and we write $a \prec b$ when one of the following cases holds:

- a and b are par cells and an auxiliary port of a is connected to the principal port of b,
- a is a par cell, b a parsing box and an auxiliary port of a is connected to a port of b,
- a is a parsing box, b a par cell and a port of a is connected to the principal port of b.

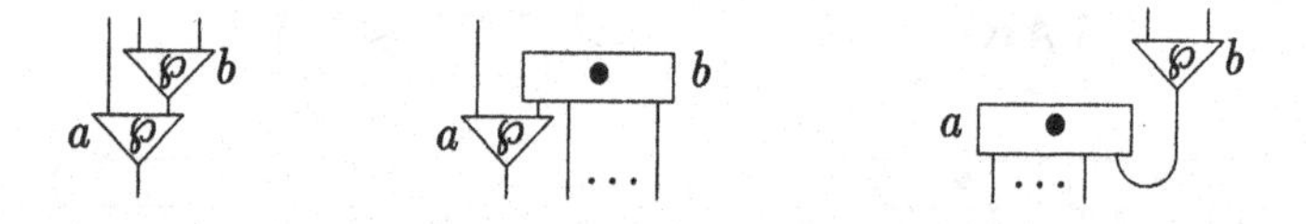

By hypothesis, π contains at least one cell, and there is no cycle $a_1 \prec a_2 \prec \cdots \prec a_{n-1} \prec a_n = a_1$, since it would give a cyclic switching. Hence there is a cell c which is dominated by no other one. Assume that c is a par cell and consider an auxiliary port of c. It cannot be connected to a free port, since it would give a non connected switching. Knowing that c is dominated by no other cell and that π is irreducible by *par*, we must have the following configuration:

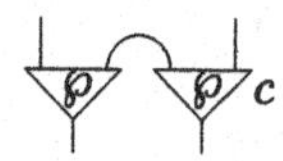

But again, this configuration gives a non connected switching, hence c is a parsing box. Knowing that π is irreducible by the parsing rules and that c is dominated by no other cell, it can only be connected to free ports and to auxiliary ports of distinct par cells:

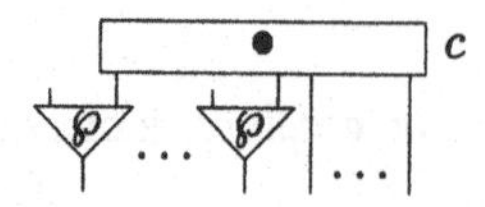

But in such a case, there is a non connected switching, unless c is only connected to free ports, which means that π is a parsing box. **Q.e.d.**

7 Units

Until now, we have not considered the multiplicative units 1 and $\bot$ (*bottom*), with the following deduction rules:

$$\frac{}{\vdash 1} \qquad \frac{\vdash \Gamma}{\vdash \bot, \Gamma}$$

All statements of section 2 to 4 extend to the full multiplicative fragment. We just need a *one* cell and a *bottom* cell:

The reduction is very simple, with an empty right member:

Of course, it is no more the case that well-formed nets are connected. Moreover, the parsing rule for the bottom cell is strongly non-deterministic:

By this rule indeed, the bottom cell is allowed to interact with any parsing box in the net! In fact, there is no hope to find a good parsing algorithm for the full multiplicative fragment. In [LW94], it is shown indeed that the multiplicative fragment of linear logic with units but without proper atoms is NP-complete. But in the absence of proper atoms, a sequent $\vdash A_1, \ldots, A_n$ is provable if and only the sequence of trees $\langle A_1 \rangle, \ldots, \langle A_n \rangle$ defines a well-formed net. Therefore, parsing with units is NP-complete!

Moreover, there is no hope to extend the Danos-Regnier criterion to the full fragment. Indeed, a condition of the kind "for every switching ..." is co-NP, and by the previous remark, the existence of a co-NP condition would mean that NP = co-NP, which is very unlikely. For this reason, it is usually thought that something goes wrong with units. Now, there are two ways to escape: change the logic, or change the notion of net.

One can argue indeed that, in the criterion, acyclicity is more important than connectedness. For example, it is enough for preventing vicious circles.

This suggests a weak version of the criterion, corresponding to a variant of linear logic with two extra deduction rules (*mix* and *empty*):

$$\frac{\vdash \Gamma \quad \vdash \Delta}{\vdash \Gamma, \Delta} \qquad \frac{}{\vdash}$$

This logic satisfies cut-elimination, but it looks a bit degenerate. For instance, $A \otimes B$ implies $A \wp B$ and 1 is equivalent to $\perp$ (see [FR90] for a more elaborate system). On the other side, the proof-boxes of [Gir87] introduce a kind of synchronization which is not justified from a purely computational viewpoint. Maybe, a better understanding of linear logic is needed to settle this question.

8 Exponentials

We have seen that the process of cut-elimination in the sequent calculus for multiplicative linear logic can be completely *localized.* The contexts Γ, Δ and Λ play indeed no active role in the reduction rule. Unfortunately, the multiplicative fragment has very little computational power: The reduction terminates in linear time because of the absence of a contraction rule.

To gain expressiveness, we must also consider the *exponential* connectives: ! (*of course*) and its dual ? (*why not*). There are four deduction rules (*promotion, dereliction, contraction* and *weakening*):

$$\frac{\vdash A, ?\Gamma}{\vdash !A, ?\Gamma} \qquad \frac{\vdash A, \Gamma}{\vdash ?A, \Gamma} \qquad \frac{\vdash ?A, ?A, \Gamma}{\vdash ?A, \Gamma} \qquad \frac{\vdash \Gamma}{\vdash ?A, \Gamma}$$

In the promotion rule, $?\Gamma$ stands for a multiset $?C_1, \ldots, ?C_n$. Intuitively, the formula $!A$ represents an inexhaustible resource: It can be accessed (by dereliction), duplicated (by contraction) or erased (by weakening). The promotion rule expresses that a resource coming from inexhaustible resources is also inexhaustible.

This extended calculus satisfies the cut-elimination property, but the reduction of a proof may be quite long (hyperexponential in fact) because of the following reduction rule which duplicates proofs:

$$\frac{\dfrac{\dfrac{\vdots}{\vdash A, ?\Gamma}}{\vdash !A, ?\Gamma} \quad \dfrac{\dfrac{\vdots}{\vdash ?A^{\perp}, ?A^{\perp}, \Delta}}{\vdash ?A^{\perp}, \Delta}}{\vdash ?\Gamma, \Delta} \longrightarrow \frac{\dfrac{\dfrac{\dfrac{\vdots}{\vdash A, ?\Gamma}}{\vdash !A, ?\Gamma} \quad \dfrac{\dfrac{\dfrac{\vdots}{\vdash A, ?\Gamma}}{\vdash !A, ?\Gamma} \quad \dfrac{\vdots}{\vdash ?A^{\perp}, ?A^{\perp}, \Delta}}{\vdash ?A^{\perp}, ?\Gamma, \Delta}}{\vdash ?\Gamma, ?\Gamma, \Delta}}{\vdash ?\Gamma, \Delta}$$

Here, the context $?\Gamma$ of the promotion rule plays an active role in the reduction. For that reason, we introduce an *exponential box* for promotion and three ordinary cells for dereliction, contraction and weakening:

The exponential box is a special kind of cell parametrized by a net: If π is a net with $n+1$ free ports, then $!\pi$ is a cell with one principal port (the leftmost one) and n auxiliary ports. The net π may itself contain exponential boxes.

Following the rules of sequent calculus, it is easy to define an appropriate notion of well-formed net. For example, the exponential box $!\pi$ is well-formed when the net π is well-formed. The contraction behaves like a *par* rule, and the weakening like a *bottom* rule. In particular, the Danos-Regnier criterion works only in the absence of weakening.[2] Finally, the typing constraints are the following:

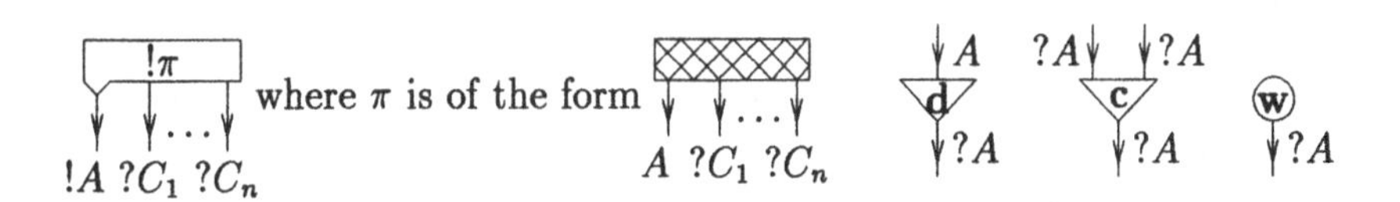

Now we can express cut-elimination directly at the level of nets. Three kinds of reductions are needed:

- *external reductions* (when two cells or boxes are connected through their principal ports):

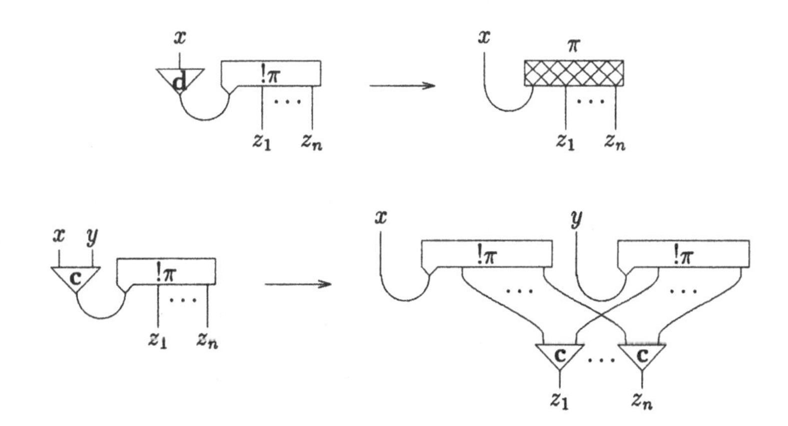

[2]For that reason, there is a box for weakening in [Gir87].

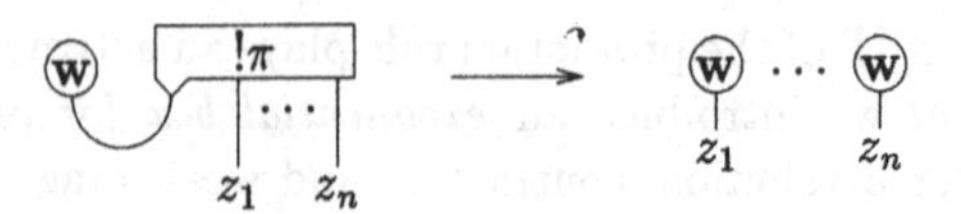

Intuitively, a dereliction opens a box, a contraction duplicates it and a weakening erases it.

- *commutative reductions* (when an auxiliary port of a box is connected to the principal port of another box):

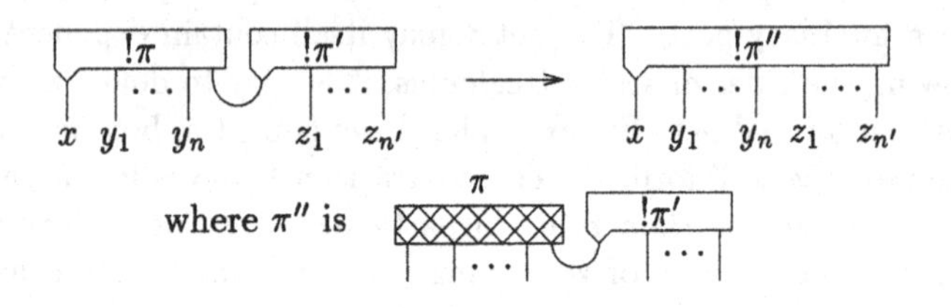

Intuitively, the second box enters the first one.

- *internal reductions* (inside a box):

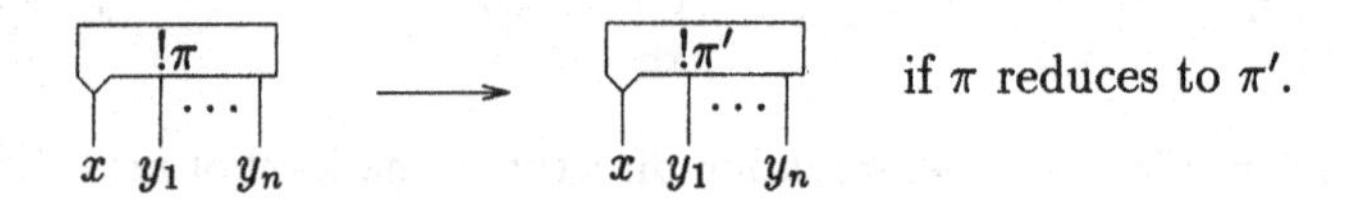

Of course, we must not forget the multiplicative reduction, which is of the external kind. This system terminates in the case of proof-nets (see [Gir87]). Furthermore, there are a lot of conflicts between the various kinds of rules, but the property of confluence holds, even in the untyped case (see [Dan90] or [Reg92]).[3]

External reduction suffices if we content ourselves with a weak notion of normal form. Here is another way of expressing this fact:

Lemma 7 *If* Π *is a proof of* $\vdash \Gamma$, *where* Γ *has no occurrence of the connective* $!$, *then the proof-net* $\langle \Pi \rangle$ *reduces to its normal form by external reduction.*

Indeed, if Γ has no occurrence of $!$ and $\langle \Pi \rangle$ is irreducible by external reduction, then it contains no box at all: This is easily proved by induction on Π. There is a similar situation in functional programming, where nobody wants to see actual values of functional type. In other words, external reduction is a kind

[3]This is not obvious at all, since for instance, the system for additive connectives is not confluent!

of lazy reduction (see [Abr93]). Note that external reduction is confluent in the strong sense of lemma 1.

From the viewpoint of cut-elimination, this fragment is much more expressive than the multiplicative one. There is indeed a translation of the implicative fragment of intuitionistic logic into the multiplicative exponential fragment of linear logic (see [Gir87]). It transforms an intuitionistic proof of $\Gamma \vdash A$ into a linear proof of $\vdash ?\Gamma, A$ with the following interpretation of the implication:

$$A \Rightarrow B = ?A^{\perp} \wp B.$$

This translation extends to untyped λ-terms (see [Dan90] or [Reg92]) so that β-reduction is interpreted in terms of reduction of well-formed nets[4]. There is another well-known translation which transforms an intuitionistic proof of $\Gamma \vdash A$ into a linear proof of $\vdash ?\Gamma, !A$ with the following interpretation of the implication:

$$A \Rightarrow B = ?A^{\perp} \wp !B.$$

The difference is that, in the first case, external reduction corresponds to the call-by-name strategy, whereas in the second case, it corresponds to the call-by-value strategy. In both cases, an unsolvable term such as $(\lambda x \cdot xx)(\lambda x \cdot xx)$ gives a (well-formed but untyped) net with no normal form, even for external reduction.

9 Interaction nets

The previous sections suggest a natural generalization of proof-nets. Consider indeed an alphabet Σ of symbols with arities. For each symbol α of arity p in Σ, we can introduce a cell with one principal port and p auxiliary ones:

If α and β are symbols of respective arities p and q, an *interaction rule* for (α, β) is a reduction of the following form:

[4]Strictly speaking, this interpretation is only defined modulo associativity and unit axioms for contraction and weakening.

where $\rho_{\alpha,\beta}$ is a net with $p+q$ free ports, built with symbols of Σ. Obviously, this rule is equivalent to the following one:

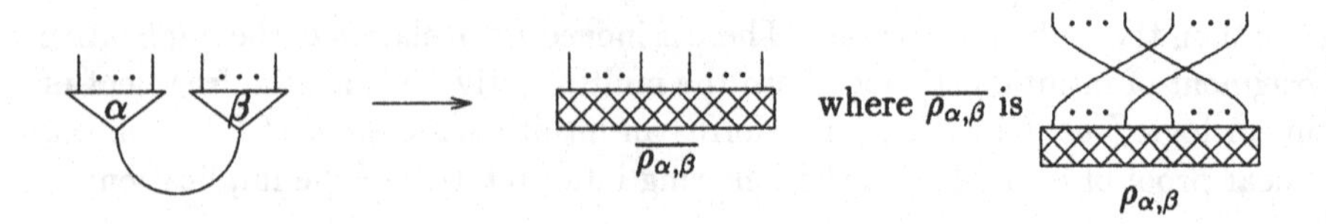

An *interaction system* is a non-ambiguous set of interaction rules. This means that the following conditions are satisfied:

- If a rule is given for (α, β), then no other rule is given for (α, β) or (β, α).
- If a rule is given for (α, α) then $\rho_{\alpha,\alpha} = \overline{\rho_{\alpha,\alpha}}$ (symmetry condition).[5]

Such a system satisfies the strong confluence property of lemma 1.

The multiplicative reduction of section 4 is a very simple kind of interaction rule where the net $\rho_{\alpha,\beta}$ contains no cell. A much more elaborate example is given by the external reduction of the previous section. In that case, the alphabet is infinite because there are infinitely many exponential boxes, but the system is *locally finite* in the following sense: If we start from a net π, the only cells that may appear in the reduction are those which occur in the construction of π (possibly inside boxes), and they are finitely many. In fact, it is possible to replace this infinite system by a finite one at the expense of some encoding (see [Mac94]).[6] Here is a completely different kind of example:

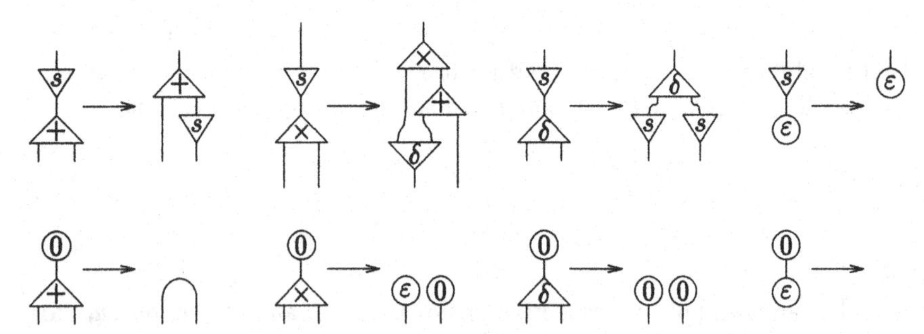

It is a program for unary arithmetic, based on the classical definitions of addition and multiplication:

$$s\,x + y = s\,(x + y), \qquad 0 + y = y, \qquad s\,x \times y = (x \times y) + y, \qquad 0 \times y = 0.$$

[5]This case was not considered in [Laf90].

[6]In [GAL92b], another encoding of exponential boxes is proposed, which is inspired by the geometry of interaction and the optimal reduction of lambda-calculus (see [GAL92a]).

It is essential to understand that the principal port of $+$ is not the output, but the first input. This comes from the fact that $+$ is defined by induction on its first argument. Of course, the same holds for $\times$, and the general interpretation of ports is the following:

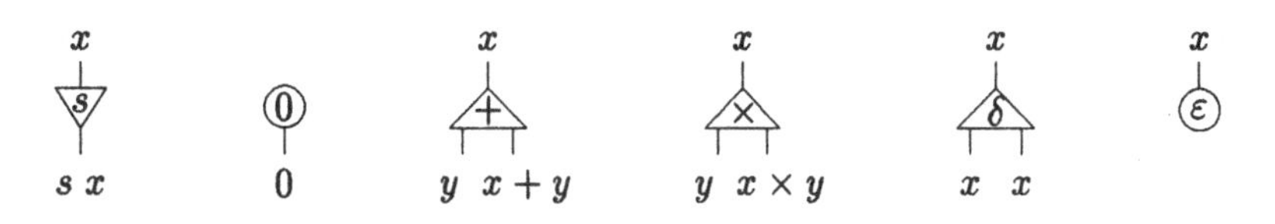

The *duplicator* δ and the *eraser* ε are needed for the multiplication: They are reminiscent of the contraction and the weakening of the previous example.

Our interaction nets provide a simple model of parallel computation. A cell is an agent which is activated only when its principal port is connected to the principal port of another agent. Of course, the two agents must interact simultaneously, but no synchronization is needed between interactions. It is not the case of other models such as cellular automata, where a global synchronization is needed. It is also interesting to notice that our interaction nets are an instance of the *connection graphs*, which were designed as an assembly language for parallel machines (see [Baw86]). Moreover, the sequential models of computation, such as Turing machines, can be revisited in the light of this new paradigm which abolishes the traditional distinction between processor and memory, or between functions and data (see [Laf94]).

Now, we could forget the logical origin of these interaction nets, but it happens that the notions introduced for proof-theoretical purposes have great computational import. Indeed, if the reduction of an interaction net terminates, and no clash or vicious circle appears, then its reduced form is *ready to interact* in the following sense: At least one free port is connected to a principal port of a cell or to another free port. On the other hand, clashes and vicious circles are configurations of deadlock which must be avoided in any computation. A typing discipline is clearly suitable for clashes, but some geometrical criterion is also needed for vicious circles. In the case of our interaction system for unary arithmetic, we can introduce a single type N and the following typing constraints:

It is easy to check that typed nets are closed under reduction. Since a clash is not typable, reducing a typed net will never create clashes. We can also introduce a switch for δ, and say that a net π is *weakly well-formed* if every switching of π is an acyclic (not necessarily connected) graph. A case analysis shows that this condition is preserved by reduction, and since a vicious circle induces a cyclic switching, reducing a weakly well-formed net will never create vicious circles. In fact, this notion of weakly well-formed net, corresponds to the variant of linear logic considered in section 7. It is equivalent to the notion of *semi-simple net* of [Laf90].

10 Conclusion

The notion of exponential proof-net is not completely convincing because of its hybrid character. The exponential box is indeed a direct translation of the promotion rule of sequent calculus, which is essentially nonlocal. There are two ways to cope with this problem. The first one is the *geometry of interaction*, which consists in replacing nets by more abstract objects (*e.g.* operators in an Hilbert space) in such a way that the cut-elimination can be localized in all cases. The second one has been sketched in the previous section. It consists in replacing the deduction rules of linear logic by interaction rules, in such a way that the concrete character of the computation is preserved, even if its logical status becomes less obvious. In fact, it is not necessary to consider arbitrary systems: In [Laf94], we prove that any interaction system can be translated into a very simple one with only three symbols that we call *interaction combinators.* This means that any computation can be decomposed into elementary steps using a very small number of interaction rules.

There is much more to be done with interaction nets, concerning the theory as well as its applications to computer science. We are thinking of applications to the following areas: programming languages for parallel computation, protocols of communication between interactive softwares, partial evaluation of programs and design of parallel computers.

References

[Abr93] **S. Abramsky**. *Computational interpretations of Linear Logic.* Theoretical Computer Science **111** 3–57. 1993.

[Baw86] **A. Bawden**. *Connection Graphs.* Proceedings of ACM Conference on Lisp and Functional Programming, 258–265. 1986.

[Dan90] **V. Danos**. *La Logique Linéaire appliquée à l'étude de divers processus de normalisation (principalement du λ-calcul).* Thèse de doctorat, Université de Paris 7. 1990.

[DR89] **V. Danos & L. Regnier**. *The structure of multiplicatives.* Archive for Mathematical Logic **28**, 181–203. 1989.

[FR90] **A. Fleury & C. Retoré**. *The Mix rule.* Prébublication de l'équipe de logique 11, Université de Paris 7, to appear in Mathematical Structures in Computer Science. 1990.

[Gal91] **J. Gallier**. *Constructive Logics. Part II: Linear Logic and Proof Nets.* Digital PRL research report 9. 1991.

[GAL92a] **G. Gonthier, M. Abadi & J.-J. Levy**. *The geometry of optimal lambda reduction.* Proceedings of 19th ACM Symposium on Principles of Programming Languages (POPL 92). 1992.

[GAL92b] **G. Gonthier, M. Abadi & J.-J. Levy**. *Linear Logic without boxes.* Proceedings of 7th Annual Symposium on Logic in Computer Science (LICS 92). 1992.

[Gir87] **J.Y. Girard**. *Linear logic.* Theoretical Computer Science **50**, 1–102. 1987.

[GLT89] **J.Y. Girard, Y. Lafont & P. Taylor**. *Proofs and Types.* Cambridge Tracts in Theoretical Computer Science **7**, Cambridge University Press. 1989.

[Laf90] **Y. Lafont**. *Interaction Nets.* Proceedings of 17th ACM Symposium on Principles of Programming Languages (POPL 90) 95–108. 1990.

[Laf94] **Y. Lafont**. *The Paradigm of Interaction.* In preparation.

[LW94] **P. Lincoln & T. Winkler**. *Constant-Only Multiplicative Linear Logic is NP-Complete.* To appear in Theoretical Computer Science. 1994.

[Mac94] **I. Mackie**. *The Geometry of Implementation (an investigation into using the Geometry of Interaction for language implementation).* Thesis, Imperial College, London. In preparation.

[Reg92] **L. Regnier**. *Lambda-calcul et réseaux.* Thèse de doctorat, Université de Paris 7. 1992.

Subnets of Proof-nets in $\mathbf{MLL}^-$

G. Bellin *   J. van de Wiele

December 17, 1994

Abstract

The paper studies the properties of the subnets of proof-nets. Very simple proofs are obtained of known results on proof-nets for $\mathbf{MLL}^-$, Multiplicative Linear Logic without propositional constants.

1 Preface

The theory of proof-nets for $\mathbf{MLL}^-$, multiplicative linear logic without the propositional constants $\mathbf{1}$ and $\bot$, has been extensively studied since Girard's fundamental paper [5]. The improved presentation of the subject given by Danos and Regnier [3] for propositional $\mathbf{MLL}^-$ and by Girard [7] for the first-order case has become canonical: the notions are defined of an arbitrary proof-structure and of a 'contex-forgetting' map $(\, . \,)^-$ from sequent derivations to proof-structures which preserves cut-elimination; correctness conditions are given that characterize proof-nets, the proof-structures $\mathcal{R}$ such that $\mathcal{R} = (\mathcal{D})^-$, for some sequent calculus derivation $\mathcal{D}$. Although Girard's original correctness condition is of an exponential computational complexity over the size of the proof-structure, other correctness conditions are known of quadratic computational complexity.

A further simplification of the canonical theory of proof-nets has been obtained by a more general classification of the subnet of a proof-net. Given a proof-net $\mathcal{R}$ and a formula A in $\mathcal{R}$, consider the set of subnets that have A among their conclusions, in particular the *largest* and the *smallest* subnet in this set, called the *empire* and the *kingdom* of A, respectively. One must give a construction proving that such a set is not empty: in Girard's fundamental paper a construction of the empires is given which is linear in the size of the

*Research supported by EC Individual Fellowship Human Capital and Mobility, contract n. 930142

proof-net. When the notion of kingdom is introduced, the essential properties of proof-nets – including the existence of a sequent derivation $\mathcal{D}$ such that $\mathcal{R} = (\mathcal{D})^-$ (Theorem 1, *sequentialization theorem*) – can be easily proved using simple properties of the kingdoms and empires, in particular the fact that the relation X *is in the kingdom of* Y is a strict ordering. [1]

Moreover the map $(\; . \;)^-$ identifies equivalence classes of sequent derivations, where $\mathcal{D}_i$ and $\mathcal{D}_j$ are equivalent if they differ only for permutations of inferences. Now consider the set of derivations $\mathcal{B}$ which have A as a conclusion, and that are subderivations of some derivation $\mathcal{D}_i$ in an equivalence class. The kingdom and the empire of a formula A in the proof-net $(\mathcal{D}_i)^-$ yield the notions of the minimum and the maximum, respectively, in such a set of subderivations (Theorem 2). This fact gives evidence that the notions is question do not depend on accidental features of the representation; therefore satisfactory generalizations of our results to larger fragments or to other logics should include Theorem 2.

Such a generalization is impossible in any logic with any form of Weakening, e.g., in the fragment **MLL** of multiplicative linear logic with the rule for the constant $\perp$. Indeed a minimal subderivation in which a formula A may be introduced by Weakening is an axiom; but the process of permuting Weakening upwards in a derivation is non-deterministic and does not always identify a unique axiom as the minimum in our set of subderivations; hence in such a logic we cannot have a meaningful notion of *kingdom.*

2 Proof Nets for Propositional MLL$^-$

We give a simple presentation of the well-known basic theory of proof nets for Multiplicative Linear Logic without propositional constants (**MLL**$^-$). The main novelty is the use of the structural properties of subnets of a proof-net, in particular the tight relations between *kingdoms* and *empires.* A pay-off is a simple and elegant proof of the following theorems:[2]

Theorem 1. *There exists a* "context-forgetting" *map* $(\; . \;)^-$ *from sequent derivations in* ***MLL***$^-$ *to proof nets for* ***MLL***$^-$ *with the following properties:*
(a) *Let* $\mathcal{D}$ *be a derivation of* Γ *in the sequent calculus for* ***MLL***$^-$*; then* $(\mathcal{D})^-$ *is a proof net with conclusions* Γ.

[1] The notion of kingdom and the discovery of its properties originated in the Équipe de Logique in the winter 1991-92 and appeared in discussions through electronic mail involving Danos, Girard, Gonthier, Joinet, Regnier, (Paris VII), Gallier and de Groote (University of Pennsylvania) and the author (University of Edinburgh).

[2] Here we prove part (a) and (b) of Theorem 1; the proof of parts (c) and (d) are clear from [5, 7].

(b) (Sequentialization) *If $\mathcal{R}$ is a proof net with conclusions Γ for $\boldsymbol{MLL}^-$, then there is a sequent calculus derivation $\mathcal{D}$ of Γ such that $\mathcal{R} = (\mathcal{D})^-$.*
(c) *If $\mathcal{D}$ reduces to $\mathcal{D}'$, then $\mathcal{D}^-$ reduces to $(\mathcal{D}')^-$.*
(d) *If $\mathcal{D}^-$ reduces to $\mathcal{R}'$ then there is a $\mathcal{D}'$ such that $\mathcal{D}$ reduces to $\mathcal{D}'$ and $\mathcal{R}' = (\mathcal{D}')^-$.*

Theorem 2. (Permutability of Inferences) (i) *Let $\mathcal{D}$ and $\mathcal{D}'$ be a pair of derivation of the same sequent $\vdash \Gamma$ in propositional $\boldsymbol{MLL}^-$. Then $(\mathcal{D})^- = (\mathcal{D}')^-$ if and only if there exists a sequence of derivations $\mathcal{D} = \mathcal{D}_1, \mathcal{D}_2, \ldots, \mathcal{D}_n = \mathcal{D}'$ such that $\mathcal{D}_i$ and $\mathcal{D}_{i+1}$ differ only for a permutation of two consecutive inferences.*

(ii) *Let $\mathcal{R}$ be a proof-net and let A be a formula occurrence in $\mathcal{R}$. Then there exists a derivation $\mathcal{D}$ with $(\mathcal{D})^- = \mathcal{R}$ and a subderivation $\mathcal{B}$ of $\mathcal{D}$ such that $(\mathcal{B})^- = eA$. A similar statement holds for kA.*

2.1 Propositional Proof Structures and Proof Nets

A *link* is an $m+n$-ary relation between formula occurrences, for some $m, n \geq 0$, $m + n \neq 0$. Suppose $X_1, \ldots, X_{m+n}$ are in a link: if $m > 0$, then $X_1, \ldots, X_m$ are called the *premises* of the link; if $n > 0$, then $X_{m+1}, \ldots, X_{m+n}$ are called the *conclusions* of the link. If $m = 0$, the link is called an *axiom* link.
Links are graphically represented as

$$\frac{X_1, \ldots, X_m}{X_{m+1}, \ldots, X_{m+n}}$$

We consider links of the following forms:

Identity Links:

axiom links: $\overline{A \quad A^\perp}$ *cut links:* $\dfrac{A \quad A^\perp}{cut}$

Multiplicative Links:

times links: $\dfrac{A \quad B}{A \otimes B}$ *par links:* $\dfrac{A \quad B}{A \wp B}$

Convention. We assume that the logical axioms and cut links are *symmetric* relations. Other links are *not* regarded as symmetric. The word "*cut*" in a cut link is not a formula, but a place-holder; following common practice, we may sometimes omit it.

Definitions 1. (i) A *proof structure* $\mathcal{S}$ for propositional $\mathbf{MLL}^-$ consists of (i) a nonempty set of *formula-occurrences* together with (ii) a set of identity links, multiplicative links satisfying the properties:

1. Every formula-occurrence in $\mathcal{S}$ is the conclusion of one and only one link;

2. Every formula-occurrence in $\mathcal{S}$ is the premise of at most one link.

We write $X \prec Y$ if X is a *hereditary premise* of Y; in this case we also say that 'X is above Y'. We shall draw proof structures in the familiar way as non-empty, not necessarily planar, graphs.

(ii) We define the following reductions on propositional $\mathbf{MLL}^-$ proof structures:

Axiom Reductions

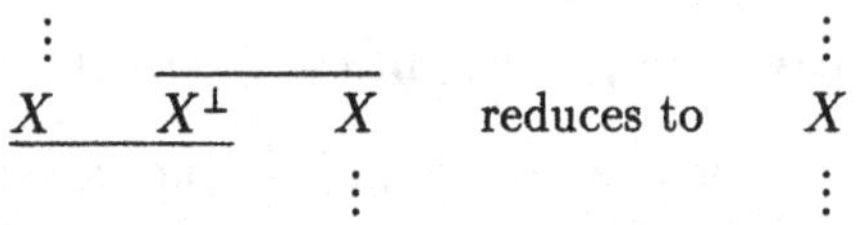

$$\dfrac{\begin{array}{c}\vdots\\ X\end{array} \quad \overline{X^\perp \qquad \begin{array}{c} X \\ \vdots \end{array}}}{} \quad \text{reduces to} \quad \begin{array}{c}\vdots\\ X\\ \vdots\end{array}$$

Symmetric Reductions

$$\dfrac{\dfrac{\vdots_1 \; X \quad \vdots_2 \; Y}{X \otimes Y} \qquad \dfrac{\vdots_3 \; X^\perp \quad \vdots_4 \; Y^\perp}{X^\perp \wp Y^\perp}}{} \quad \text{reduces to} \quad \dfrac{\vdots_1 \; X \quad \vdots_3 \; X^\perp}{} \qquad \dfrac{\vdots_2 \; Y \quad \vdots_4 \; Y^\perp}{}$$

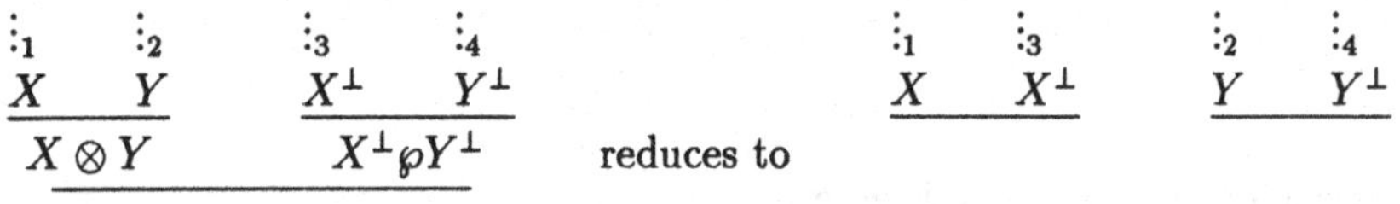

Definitions 2. Let $\mathcal{R}$ be a propositional proof structure for $\mathbf{MLL}^-$.

(i) A *Danos-Regnier switching* s for $\mathcal{R}$ consists in the choice for each *par* link $\mathcal{L}$ in $\mathcal{R}$ of one of the premises of $\mathcal{L}$.

(ii) Given a switching s for $\mathcal{R}$, we define the undirected *Danos-Regnier graph* $s(\mathcal{R})$ as follows:

- the vertices of $s(\mathcal{R})$ are the formulas of $\mathcal{R}$;
- there is an edge between vertices X and Y exactly when:

1. X and Y are the conclusions of a logical axioms or the premises of a cut link; or

2. X is a premise and Y the conclusion of a *times* link; or else

3. Y is the conclusion of a *par* and X is the occurrence selected by the switching s.

Definition 3. Let $\mathcal{R}$ be a multiplicative proof-structure. $\mathcal{R}$ is a *proof-net* for propositional $\mathbf{MLL}^-$ if for every switching s of $\mathcal{R}$, the graph $s(\mathcal{R})$ is *acyclic* and *connected* (i.e., an undirected *tree*).

2.2 Subnets

Definitions 4. Let $m : \mathcal{S} \to \mathcal{R}$ be any injective map of $\mathbf{MLL}^-$ proof structures (regarded as sets of formula occurrences) such that X and $m(X)$ are occurrences of the same formula.

(i) We say that m *preserves the links* if for every $\mathcal{L}$ in $\mathcal{S}$ there is a link $\mathcal{L}'$ in $\mathcal{R}$ of the same kind such that

$$\mathcal{L} : \frac{X_1, \ldots, X_k}{X_{k+1}, \ldots X_{k+n}} \qquad \mapsto \qquad \mathcal{L}' : \frac{mX_1, \ldots, mX_k}{mX_{k+1}, \ldots mX_{k+n}}$$

(ii) A proof-structure $\mathcal{S}$ is a *substructure* of a proof-structure $\mathcal{R}$ if there is an injective map $\iota : \mathcal{S} \to \mathcal{R}$ preserving links. If $\mathcal{S}$ is a substructure of $\mathcal{R}$, then the lowermost formula occurrences of $\mathcal{S}$ are also called the *doors* of $\mathcal{S}$.

(iii) We write $st\Sigma$ for the *smallest substructure* of $\mathcal{R}$ containing Σ.

(iv) A *subnet* is a substructure which satisfies the condition of proof-nets.

Remark. In definition 4.(ii) let ι be the identity map. A subset $\mathcal{S}$ of $\mathcal{R}$ (with the links of $\mathcal{R}$ holding among the occurrences in $\mathcal{S}$) is a substructure if and only if

(1) $\mathcal{S}$ is closed under hereditary premises and
(2) if $\overline{X_0 \quad X_1}$ is an axiom and $X_i \in \mathcal{S}$ then $X_{1-i} \in \mathcal{S}$.

In particular, the set of formula occurrences in $st(\Sigma)$ consists of Σ, of all the hereditary premises of Σ and of the axioms above them:

$$st(\Sigma) = \bigcup_{Z \in \Sigma} \{X : X \preceq Z\} \cup \bigcup_{Z \in \Sigma} \{X \in \overline{X \quad Y} : Y \preceq \Sigma\}.$$

Lemma 1. *Let $\mathcal{R}_1$ and $\mathcal{R}_2$ be subnets of the proof net $\mathcal{R}$. Then*
(i) *$\mathcal{S} = \mathcal{R}_1 \cup \mathcal{R}_2$ is a subnet if and only if $\mathcal{R}_1 \cap \mathcal{R}_2 \neq \emptyset$.*

(ii) *If $\mathcal{R}_1 \cap \mathcal{R}_2 \neq \emptyset$ then $\mathcal{R}_0 = \mathcal{R}_1 \cap \mathcal{R}_2$ is a subnet.*

Proof. Let $\mathcal{R}$ be a proof net and $\mathcal{R}'$ any substructure. Given a switching s' for $\mathcal{R}'$, extend s' to a switching s for $\mathcal{R}$; then $s'\mathcal{R}'$ is a subgraph of $s\mathcal{R}$, hence $s'\mathcal{R}'$

is acyclic, since $s\mathcal{R}$ is. Therefore we need only to consider the connectedness of $s\mathcal{S}$ and $s\mathcal{R}_0$.

To prove (i), assume $\mathcal{R}_1$ and $\mathcal{R}_2$ are subnets with nonempty intersection and fix a switching s for $\mathcal{S} = \mathcal{R}_1 \cup \mathcal{R}_2$. For $i = 1, 2$ let $s\mathcal{R}_i$ be the restriction of $s\mathcal{R}$ to $\mathcal{R}_i$; then $s\mathcal{R}_i$ is connected since $\mathcal{R}_i$ is a subnet. Let A be in $\mathcal{R}_1$ and B in $\mathcal{R}_2$; if $C \in \mathcal{R}_1 \cap \mathcal{R}_2$, then A is connected with C since $s\mathcal{R}_1$ is connected and B is connected with C since $s\mathcal{R}_2$ is connected, hence A is connected with B as required. The converse is immediate, namely, if $\mathcal{R}_1 \cap \mathcal{R}_2 = \emptyset$, then any Danos-Regnier graph on $\mathcal{R}_1 \cup \mathcal{R}_2$ is disconnected.

To prove (ii), let s_0 be a switching for $\mathcal{R}_0 = \mathcal{R}_1 \cap \mathcal{R}_2$; let s_1, s_2 be extensions of s_0 to $\mathcal{R}_1$, $\mathcal{R}_2$, respectively; then $s = s_1 \cup s_2$ is a switching of $\mathcal{R}_1 \cup \mathcal{R}_2$. If A and B occur in $\mathcal{R}_0$, then they are connected by a path π_1 in $s_1\mathcal{R}_1$ and by a path π_2 in $s_2\mathcal{R}_2$; if $\pi_1 \neq \pi_2$, then there is a cycle in $s\mathcal{S}$, which is impossible. But $\pi_1 = \pi_2$ means that A and B are connected in $s_0\mathcal{R}_0$. ■

Proposition 1: (i) *Let $\mathcal{R}_1$ and $\mathcal{R}_2$ be proof nets and let*

$$\mathcal{S} = \mathit{Times}\,(\mathcal{R}_1, \mathcal{R}_2) = \frac{\begin{matrix}\mathcal{R}_1 & \mathcal{R}_2 \\ A & B\end{matrix}}{A \otimes B} \quad \text{or} \quad \mathcal{S} = \mathit{Cut}\,(\mathcal{R}_1, \mathcal{R}_2) = \frac{\begin{matrix}\mathcal{R}_1 & \mathcal{R}_2 \\ A & A^{\perp}\end{matrix}}{\mathit{cut}}$$

Then $\mathcal{S}$ is a proof net if and only if $\mathcal{R}_1 \cap \mathcal{R}_2 = \emptyset$.

(ii) *Let $\mathcal{R}_0$ be a substructure of the proof net $\mathcal{R}$ and let*

$$\mathcal{S} = \mathit{Par}\,(\mathcal{R}_0) = \frac{\begin{matrix} & \mathcal{R}_0 & \\ A_1 & & A_2\end{matrix}}{A \wp B}$$

Then $\mathcal{S}$ is a subnet if and only if $\mathcal{R}_0$ is a subnet.

Proof. (i) Let s be a switching of $\mathcal{S} = \mathit{Times}\,(\mathcal{R}_1, \mathcal{R}_2)$; since $\mathcal{R}_1$ and $\mathcal{R}_2$ are proof nets, each of the graphs $s\mathcal{R}_1$ and $s\mathcal{R}_2$ are acyclic and connected; in addition to $s\mathcal{R}_1 \cup s\mathcal{R}_2$, $s\mathcal{S}$ has the vertex $A \otimes B$ and two edges $(A, A \otimes B)$ and $(B, A \otimes B)$, which establish a connection between $s\mathcal{R}_1$ and $s\mathcal{R}_2$; this is the only connection since $\mathcal{R}_1$ and $\mathcal{R}_1$ are disjoint.
Conversely, if $\mathcal{R}_1 \cap \mathcal{R}_2 \neq \emptyset$, then by lemma 1.(i) $\mathcal{R}_1 \cup \mathcal{R}_2$ is a subnet. Therefore given any switching s of $\mathcal{S}$, the nodes A and B in are connected already in $s(\mathcal{R}_1 \cup \mathcal{R}_2)$; also the edges along link $\frac{A \quad B}{A \otimes B}$ yield another connection between the vertices A and B, hence there is a cycle in $s\mathcal{S}$. ■
Part (ii) is immediate: for any switching s of $\mathcal{R}$, $s\mathcal{S}$ comes from $s\mathcal{R}_0$ by introducing an additional edge $(A_i, A_1 \wp A_2)$ to a leaf A_i, where $A \wp B$ is a new leaf. ■

By induction on the definition of a sequent derivation in $\mathbf{MLL}^-$ we define the map $(\, . \,)^-$ from sequent derivations to proof structures ("forgetting the context").

Theorem 1.(a) *Let $\mathcal{D}$ be a derivation in the sequent calculus for $\mathbf{MLL}^-$; then $(\mathcal{D})^-$ is a proof net.*

Proof. Axioms are proof nets, and the property of being a net is preserved under the *times*, *cut* and *par* rules by Proposition 1. ■

Definitions 5. Let Σ be a set of formula-occurrences in a proof-net $\mathcal{R}$.
(i) The *territory* $t\Sigma$ of is the smallest subnet of $\mathcal{R}$ including Σ (*not necessarily as doors*).
(ii) The *kingdom* kA [the *empire* eA] of a formula-occurrence A in a proof-net $\mathcal{R}$ is the smallest [the largest] subnet of $\mathcal{R}$ having A as a door.
(iii) Let $X \ll Y =_{df} X \in kY$.

Remarks. (i) Given a proof-net $\mathcal{R}$ and formula occurrences Σ in $\mathcal{R}$, the subnet $t\Sigma$ always exists by Lemma 1.
(ii) Suppose for no X, Y in Σ we have that X is a hereditary premise of Y ($X \prec Y$). Then $st\Sigma$, the smallest substructure containing Σ, has all the occurrences in Σ among its doors. On the other hand, there may not be a *subnet* having all of Σ among its doors.
(iii) The existence of kA and eA is immediate by Lemma 1 once we prove there exists a subnet having A as a door. This can be done by giving an explicit construction of eA as in [5, 7] and in the following section.

2.3 Empires and Kingdoms: Existence and Properties

Among the results in this section, for the proof of the Sequentialization theorem we need only the fact that for each formula occurrence A in a proof-net $\mathcal{R}$ there exists a subnet having A as a door.

Definition 6. Let A be a formula occurrence in the proof net $\mathcal{R}$. For a given D-R-switching s, let $s(\mathcal{R}, A)$ be (the set of formula occurrences and of links occurring in) the connected component of the graph $s\mathcal{R}$ which is obtained as follows:

- if A is a premise of a link in $\mathcal{R}$ with conclusion Z and there is an edge (A, Z) in the D-R-graph $s\mathcal{R}$, then remove (A, Z) and let $s(\mathcal{R}, A)$ be the component containing the vertex A.
- otherwise, let $s(\mathcal{R}, A)$ be $s\mathcal{R}$.

We write $\overline{s(\mathcal{R}, A)}$ for the connected component not containing A after the removal of the edge (A, Z) from $s\mathcal{R}$, if such an edge exists; $\overline{s(\mathcal{R}, A)}$ is empty otherwise.

Definition 7. Let $\mathcal{R}$ be a proof-net and let Σ be a set of formula-occurrences in $\mathcal{R}$. We write $path_s(\Sigma)$ for the smallest subgraph of $s\mathcal{R}$ connecting all formula-occurrences in Σ. Clearly $path_s(A, B)$ is a path of $s\mathcal{R}$, for every A, B in $\mathcal{R}$ and every switching s for $\mathcal{R}$.

Proposition 2. (Characterizartions of empires; cf. [3, 5, 7]) *Let $\mathcal{R}$ be a proof net. Then $e(A)$ (the largest subnet of $\mathcal{R}$ containing A as a conclusion) exists and is characterized by the following equivalent conditions:*

(a) $\bigcap_s s(\mathcal{R}, A)$, *where s varies over all possible switchings;*

(b) *the smallest set of formula occurrences in $\mathcal{R}$ closed under the following conditions:*

(i) $A \in e(A)$;

(ii) if $\frac{X_1 \quad X_2}{Y}$ is a link in $\mathcal{S}$ and $Y \in e(A)$, then $X_1, X_2 \in e(A)$, ($\uparrow$-*step*);

(iii) if $\overline{X_0 \quad X_1}$ is an axiom in $\mathcal{S}$ and $X_i \in e(A)$, then $X_{1-i} \in e(A)$ ($\rightarrow$-*step*);

(iv) if $\frac{X_1 \quad X_2}{X_1 \otimes X_2}$ is a link in $\mathcal{S}$, and for $i = 1$ or 2 $X_i \neq A$ and $X_i \in e(A)$, then $X_1 \otimes X_2 \in e(A)$ ($\downarrow$-*step*);

(v) if $\frac{X_1 \quad X_2}{X_1 \wp X_2}$ is a link in $\mathcal{S}$, $X_1 \neq A \neq X_2$ and $\{X_1, X_2\} \subset e(A)$, then $X_1 \wp X_2 \in e(A)$ ($\Downarrow$-*step*).

(According to our conventions, $X_i \neq A$ means that X_i and A are different formula *occurrences*.)

Proof. The following proof of $(a) = (b)$ follows the argument in [7]. To show that $(b) \subseteq (a)$ we show that the set (a) is closed under the conditions $(i) - (v)$ defining (b). This is easy for clauses (i), (iii), (iv) and (v) of (b), and also for clause (ii), if the link in question is a *times* link. Now suppose that for some *par* link $\mathcal{L}$ the conclusion $X_1 \wp X_2 \in \bigcap_s s(\mathcal{R}, A)$, but, say, for the premise X_2 we have $X_2 \notin \bigcap_s s(\mathcal{R}, A)$. Then for some s we have that $X_1 \wp X_2$ belongs to $s(\mathcal{R}, A)$ and X_2 does not. Therefore A is premise of a link with conclusion Z and X_2 belongs to the same connected component as Z, i.e., to $\overline{s(\mathcal{R}, A)}$; let π be $path_s(X_2, Z)$, the path connecting X_2 and Z in $\overline{s(\mathcal{R}, A)}$. Since the switching s in $\mathcal{L}$ is Left and the edge $(X_1, X_1 \wp X_2)$ belongs to $s(\mathcal{R}, A)$, it plays no role in the connections π between X_2 and Z. Therefore if s' is like s, except that the switch on $\mathcal{L}$ is changed from Left to Right, then we still have a connection π between X_2 and Z; since $X_1 \wp X_2 \in \bigcap_s s(\mathcal{R}, A)$, π can be extended to a connection $path_{s'}(A, Z)$, between A and Z in $s'(\mathcal{R}, A)$; but then in $s'A$ we have a cycle, and this is a contradiction. Therefore $\{X_1, X_2\} \subset eA$.

To show that $(a) \subseteq (b)$ we consider a *principal switching* s for A: this is a switching such that for every *par* link $\mathcal{L}$, if a premise X_i of $\mathcal{L}$ is in (b), but the conclusion $X_0 \wp X_1$ is not, then s chooses X_{1-i}. We claim that if s is a principal switching, then $s(\mathcal{R}, A)$ is precisely (b).

Notice that any set $\mathcal{S}$ closed under clauses $(i) - (v)$ has the property that if $\mathcal{S}$ contains X, then it contains also every formula occurrence Z such that X and Z are in a link $\mathcal{L}$, in all cases *except perhaps the following*:

(1) X is A and a premise of $\mathcal{L}$, while Z is the conclusion of $\mathcal{L}$;

(2) $\mathcal{L}$ is a *par* link, X is a premise and Z the conclusion of $\mathcal{L}$, and the other premise Y is not it $\mathcal{S}$.

It follows that the set (b) is a substructure of $\mathcal{R}$ whose doors can only be conclusions of $\mathcal{R}$, or cuts, or occurrences X as in (1) or (2).

Now suppose a formula-occurrence W is in (a) but not in (b); choose a switching s principal for A. Since $s(\mathcal{R}, A)$ is connected and (b) is a substructure, the path π connecting A with W in $s(\mathcal{R}, A)$ must exit (b) from a door X as in cases (1) or (2). But this is impossible by the definition of principal switching and of $s(\mathcal{R}, A)$. Hence $(a) \subseteq (b)$ as claimed.

We must show that A is a door of the substructure equivalently defined by (a) and (b). Let $Z \in \bigcap_s s(\mathcal{R}, A)$ and suppose $A \prec Z$. Choose a switching s such that if $\dfrac{X_0 \quad X_1}{X_0 \wp X_1}$ is a link such that $A \preceq X_i \prec Z$, then s chooses X_{1-i}. We claim that there must be a *times* link $\dfrac{B \quad C}{B \otimes C}$ in $s(\mathcal{R}, A)$ such that, say, $A \preceq C \prec Z$: otherwise, $Z \notin s(\mathcal{R}, A)$, by the choice of s and the definition of $s(\mathcal{R}, A)$. Thus let $\mathcal{L}$ be the uppermost such link: then the path π connecting A and B in $s(\mathcal{R}, A)$ does not pass through C; but then in $s\mathcal{R}$ we have two distinct paths connecting A and B, and this contradicts the acyclicity of $s\mathcal{R}$.

Since (b) is a substructure satisfying the condition (a), for each s the restriction of $s(\mathcal{R}, A)$ to (b) is acyclic and connected, hence (b) is a subnet. We have proved that given a proof-net $\mathcal{R}$ and a formula-occurrence A in $\mathcal{R}$, a subnet with conclusion A always exists.

But (a) is also the largest among such subnets: let $\mathcal{S}$ be a substructure of $\mathcal{R}$ with A as a door and suppose $Z \in \mathcal{S} \setminus (a)$; then for some s, we have $Z \notin s(\mathcal{R}, A)$, from which it follows that no path connects A and Z in $s\mathcal{S}$; hence $\mathcal{S}$ is not a subnet. We conclude that $e(A) = (a) = (b)$. ∎

The construction of a principal switching was given first in Girard's *Trip Theorem* (cf. [5], 2.9.5.); using Girard's notion of a *trip* the principal switching constructed 'dynamically', by making the following choices during a trip.

Starting from A, the trip proceed upwards in $\mathcal{R}$, and at a branching point, i.e., at *times* link, we choose arbitrarily;

- if the trip reaches a *par* link for the first time from below, then we fix s arbitrarily and the trip continues to the chosen premise;
- if the trip reach a *par* link *for the first time* from a premise, then we let s choose the *other* premise.

The Trip Theorem shows that eA is exactly the set of occurrences visited between the first and the second visit to A. The algorithm is transfered to our setting using the correspondence between trips and D-R-graphs established by Danos and Regnier [3]. One advantage of such a formulation is that the following corollary becomes completely obvious.

Corollary. *The complexity of the computation of eA is linear on the size of the proof-net.* ∎

Proposition 3.(I) (properties of territories). *Let $\mathcal{R}$ be a proof-net and let Σ be a set of occurrences in $\mathcal{R}$. Then the territory $t\Sigma$ satisfies*

$$t\Sigma = t(path_s(\Sigma)) = \bigcup_{X \in path_s(\Sigma)} tX$$

for any switching s. ∎

Proposition 3.(II) (characterizations of kingdoms).[3] *Let $\mathcal{R}$ be a proof net. Then the kingdom kA of A in $\mathcal{R}$ (the smallest subnet of $\mathcal{R}$ having A as a conclusion), exists and is characterized by the following equivalent conditions:*

(a) tA;

(b) *the smallest set satisfying the following conditions (Danos et al.):*

(o) $A \in kA$.

(i) Let $\overline{X \quad X^{\perp}}$ occur in $\mathcal{R}$. Then

$$\overline{X \quad X^{\perp}} = kX = t(X, X^{\perp}) = kX^{\perp}.$$

(ii) Let $\mathcal{L} : \dfrac{A \quad B}{A \otimes B}$ be a link in $\mathcal{R}$. Then

$$kX \otimes Y = kX \cup kY \cup \{X \otimes Y\}.$$

[3]Characterization (*b*) is due to Danos and others, as specified in footnote 1. Characterization (*c*) was suggested to us by J-Y. Girard.

(iii) Let $\mathcal{L} : \dfrac{X \quad Y}{X \wp Y}$ be a link in $\mathcal{R}$. Then

$$kX\wp Y = \bigcup \{kC | C \in path_s(X, Y)\} \cup \{X \wp Y\}$$

for any switching s.

(c) *the smallest set of formula occurrences closed under the following conditions:*

(i) $A \in k(A)$;

(ii) if $\frac{X_1 \quad X_2}{Y}$ is a link in $\mathcal{S}$ and $Y \in k(A)$, then $X_1, X_2 \in k(A)$ [similarly, if $\frac{X[t/y]}{\exists y.X}$ is a link in $\mathcal{S}$ and $\exists y.X \in k(A)$, then $X[t/y] \in k(A)$] ($\uparrow$*-step*);

(iii) if $\overline{X_0 \quad X_1}$ is an axiom in $\mathcal{S}$ and $X_i \in k(A)$, then $X_{1-i} \in k(A)$ ($\rightarrow$*-step*);

(iv) if $\frac{...X...}{Y}$ is a link in $\mathcal{S}$ $X \neq A \neq Y$, $X \in k(A)$, then $Y \in kA$ iff $A \notin eX$ ($\downarrow$*-step*). ∎

•The proof is left to the reader; for case $(c)(iv)$, see the following Lemma 2.

2.4 Sequentialization Theorem

Lemma 2. (Empire-Kingdom Nesting) *Let $\mathcal{L}_1 : \frac{...A...}{C}$ and $\mathcal{L}_2 : \frac{...B...}{D}$ be distinct links in a proof net $\mathcal{R}$ for **MLL$^-$**. Suppose $B \in eA$; then $D \notin eA$ if and only if $C \in kD$.*

Proof. Clearly $B \in eA \cap kD$, hence $\mathcal{R}_0 = eA \cap kD$ and $\mathcal{S} = eA \cup kD$ are subnets of $\mathcal{R}$. If $C \notin kD$ and $D \notin eA$, then $\mathcal{S}$ is a subnet with conclusion A, which is larger than eA, since it contains D: this contradicts the definition of the empire of A. If $C \in kD$ and $D \in eA$, then $\mathcal{R}_0$ is a subnet with conclusion D, which is smaller than kD since it does not contain C: this contradicts the definition of the kingdom of D. ∎

Lemma 3. (Kingdom Ordering) (i) *Let $\mathcal{R}$ be a proof net and let X, Y occur in $\mathcal{R}$. If $X \ll Y$ and $Y \ll X$ then either X and Y are the same occurrence or they occur in an axiom $\overline{X \quad Y}$ of $\mathcal{R}$.* (ii) *Hence $\ll$ is an ordering* of the conclusions of non-axiom links.

Proof. For an axiom $\mathcal{A} = \overline{X\ X^\perp}$ we have $kX = \mathcal{A} = kX^\perp$. Otherwise, let $X \in kY$, with X and Y distinct; if also $Y \in kX$, then $kY \cap kX$ is a subnet, and necessarily $kX = kX \cap kY = kY$.

If X is $X_1 \wp X_2$ in a link $\mathcal{L}$: $\dfrac{X_1 \quad X_2}{X_1 \wp X_2}$ then the result of removing X and $\mathcal{L}$ from kY is still a subnet, and this contradicts the definition of kY.

If X is $X_1 \otimes X_2$ in a link $\dfrac{X_1 \quad X_2}{X_1 \otimes X_2}$ then clearly $kX = k(X_1) \cup k(X_2) \cup \{X\}$, hence for $i = 1$ or 2, $Y \in k(X_i)$; but by Lemma 2, Y is not even in $e(X_i)$. ■

Theorem 1.(b) (Sequentialization) *If $\mathcal{R}$ is a proof net with conclusions Γ, then there is a sequent calculus derivation $\mathcal{D}$ of Γ such that $\mathcal{R} = (\mathcal{D})^{-}$.*

Proof. By induction on the size of $\mathcal{R}$. If $\mathcal{R}$ is an axiom, then $\mathcal{D}$ is an axiom sequent. If one of the lowermost links is a *par* or *for all* link, then we remove such a link, we apply the induction hypothesis to the resulting subnet and we conclude by applying a suitable *par* inference. Now suppose that all the conclusions of $\mathcal{R}$ are conclusions either of an axiom or of a *times* link: we choose a terminal *times* link $\mathcal{L}$ whose conclusion $X = A_i \otimes B_i$ is maximal w.r.t. $\ll$. In this case eA_i and eB_i split $\mathcal{R} \setminus \{A_i \otimes B_i\}$. Suppose not; then there is a link $\mathcal{L} : \frac{\ldots D \ldots}{C}$ such that, say, $D \in eB_i$ and $C \notin eB_i$. But C occurs at or above another conclusion $Y = A_j \otimes B_j$. By the lemma 2 $X = A_i \otimes B_i \in kC$; also $C \in kY$ hence $kC \subset kY$; thus we obtain $X \in kY$, contradicting the choice of X. ■

Remark. The computational complexity of Girard's *no-short-trip* condition and of Danos-Regnier's requirement that all D-R-graphs be acyclic and connected is clearly exponential on the size of the given proof-structure. It is known (see, e.g., [3, 4, 1]) that there are procedures to decide whether or not a proof-structure $\mathcal{R}$ for $\mathbf{MLL}^{-}$ is a proof-net in time quadratic over the cardinality of $\mathcal{R}$.

2.5 Permutability of Inferences in the Sequent Calculus

Given a derivation $\mathcal{D}$ and two formula-occurrences X_1 and X_2 in some sequents of $\mathcal{D}$, if X_1 is an ancestor of X_2 then certainly the inference introducing X_1 must occur above the inference introducing X_2. We are concerned with occurrences X_1 and X_2 in $\mathcal{D}$ such that neither one is an ancestor of the other. Suppose X_1 is introduced above X_2 in $\mathcal{D}$, we ask whether there is a derivation $\mathcal{D}'$ which is obtained from $\mathcal{D}$ by successive permutation of the inferences and such that X_1 is introduced below X_2 in $\mathcal{D}'$.

Counterexample. The following is a derivation in $\mathbf{MLL}^{-}$ in which the applications of the $\otimes$-rule and of the $\wp$-rule cannot be permuted.

$$\dfrac{\dfrac{\dfrac{\vdash P^{\perp}, P \qquad \vdash Q, Q^{\perp}}{\vdash P^{\perp}, P\otimes Q, Q^{\perp}}\otimes}{\vdash Q^{\perp}, P^{\perp}, P\otimes Q}\,exchange}{\vdash Q^{\perp}\wp P^{\perp}, P\otimes Q}\wp$$

Remark. In the sequent calculus for propositinal $\mathbf{MLL}^-$ $\otimes/\wp$, $cut/\wp$ and $\exists/\forall$ are the only exceptions to the permutability of inferences where neither one of the principal formulas is an ancestor of the other.

A full characterization of permutability of inference in $\mathbf{MLL}^-$ is obtained using the 'context-forgetting' map $(\,.\,)^-$ of derivations into proof-nets and the notions of empire and kingdom. Such a map uniquely associates each inference $\mathcal{I}$ in $\mathcal{D}$ other than Exchange with a link $\mathcal{L}$ in $(\mathcal{D})^-$ and the principal formula(s) of $\mathcal{I}$ with the conclusion(s) of $\mathcal{L}$.

Theorem 2. (i) *Let $\mathcal{D}$ and $\mathcal{D}'$ be a pair of derivation of the same sequent $\vdash \Gamma$ in propositional $\mathbf{MLL}^-$. Then $(\mathcal{D})^- = (\mathcal{D}')^-$ if and only if there exists a sequence of derivations $\mathcal{D} = \mathcal{D}_1, \mathcal{D}_2, \ldots, \mathcal{D}_n = \mathcal{D}'$ such that $\mathcal{D}_i$ and $\mathcal{D}_{i+1}$ differ only for a permutation of two consecutive inferences.*

(ii) *Let $\mathcal{R}$ be a proof-net and let A be a formula occurrence in $\mathcal{R}$. Then there exists a derivation $\mathcal{D}$ with $(\mathcal{D})^- = \mathcal{R}$ and a subderivation $\mathcal{B}$ of $\mathcal{D}$ such that $(\mathcal{B})^- = eA$. A similar statement holds for kA.*

Proof. (i) The "if" part is clear. To prove the "only if" part, let $(\mathcal{D})^- = \mathcal{R} = (\mathcal{D}')^-$; consider a branch of $\mathcal{D}$ and let $\mathcal{I}_0$ the last inference from bottom up where $\mathcal{D}$ agrees with $\mathcal{D}'$. If $\mathcal{I}_0$ is an axiom, then $\mathcal{D}$ and $\mathcal{D}'$ entirely agree in the order of inferences in this branch. Otherwise, let $\mathcal{I}_A$ be the inference immediately above $\mathcal{I}_0$ in the branch of $\mathcal{D}$ under consideration, and let $\mathcal{I}'_A$ be the inference of $\mathcal{D}'$ such that the principal formulas of $\mathcal{I}_A$ and $\mathcal{I}'_A$ are mapped to the same formula occurrence A of $\mathcal{R}$: such an $\mathcal{I}'_A$ exists, since $(\mathcal{D})^- = (\mathcal{D}')^-$.

Moreover, let $\mathcal{I}'_1, \ldots, \mathcal{I}'_k$ be the inferences which occur in $\mathcal{D}'$ between $\mathcal{I}'_A$ and $\mathcal{I}_0$ (proceeding downwards). Notice that if the principal formula of any $\mathcal{I}'_i$ for $i \leq k$ is mapped to a formula B of $\mathcal{R}$, then the inference $\mathcal{I}_B$ of $\mathcal{D}$ whose principal formula is mapped to B also occurs *above* the inference $\mathcal{I}_0$, by our assumption that $\mathcal{D}$ and $\mathcal{D}'$ agree in the given branch up to $\mathcal{I}_0$. It follows that no descendant of A is active in $\mathcal{I}'_1, \ldots, \mathcal{I}'_k$.

If the inference $\mathcal{I}'_A$ is an instance of the *par* rule, then clearly it can be permuted below $\mathcal{I}'_1, \ldots, \mathcal{I}'_k$. If $\mathcal{I}'_A$ is a *times* rule, say, A is $A_1 \otimes A_2$, then we have

$$\mathcal{I}_A : \dfrac{\begin{array}{cc} \mathcal{B}_1 & \mathcal{B}_2 \\ \vdots & \vdots \\ \vdash \Gamma_1, A_1 & \vdash \Gamma_2, A_2 \end{array}}{\vdash \Gamma_1, \Gamma_2, A_1 \otimes A_2} \qquad \mathcal{I}'_A : \dfrac{\begin{array}{cc} \mathcal{B}'_1 & \mathcal{B}'_2 \\ \vdots & \vdots \\ \vdash \Delta_1, A_1 & \vdash \Delta_2, A_2 \end{array}}{\vdash \Delta_1, \Delta_2, A_1 \otimes A_2}$$

If $\mathcal{I}'_1$ is another *times* rule, then clearly it can be permuted above $\mathcal{I}'_A$. If $\mathcal{I}'_1$ is a *par* rule, then consider the inference $\mathcal{I}_C$ of $\mathcal{D}$ such that the principal formulas of $\mathcal{I}'_1$ and $\mathcal{I}_C$ are mapped to the same formula occurrence $C = C_0 \wp C_1$ of $\mathcal{R}$. Now $(\mathcal{B}_j)^-$ is a subnet of $\mathcal{R}$ with A_j as a conclusion, hence $(\mathcal{B}_j)^- \subseteq e(A_j)$;

similarly $(\mathcal{B}'_j)^- \subseteq e(A_j)$. Since $\mathcal{I}_C$ occurs above $\mathcal{I}_A$, the link $\dfrac{C_0 \quad C_1}{C_0 \wp C_1}$ occurs in $e(A_j)$; moreover, $e(A_j) \cap e(A_{1-j}) = \emptyset$, hence the active formulas C_0 and C_1 of $\mathcal{I}_1$ are both in the same branch $\mathcal{B}'_j$ of $\mathcal{D}'$. It follows that $\mathcal{I}_1$ can be permuted above $\mathcal{I}'_A$.

(ii) Let $(\mathcal{D})^- = \mathcal{R}$; let $\mathcal{I}_A$ be the inference in $\mathcal{D}$ whose principal formula is mapped to A in $\mathcal{R}$; let $\mathcal{B}_A$ be the subderivation of $\mathcal{D}$ ending with $\mathcal{I}_A$. To find a derivation $\mathcal{D}'$ and a subderivation $\mathcal{B}'$ such that $(\mathcal{B}') = eA$, let k be the number of formula-occurrences in $eA \setminus (\mathcal{B}_A)^-$: then there are also k inferences in $\mathcal{D}$ which must be successively permuted above $\mathcal{I}_A$. We proceed by induction on eA, as characterized by Proposition 2. We need to consider only the following cases:

$\downarrow$-step for *times*, clause (iv): $X \in eA$ and $X \neq A$ implies $X \otimes Y \in eA$. By induction hypothesis we may assume that X is introduced above $\mathcal{I}_A$. If $\mathcal{I}'$ introduces $X \otimes Y$ and occurs below $\mathcal{I}_A$, then X is a passive formula of every sequent between $\mathcal{I}_A$ and $\mathcal{I}'$. If we permute $\mathcal{I}'$ with the inference $\mathcal{I}''$ immediately above it, we do not increase the number of formulas in $eA \setminus (\mathcal{B})^-$, After a finite number of steps, the inference introducing $X \otimes Y$ is permuted above $\mathcal{I}_A$ and we have reduced k.

$\Downarrow$-step for *par* links, clause (v): $X \in eA, Y \in eA$ and $X \neq A \neq Y$ imply $X \wp Y \in eA$. By induction hypothesis we assume that both X and Y are introduced above $\mathcal{I}_A$, and let $\mathcal{I}'$ be the inference introducing $X \wp Y$ below $\mathcal{I}_A$. It follows that for each application of the $\otimes$-rule between $\mathcal{I}_A$ and $\mathcal{I}'$ the ancestors of $X \wp Y$ occur in one branch only, namely that containing $\mathcal{I}_A$. Therefore the inference $\mathcal{I}'$ can always be permuted with the inference $\mathcal{I}''$ immediately above it, even in the case when $\mathcal{I}''$ is a $\otimes$-rule. After a finite number of steps we reduce k.

Finally, to find a derivation $\mathcal{D}''$ and a subderivation $\mathcal{B}''$ such that $(\mathcal{B}'') = kA$, consider the doors of $k(A)$ which are premises of some link; let $X_1, \ldots, X_n$ be the conclusions of such links. Since $X_i \notin kA$ by Lemma 2, we have $A \in e(X_i)$ and by the above argument, the inference $\mathcal{I}_A$ can be permuted above the inference $\mathcal{I}_i$ introducing X_i in $\mathcal{D}$. The argument can be repeated for all $i \leq n$, without permuting $\mathcal{I}_A$ below a previously considered $\mathcal{I}_j$; the result follows. ∎

3 Proof Nets for First Order MLL$^-$

This section is essentially based on Girard [7].

3.1 First-Order Proof-Structures

We work with a *first-order* language for $\mathbf{MLL}^-$ and consider multiplicative proof-structures with the addition of the following links.

First-order links:

$$\textit{for all:}\quad \frac{A}{\forall x.A} \qquad \textit{exists:}\quad \frac{A[t/x]}{\exists x.A}$$

Definition 8. The variable x (possibly) occurring free in the premise of a *for all* link $\mathcal{L} : \dfrac{A}{\forall x.A}$ is called the *eigenvariable* associated with the link $\mathcal{L}$. Notice that the same variable x occurs free in the premise and bound in the conclusion of $\mathcal{L}$. We associate with each eigenvariable x a constant $\underline{x}$. Obviously, a link of the form $\dfrac{A[\underline{x}/x]}{\forall x.A}$ is *incorrect.*

Definitions 9. (i) A proof structure for *first order* $\mathbf{MLL}^-$ is defined as before with the addition of the following conditions:

3. (a) Each occurrence of a quantifier link uses a distinct bound variable.

   (b) If a variable occurs freely in some formula of the structure, then the variable is the eigenvariable of exactly one $\forall$-link.

   (c) The conclusions of the proof structure are closed formulas.

4. We say that in a first-order proof-structure $\mathcal{S}$ eigenvariables are used *strictly* if no substitution of any set of occurrences of an eigenvariable x with the constant $\underline{x}$ yields a correct proof structure with the same conclusions as $\mathcal{R}$. We require also that in first-order proof-structures eigenvariable are used strictly.[4]

(ii) Let $\mathcal{R}$ be a proof structure for $\mathbf{MLL}^-$ and let x be an eigenvariable in $\mathcal{R}$. The *free range of x in $\mathcal{S}$* is the set of all formula occurrences in which the eigenvariable x occurs freely. The *existential border* of x is the set of all the formula occurrences which are the conclusion of a link $\mathcal{L} : \dfrac{B[t/y]}{\exists y.B}$ where x occurs in the premise but not conclusion of $\mathcal{L}$. We say also that the link $\mathcal{L}$ is in the existential border of x.

[4]We modify the setting of Girard [7] only with the condition of a strict use of the eigenvariables; this is enough to give a smooth tratment of kingdom and empires.

(iii) We define the following additional reductions.

Symmetric Reductions

$$\dfrac{\dfrac{\begin{array}{c}\vdots\\ A[t/x]\end{array}}{\exists x.A} \quad \dfrac{\begin{array}{c}\mathcal{R}(x)\\ A^{\perp}\end{array}}{\forall x.A^{\perp}}}{cut} \qquad \text{reduces to} \qquad \dfrac{\begin{array}{c}\vdots\\ A[t/x]\end{array} \quad \begin{array}{c}\mathcal{R}[t/x]\\ A^{\perp}[t/x]\end{array}}{cut}$$

where $\mathcal{R}(x)$ is the smallest substructure containing all occurrences of the eigenvariable x and $\mathcal{R}[t/x]$ results from $\mathcal{R}(x)$ by replacing t for x everywhere.

The definition of Danos-Regnier graph for first order proof structures is extended as follows.

Definitions 10. Let $\mathcal{R}$ be a proof structure for first order $\mathbf{MLL}^{-}$.

(i) A *Danos-Regnier switching* s in a first order proof structure $\mathcal{R}$ for $\mathbf{MLL}^{-}$ consists in a switch for each *par* and *for all* link of $\mathcal{R}$, where

- a switch for a *par* link is the choice of one of the premises of the link and
- a switch for a *for all* link with associate eigenvariable x is a choice of either (1) the premise of the link or of a formula occurrence in (2) the free range or in (3) the existential border of x (case (1) is needed if x does not occur free in $\mathcal{R}$).

(ii) Given a switching s for $\mathcal{R}$, we define the undirected *Danos-Regnier graph* $s(\mathcal{R})$ as follows:

- the vertices of $s(\mathcal{R})$ are the formulas of $\mathcal{R}$;
- there is an edge between vertices X and Y exactly when:
  (a) X and Y are the conclusions of a logical axioms or the premises of a cut link;
  (b) X is a premise and Y the conclusion of a *times* or *exists* link;
  (c) Y is the conclusion of a *par* or *for all* link and X is the occurrence selected by the switching s.

(iii) $\mathcal{R}$ is a *proof net* for first order $\mathbf{DL}$ [$\mathbf{MLL}^{-}$] if for every switching s of $\mathcal{R}$, the graph $s(\mathcal{R})$ is acyclic [and connected].

The requirement that eigenvariable should be used strictly guarantees that the following structure is incorrect:

$$\frac{A(x)}{\forall x.A} \qquad \frac{\dfrac{A^\perp(x)}{\exists x.A^\perp} \qquad \dfrac{B(x)}{\exists x.B}}{\exists x.A^\perp \otimes \exists x.B} \qquad \frac{B^\perp(x)}{\exists x.B^\perp}$$

and must be rewritten as

$$\frac{A(x)}{\forall x.A} \qquad \frac{\dfrac{A^\perp(x)}{\exists x.A^\perp} \qquad \dfrac{B(c)}{\exists x.B}}{\exists x.A^\perp \otimes \exists x.B} \qquad \frac{B^\perp(c)}{\exists x.B^\perp}$$

where c is a new constant.

The following is an equivalent way of characterizing the same property.

Definition 11. An *x-tread* in a proof-structure $\mathcal{R}$ is a sequence $C_1, \ldots, C_n$ of formula occurrences which contain the free variable x and such that for each $i < n$ there is a link $\mathcal{L}$ such that either (1) C_i is the premise and C_{i+1} is the conclusion of $\mathcal{L}$ or (2) C_i and C_{i+1} are conclusions of $\mathcal{L}$ (an axiom link) or (3) C_i is the conclusion and C_{i+1} is the premise of $\mathcal{L}$.

Fact 1. *In a proof structure eigenvariables are used strictly if and only if every occurrence of an eigenvariable x belongs to an x-thread ending with the $\forall$-link associated with x.* ■

3.2 Subnets

The definition of a *substructure* $\mathcal{S}_0$ of a proof-structure $\mathcal{S}$ must take into account the requirement that all conclusion of $\mathcal{S}_0$ should be closed formulas.

Definitions 12. (i) Let $\mathcal{S}$ be a proof structure for first order **MLL**. A set of formula occurrences and links $\mathcal{S}_0$ is a *substructure* of $\mathcal{S}$ if $\mathcal{S}_0$ is a proof structure and there is an injective map $\iota : \mathcal{S}_0 \to \mathcal{S}$ preserving links such that X and $\iota(X)$ are the same formula or X comes from $\iota(X)$ by a substitution of a free variable x with $\underline{x}$. (We will usually omit to mention the map ι.)
As before, a *subnet* is a substructure which satisfies the condition of proof-nets.

Fact 2. *If $\mathcal{S}$ is a substructure of a first order proof-structure $\mathcal{R}$ and a link $\mathcal{L} : \dfrac{A}{\forall x.A}$ occurs in $\mathcal{S}$, then the free range of x and its existential border are contained in $\mathcal{S}$.*

Proof. All eigenvariables are used strictly in $\mathcal{S}$ by definition. Suppose $\mathcal{L}$ occurs in $\mathcal{S}$ but x occurs outside $\mathcal{S}$; then there is an x-thread 'crossing the border of' $\mathcal{S}$, say at a door C. This means that any substitution of $\underline{x}$ for x in C spoils the link $\mathcal{L}$, i.e., $\mathcal{S}$ cannot be a substructure, a contradiction. ∎

Lemma 1 (first order case) *In first order* **MLL**$^{-}$, *the intersection and the union of subnets are subnets if and only if the intersection is nonempty.*

Proof. The argument for the propositional case applies here; we need only to make sure that if $\mathcal{R}_1$ and $\mathcal{R}_2$ are subnets of a proof-net $\mathcal{R}$ with $\mathcal{R}_1 \cap \mathcal{R}_2 \neq \emptyset$, then $\mathcal{S} = \mathcal{R}_1 \cup \mathcal{R}_2$ and $\mathcal{R}_0 = \mathcal{R}_1 \cap \mathcal{R}_2$ are *first-order substructures*, and in particular, the eigenvariables are used strictly and their conclusions are closed.
If a $\forall$-link of $\mathcal{R}$ does not occur in $\mathcal{S}$, then the associated eigenvariable z is replaced by $\underline{z}$ in the subnets $\mathcal{R}_1$ and in $\mathcal{R}_2$, hence in $\mathcal{S}$ too.
If a $\forall$-link with eigenvariable z occurs in $\mathcal{R}_0$, then (since eigenvariables are used strictly in $\mathcal{R}$) z also occurs inside $\mathcal{R}_0$ but not in any door of $\mathcal{R}_0$, by the Fact 2.
Finally, if a $\forall$ link with eigenvariable z occurs, say, in $\mathcal{R}_1 \setminus \mathcal{R}_2$, then any occurrence of z in the substructure $\mathcal{R}_0$ is replaced by $\underline{z}$. Moreover z does not occur in the doors of $\mathcal{S}$: indeed by the same corollary, z does not occur in the doors of $\mathcal{R}_1$, hence it does not occur in $\mathcal{R}_2 \setminus \mathcal{R}_1$ either. ∎

Proposition 1. (first order cases) *Let $\mathcal{R}_0$ be a substructure of the proof net $\mathcal{R}$. Then*
(iii)

$$\mathcal{S} = \textit{For All}\,(\mathcal{R}_0) = \begin{array}{cc} & \mathcal{R}_0[x/\underline{x}] \\ \Gamma & \dfrac{A}{\forall x.A} \end{array}$$

is a subnet if only if $\mathcal{R}_0$ is a subnet and $\underline{x}$ does not occur in Γ.
(iv) *The substructure*

$$\mathcal{S} = \textit{Exists}\,(\mathcal{R}_0) = \begin{array}{c} \mathcal{R}_0 \\ \dfrac{A[t/x]}{\exists x.A} \end{array}$$

is a subnet if $\mathcal{R}_0$ is one.

Proof. (iii) $\mathcal{S}$ is a substructure, since the substitution of x for $\underline{x}$ does not affect the conclusions of $\mathcal{S}$, which remain closed. Given a switching s for $\mathcal{S}$, $s\mathcal{S}$ differs from $s\mathcal{R}_0$ only for having a leaf $\forall x.A$ connected by an edge to some vertex of $\mathcal{R}_0$; thus $s\mathcal{S}$ is acyclic and connected, since $s\mathcal{R}_0$ is. (iv) is similar but easier. ∎

Remark. It is not true that if $\mathcal{S} = \textit{Exists}\ \mathcal{R}_0$ and $\mathcal{S}$ is a proof-net then $\mathcal{R}_0$ is a proof-net: for instance in $A[t/x]$ the term t may contain the eigenvariable of some *for all* link which occur in $\mathcal{R}_0$.

As before Theorem 1.(a) follows as a corollary. (Notice that if $\vdash \Gamma$ is the end sequent of $\mathcal{D}$ and a free variable x occurs in Γ, then $(\mathcal{D})^- = (\mathcal{D}[\underline{x}/x])^-$, a proof-structure with conclusions $\Gamma[\underline{x}/x]$.)

Theorem 1.(a) (first-order case) *Let $\mathcal{D}$ be a derivation in the sequent calculus for first order $\mathbf{MLL}^-$; then $(\mathcal{D})^-$ is a proof net.* ∎

3.3 Empires and Kingdoms: Existence and Properties

As in the propositional case, we need to prove that given a proof-net $\mathcal{R}$ and a formula A in $\mathcal{R}$, *there always exists a subnet of $\mathcal{R}$ having A among its conclusions.*

Proposition 2. (Characterization of empires, first-order case; cf. [7]) *Let $\mathcal{R}$ be a proof net for first order* $\mathbf{MLL}^-$ *and let A occur in $\mathcal{R}$. Then the empire eA of A in $\mathcal{R}$ exists and is characterized by the following equivalent conditions:*

(a) $\bigcap_s s(\mathcal{R}, A)$, *where s varies over all possible switchings;*

(b) *the smallest set of formula occurrences in $\mathcal{R}$ closed under conditions* (b)(i)-(v) *of Proposition 2 for propositional multiplicative links and moreover* (vi) if $\dfrac{X[t/y]}{\exists y.X}$ is a link in $\mathcal{S}$ and $X[t/y] \neq A$, then $\exists y.X \in e(A)$ if and only if $X[t/y] \in e(A)$, ($\uparrow$- and $\downarrow$-*steps*);

(vii) if $\dfrac{X}{\forall y.X}$ is a link in $\mathcal{S}$ and $X \neq A$, then $\forall y.X \in eA$ if and only if the free range of y and the occurrences in its existential border belong to eA ($\Uparrow$- and $\Downarrow$-*steps*).

Proof. We follow Girard [7]. (vii) Suppose $\forall y.X \in eA$, but for some C in the free range of y we have $C \notin eA$. Then A must be a premise of some link with conclusion Z, and for some s we have $\forall y.X \in s(\mathcal{R}, A)$ and $C \in \overline{s(\mathcal{R}, A)}$, where $\overline{s(\mathcal{R}, A)}$ is the connected component not containing A after removal of the edge (A, Z) from $s\mathcal{R}$. Therefore in $s(\mathcal{R}, A)$ there is a path connecting A and $\forall y.X$ and moreover in $\overline{s(\mathcal{R}, A)}$ there is a path connecting Z and C which obviously does not depend on the switch for $\forall y.X$. Now if we change the switch for $\forall y.X$ to choose C leaving all other choices unchanged, then we obtain a switch s' such that $s'\mathcal{R}$ is cyclic: indeed there still remains a connection between Z and C in $\overline{s'(\mathcal{R}, A)}$ (which lies outside eA) and there certainly is a distinct connection between A and $\forall y.X$ in $s'(\mathcal{R}, A)$ (since $\forall y.X \in eA$). But then $s'\mathcal{R}$

contains a cycle, a contradiction. ■

The example at the beginning of the present section shows that an eigenvariable x can occur outside the kingdom of $\forall x.A$, unless a strict use of eigenvariables is required. We have the following *characterization of kingdoms in first order* **MLL**$^-$ (which is not true in in the setting of [7]).

Proposition 3. (Inductive definition of kingdoms, first-order cases) *Let $\mathcal{R}$ be a proof net for first order **MLL**$^-$. Then kA, the kingdom of a in $\mathcal{R}$ exists and is characterized as the smallest set of formula occurrences closed under conditions* (i)-(iv) *of Proposition 3 for multiplicative propositional links and moreover*

(ii)' if $\dfrac{X[t/y]}{\exists y.X}$ is a link in $\mathcal{S}$ and $\exists y.X \in k(A)$, then $X[t/y] \in k(A)$, ($\uparrow$-*step*);

(v) if $\dfrac{X}{\forall y.X}$ is a link in $\mathcal{S}$ and $\forall y.X \in kA$, then the free range of y and the occurrences in its existential border belong to kA ($\Uparrow$-*step*). ■

3.4 Sequentialization

The proof of Lemma 2 extends to the first-order case without modificatons.

Lemma 2. (Empire-Kingdom Nesting) *Let $\mathcal{L}_1 : \frac{...A...}{C}$ and $\mathcal{L}_2 : \frac{...B...}{D}$ be distinct links in a proof net $\mathcal{R}$. Suppose $B \in eA$; then $D \notin eA$ if and only if $C \in kD$.* ■

Lemma 3. (Ordering of the kingdoms, first-order case) *In proof-nets for first order* **MLL**$^-$ *the relation $\ll$ is a strict ordering of formula-occurrences that are not conclusions of axiom links.*

Proof. Suppose $X \in kY$, where X and Y not the conclusions of axioms links. Two cases are to be added to the propositional proof.

Let X be the conclusion of a link $\dfrac{A[t/x]}{\exists x.A}$. It follows from the definition of kingdom and proposition 1 that $kX = k(\exists x.A) = k(A[t/x]) \cup \{\exists x.A\}$. If X and Y are distinct and also $Y \in kX$, then $Y \in k(A[t/x])$ and this is absurd, since $Y \notin e(A[t/x])$ follows from $\exists x.A \in kY$ by lemma 2.

Finally, let X be the conclusion of a link $\dfrac{A}{\forall x.A}$. If follows from proposition 1 that $kX \setminus \{\forall x.A\} \subset eA$. If X and Y are distinct and also $Y \in kX$, then $Y \in eA$, and this contradicts lemma 2. ■

Theorem 1.(b) *The Sequentialization Theorem holds in first order **MLL**$^-$.*

Proof. We consider first the lowermost *par* and *for all* links, if such links exist. Otherwise, we choose a terminal link $\mathcal{L}$ whose conclusion is maximal w.r.t. $\lll$. If $\mathcal{L}$ is an *exists* link, then the result of removing it is still a proof-net. Suppose not; then $\mathcal{L}$: $\dfrac{A[t/x]}{\exists x.A}$ is in the existential border of y, where y is associated with $\dfrac{B}{\forall y.B}$ then $\exists x.A \in k(\forall y.B)$, by Fact 2, hence $\exists x.A$ it cannot be maximal w.r.t. $\lll$. The rest of the proof is as before. ∎

3.5 Permutability of Inferences in the Sequent Calculus

Counterexample. Let x occur free in P. The following is a derivation in MLL^- in which the applications of the $\exists$-rule and of the $\forall$-rule cannot be permuted.

$$\dfrac{\dfrac{\vdash P^\perp, P}{\vdash P^\perp, \exists x.P}\,\exists}{\vdash \forall x.P^\perp, \exists x.P}\,\forall$$

Theorem 2. (first order case) *The Theorem on permutability of inferences holds in first order* $\boldsymbol{MLL}^-$.

Proof. (i) Assuming the pure parameter property, the argument is similar to the propositional case, where *for all* rules behave like *par* rules and *exists* rules like *times* rules. The nontrivial case is the following: an inference $\mathcal{I}'_A$ of $\mathcal{D}'$ has the principal formula $A = \exists x.A_1$ and must be permuted below a *for all* rule $\mathcal{I}'_1$. As before we argue that in $\mathcal{D}$ we have an inference $\mathcal{I}_B$ such that $\mathcal{I}'_1$ and $\mathcal{I}_B$ are mapped to $B = \forall y.B_1$ and that such an inference must occur above the inference $\mathcal{I}_A$ whose active formula is $A_1[t/x]$; by the pure parameter property of $\mathcal{D}$, y does not occur in t, and the permutation is permissible.

(ii) As before, the argument is by induction on $eA \setminus (\mathcal{B}_A)^-$; to the propositional cases we add the following cases (the cases of existential links being unproblematic):

($\Uparrow$-step) *for all* link, clause (*vii*): By the pure parameter property the eigenvariables occur only above the associated $\forall$-inference, which already occurs above $\mathcal{I}_A$ by induction hypothesis.

($\Downarrow$-step) *for all* links, clause (*vii*): Let $\mathcal{I}'$ be the inference introducing $\forall y.X$ below $\mathcal{I}$, where $\forall y.X \in eA$. By induction hypothesis the eigenvariable y occurs only in sequents above $\mathcal{I}_A$, except for one occurrence of a formula $X(y)$ (an ancestor of $\forall y.X$) for each sequent between $\mathcal{I}_A$ and $\mathcal{I}'$. Hence we can always permute $\mathcal{I}'$ with the inference immediately above it. ∎

References

[1] G. Bellin. Mechanizing Proof Theory: Resource-Aware Logics and Proof-Transformations to Extract Implicit Information, Phd Thesis, Stanford University. Available as: Report CST-80-91, June 1990, Dept. of Computer Science, Univ. of Edinburgh.

[2] G. Bellin. Proof Nets for Multiplicative and Additive Linear Logic, Report LFCS-91-161, May 1991, Dept. of Computer Science, Univ. of Edinburgh.

[3] V. Danos and L. Regnier. The Structure of Multiplicatives, *Arch. Math. Logic* (1989) **28**, pp. 181-203.

[4] J. Gallier. Constructive Logics. Part II: Linear Logic and Proof Nets, preprint, Dept. of Computer and Information Science, University of Pennsylvania, 200 South 33rd St., Philadelphia, PA 19104, USA, Email: `jean@saul.cis.upenn.edu`

[5] J-Y. Girard. Linear Logic, *Theoretical Computer Science* **50**, 1987, pp. 1-102.

[6] J-Y. Girard. Quantifiers in Linear Logic, Proceedings of the SILFS conference held in Cesena, January 1987.

[7] J-Y. Girard. Quantifiers in linear logic II. Preprint, Équipe de Logique Mathématique, Université Paris VII, Tour 45-55. 5ᵉ étage, 2, Place Jussieu, 75251 PARIS Cedex 05. France.

Gianluigi Bellin and Jacques van de Wiele
Équipe de Logique
Université de Paris VII
Tour 45-55, 5ᵉ étage
2 Place Jussieu
75251 Paris Cedex 05
France

NONCOMMUTATIVE PROOF NETS

V. Michele ABRUSCI

Dipartimento di Scienze Filosofiche , Università di Bari
Palazzo Ateneo, Piazza Umberto, 70121 Bari - Italy
abrusci@vm.unibari.it

Introduction

The aim of this paper is to give a purely graph-theoretical definition of *noncommutative proof nets*, i.e. graphs coming from proofs in **MNLL** (*multiplicative noncommutative linear logic,* the ($\otimes$,$\wp$)-fragment of the one-sided sequent calculus for classical noncommutative linear logic, introduced in [Abr91]). Analogously, one of the aims of [Gir87] was to give a purely graph-theoretical definition of *proof nets,* i.e. graphs coming from the proofs in **MLL** (*multiplicative linear logic,* the ($\otimes$,$\wp$)-fragment of the one-sided sequent calculus for classical linear logic - better, for classical *commutative* linear logic). - The relevance of the purely graph-theoretical definition of proof nets for the development of commutative linear logic is well-know; thus we hope the results of this paper will be useful for a similar development of noncommutative linear logic.

The *language* for **MNLL** is an extension of the language for **MLL**, obtained simply adding, as atomic formulas, propositional letters with *an arbitrary finite number of negations* written *after* the propositional letter (*linear post-negation)* or *before* the propositional letter (*linear retro-negation).* Every formula A of **MNLL** may be translated into a formula Tr(A) of **MLL** (simply by replacing each propositional letter with an even number of negations by the propositional letter without negations, and each propositional letter with an odd number of negations by the propositional letter with only one negation after the propositional letter). - If we restrict the rules of the *sequent calculus* for **MNLL** to the language for **MLL**, and we add the *exchange rule*, we get exactly a formulation of the sequent calculus for **MLL**; moreover, if $\mathfrak{D}$ is a proof in the sequent calculus for **MNLL** then $\mathrm{Tr}(\mathfrak{D})$ (i.e. what is obtained from $\mathfrak{D}$ by replacing each formula A by $\mathrm{Tr}(A)$) is a proof in the sequent calculus for **MLL**. - The language and the sequent calculus for **MNLL** are presented in the section 1.

In [Gir87]:
a) it is defined a class of graphs of occurrences of formulas of **MLL**, the class of the *proof structures* (better, *commutative proof structures*) and it is shown how to every proof $\mathfrak{D}$ of a sequent $\vdash\Gamma$ in the sequent calculus for **MLL** can be associated a commutative proof structure $\Phi(\mathfrak{D})$ whose conclusions are exactly the formulas in Γ, the *commutative proof structure coming from* $\mathfrak{D}$,
b) it is defined (by means of a purely graph-theoretical definition, independent from the sequent calculus) a subclass of commutative proof structures, the proof structures without short trips, called *proof nets* (better, *commutative proof nets*),
c) it is proved that $\pi = \Phi(\mathfrak{D})$ for some proof $\mathfrak{D}$ in the sequent calculus for **MLL** iff π is a commutative proof net.
The analogous of a), b) and c) for **MNLL** is performed in sections 2,3,4 of this paper.

In the section 2, following ideas already present in [Abr91], we introduce *noncommutative proof structures* in such a way that if π is a noncommutative proof structure then $\mathrm{Tr}(\pi)$ (i.e. what is obtained by replacing each occurrence of formula A in π by $\mathrm{Tr}(A)$) is a commutative proof structure. To every proof $\mathfrak{D}$ of a sequent $\vdash\Gamma$ in the sequent calculus for **MNLL** we associate (again, following a procedure already given in [Abr91]) a noncommutative proof structure $\Phi(\mathfrak{D})$ s.t.
a) the conclusions of $\Phi(\mathfrak{D})$ are exactly the formulas in Γ,
b) $\mathrm{Tr}(\Phi(\mathfrak{D}))$ is exactly the commutative proof structure coming from the

proof $\mathrm{Tr}(\mathfrak{D})$ in **MLL** and therefore $\mathrm{Tr}(\Phi(\mathfrak{D}))$ is a commutative proof net,
c) $\Phi(\mathfrak{D})$ is planar.
Moreover, by looking at the proof $\mathfrak{D}$ in sequent calculus, it it possible to arrange $\Phi(\mathfrak{D})$ in order to have that the conclusions (read from the left to the right) are in the order Γ.

The property of a noncommutative proof structure π "$\mathrm{Tr}(\pi)$ is a commutative proof net and π is planar" is a purely geometrical property, but it is not equivalent to "π comes from a proof in **MNLL**", as we show by means of counterexamples (moreover, the property "π is a commutative proof net and π is planar" is not equivalent to "π comes from a proof in **MLL** which is also a proof in **MNLL**").

In the section 3 we define *noncommutative proof nets.* A noncommutative proof net π with conclusion Γ is a noncommutative proof structure s.t.
(i) $\mathrm{Tr}(\pi)$ is a proof net
(ii) π is *good*
(iii) π *induces the linear order* Γ of the conclusions;
where (i)-(ii)-(iii) are properties defined independently from the sequent calculus.

In the section 4, we prove that π is a noncommutative proof net with conclusions Γ iff $\pi = \Phi(\mathfrak{D})$ for some proof $\mathfrak{D}$ of the sequent $\vdash \Gamma$ in **MNLL**.

In a forthcoming paper we shall define reduction rules on noncommutative proof nets, and we shall prove the normalization theorem for noncommutative proof nets, i.e. that every noncommutative proof net reduces in a finite number of steps into a cut-free noncommutative proof net.

Acknowledgment. The author is indebted to the referee for very useful suggestions.

1. Sequent calculus for MNLL

1.1 Language of MNLL

1.1.0 Definition. (Formulas and sequents). (cf. [Abr91] for details)

Let $\mathbb{N}$ be the set of positive integers, $\mathbb{Z}$ the set of positive and negative integers.

Formulas of **MNLL** are defined as follows: (i) if P is a propositional letter and $n \in \mathbb{Z}$, then P^n is a formula of **MNLL** ; (ii) if A and B are formulas of **MNLL**, then $A \otimes B$ and $A \wp B$ are formulas of **MNLL** ; (iii) nothing else is a formula of **MNLL** .

If P is a propositional letter and $n \in \mathbb{N}$, then:
P^0 is P ("P without linear negations") ; P^{+n} is "n linear post-negations of P" ($P^{\perp}$ for $n = 1$ and $P^{\perp \ldots \perp\,(n \text{ times})}$ for $n > 1$); P^{-n} is "n linear retro-negations of P" (${}^{\perp}P$ for $n = 1$ and ${}^{\perp \ldots \perp\,(n \text{ times})}P$ for $n > 1$).

A *sequent* of **MNLL** is any configuration $\vdash \Gamma$, where Γ is a finite sequence of formulas of **MNLL** .

1.1.1 Definition. (Metalinguistic negation of formulas). (cf. [Abr91] for details)

For each formula A of **MNLL**, we define $A^{\perp}$ (the *linear post-negation of* A) and ${}^{\perp}A$ (the *linear retro-negation of* A) as follows:
(i) if P is a propositional letter, and $n \in \mathbb{Z}$, $(P^n)^{\perp} = P^{n+1}$, ${}^{\perp}(P^n) = P^{n-1}$; (ii) $(A \otimes B)^{\perp} = B^{\perp} \wp A^{\perp}$, ${}^{\perp}(A \otimes B) = {}^{\perp}B \wp {}^{\perp}A$; (iii) $(A \wp B)^{\perp} = B^{\perp} \otimes A^{\perp}$, ${}^{\perp}(A \wp B) = {}^{\perp}B \otimes {}^{\perp}A$.

If A is a formula of **MNLL** and $n \in \mathbb{Z}$, then A^n will denote

- A, if $n = 0$
- $A^{\perp \ldots \perp\,(n \text{ times})}$, if $n > 0$
- ${}^{\perp \ldots \perp\,(-n \text{ times})}A$, if $n < 0$

The reader will verify that for each formula A of **MNLL** and for each $n, m \in \mathbb{Z}$

$$(A^n)^m = A^{n+m}$$

and in particular $\quad ({}^{\perp}A)^{\perp} = {}^{\perp}(A^{\perp}) = A$.

1.1.2. Definition. (Translation of formulas of **MNLL** into formulas of **MLL**).

Every formula A of **MNLL** may be translated in a formula $\mathrm{Tr}(A)$ of **MLL**:

(i) if $n \in \mathbb{Z}$, $\mathrm{Tr}(P^n) = P$ if n is even, $\mathrm{Tr}(P^n) = P^{\perp}$ if n is odd;

(ii) $\mathrm{Tr}(A \otimes B) = \mathrm{Tr}(A) \otimes \mathrm{Tr}(B)$, $\mathrm{Tr}(A \wp B) = \mathrm{Tr}(A) \wp \mathrm{Tr}(B)$.

If Γ is a finite sequence of formulas of **MNLL**, $\mathrm{Tr}(\Gamma)$ is defined in an obvious way.

The reader will verify that, for every formula A of **MNLL**,

$$\mathrm{Tr}(A^{\perp}) = \mathrm{Tr}({}^{\perp}A) = (\mathrm{Tr}(A))^{\perp}$$

where $(\mathrm{Tr}(A))^{\perp}$ is the metalinguistic linear negation of $\mathrm{Tr}(A)$ in **MLL**, defined as follows:

$$(P)^{\perp} = P^{\perp},\ (P^{\perp})^{\perp} = P$$

$$(A \otimes B)^{\perp} = B^{\perp} \wp A^{\perp},\ (A \wp B)^{\perp} = B^{\perp} \otimes A^{\perp}.$$

1.2 Sequent calculus for MNLL

1.2.0. Definition. (Rules of the one-sided sequent calculus).

The *rules of the one-sided sequent calculus* for **MNLL** are the following ones:

$$\frac{}{\vdash A^{\perp}, A}\ (\mathrm{id})$$

$$\frac{\vdash \Gamma, A, \Gamma' \qquad \vdash A^{\perp}, \Delta}{\vdash \Gamma, \Delta, \Gamma'}\ (\mathrm{cut1}) \qquad \frac{\vdash \Gamma, A \qquad \vdash \Delta, A^{\perp}, \Delta'}{\vdash \Delta, \Gamma, \Delta'}\ (\mathrm{cut2})$$

$$\frac{\vdash \Gamma, A, \Gamma' \qquad \vdash B, \Delta}{\vdash \Gamma, A \otimes B, \Delta, \Gamma'}\ (\otimes 1) \qquad \frac{\vdash \Gamma, A \qquad \vdash \Delta, B, \Delta'}{\vdash \Delta, \Gamma, A \otimes B, \Delta'}\ (\otimes 2)$$

$$\frac{\vdash \Gamma, A, B, \Gamma'}{\vdash \Gamma, A \wp B, \Gamma'} (\wp)$$

1.2.1. Remarks.

(i) Since $A = ({}^{\perp}A)^{\perp}$, the followings rules:

$$\overline{\vdash A, {}^{\perp}A} \qquad \frac{\vdash \Gamma, {}^{\perp}A, \Gamma' \quad \vdash A, \Delta}{\vdash \Gamma, \Delta, \Gamma'} \qquad \frac{\vdash \Gamma, {}^{\perp}A \quad \vdash \Delta, A, \Delta'}{\vdash \Delta, \Gamma, \Delta'}$$

are just another way to write the rules (Id), (cut1) and (cut2).

(ii) The sequent calculus for **MNLL** is *not degenerate*, i.e. for every formula A

$\vdash A$ is not provable in **MNLL** or $\vdash A^{\perp}$ is not provable in **MNLL**

Therefore, in (cut1) and (cut2) one of the three contexts involved is not empty.

1.2.2 Remarks. (Relation to the sequent calculus for **MLL**)

(ii) By restricting the rules of the one-sided sequent calculus for **MNLL** to formulas of **MLL**, by considering $A^{\perp}$ in these rules as the metalinguistic linear negation of A in **MLL**, and by adding the *exchange rule*, we get an alternative formulation of the sequent calculus for **MLL** equivalent to the one given in [Gir87].

(ii) If $\mathfrak{D}$ is a proof of $\vdash \Gamma$ in **MNLL**, and we replace every formula A in $\mathfrak{D}$ by $\mathrm{Tr}(A)$, then we get a proof $\mathrm{Tr}(\mathfrak{D})$ in **MLL** of $\vdash \mathrm{Tr}(\Gamma)$ without exchange rule.

(iii) The following rules are derived rules in **MLL** but not in **MNLL**:

(1) The variant of (Id) with the exchange of the order of the formulas

$$\overline{\vdash A, A^{\perp}}$$

(2) The following general cut-rule including (cut1) and (cut2)

$$\frac{\vdash \Gamma, A, \Gamma' \quad \vdash \Delta, A^{\perp}, \Delta'}{\vdash \Delta, \Gamma, \Delta', \Gamma'} (\mathrm{cut}) \quad \text{with } \Gamma' \neq \emptyset \text{ and } \Delta \neq \emptyset$$

(3) The following general $\otimes$-rule including ($\otimes$1) and ($\otimes$2)

$$\frac{\vdash \Gamma, A, \Gamma' \qquad \vdash \Delta, B, \Delta'}{\vdash \Delta, \Gamma, A \otimes B, \Delta', \Gamma'} (\otimes) \text{ with } \Gamma' \neq \emptyset \text{ and } \Delta \neq \emptyset$$

(4) The following general $\wp$-rule

$$\frac{\vdash \Gamma, A, \Delta, B, \Gamma'}{\vdash \Gamma, \Delta, A \wp B, \Gamma'} (\wp) \text{ with } \Delta \neq \emptyset$$

2. Noncommutative proof structures

2.3. Definition. (Noncommutative proof structures).

(i) π is a *noncommutative proof structure* iff π is a graph of occurrences of formulas of **MNLL** (called *vertices of* π), s.t.

- each vertex in π is conclusion of exactly one link in π
- each vertex in π is premise of at most one link in π
- the links in π are the following ones:

Ax-link: $A^{\perp} \quad A$

(no premise, $A^{\perp}$ first conclusion, A second conclusion)

Cut-link: $A \quad A^{\perp}$

(A first premise, $A^{\perp}$ second premise, no conclusion)

$\otimes$-link: $A \quad B$

$A \otimes B$

A first premise, B second premise, $A \otimes B$ conclusion

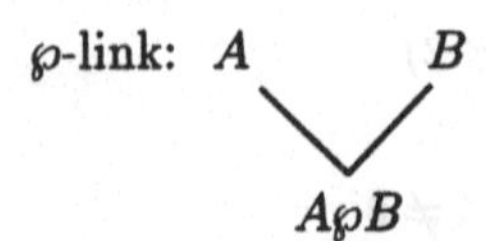

A first premise, B second premise, $A\wp B$ conclusion

Since $A^{\perp}$ is not $A^{\perp\perp}$, it is not ambiguous to determine in an Ax-link (or in a Cut-link) what is the first conclusion (the first premise) and what is the second conclusion.

The *commutative translation of* a noncommutative proof structure π is the graph obtained from π by replacing each vertex A by Tr(A). Tr(π) is obviously a commutative proof structure, in the sense of [Gir87].

A is a *conclusion* of a noncommutative proof structure π iff A is an vertex in π and A is not a premise of a link in π.

2.4. Definition. (Noncommutative proof structures coming from proofs in **MNLL**)

Let $\mathfrak{D}$ be a proof of a sequent $\vdash\Gamma$ in **MNLL**. The *noncommutative proof structure* $\Phi(\mathfrak{D})$ *coming from* $\mathfrak{D}$ is defined as usual, by induction on $\mathfrak{D}$, with the property that an occurrence of formula A is a conclusion of $\Phi(\mathfrak{D})$ iff A is in Γ.

(i) If $\mathfrak{D} = \vdash A^{\perp}, A$ then $\Phi(\mathfrak{D}) = A^{\perp} \quad A$

(ii) if $$\mathfrak{D} = \cfrac{\begin{matrix}\mathfrak{D}_1 \\ \vdots \\ \vdash \Gamma, A, \Gamma'\end{matrix} \qquad \begin{matrix}\mathfrak{D}_2 \\ \vdots \\ \vdash A^{\perp}, \Delta\end{matrix}}{\vdash \Gamma, \Delta, \Gamma'}\text{(cut1)}$$

or $$\mathfrak{D} = \cfrac{\begin{matrix}\mathfrak{D}_1 \\ \vdots \\ \vdash \Gamma, A\end{matrix} \qquad \begin{matrix}\mathfrak{D}_2 \\ \vdots \\ \vdash \Delta, A^{\perp}, \Delta'\end{matrix}}{\vdash \Delta, \Gamma, \Delta'}\text{(cut2)}$$

then $\Phi(\mathfrak{D}) =$ $\Phi(\mathfrak{D}_1)$ … A … $\Phi(\mathfrak{D}_2)$ … $A^\perp$ …

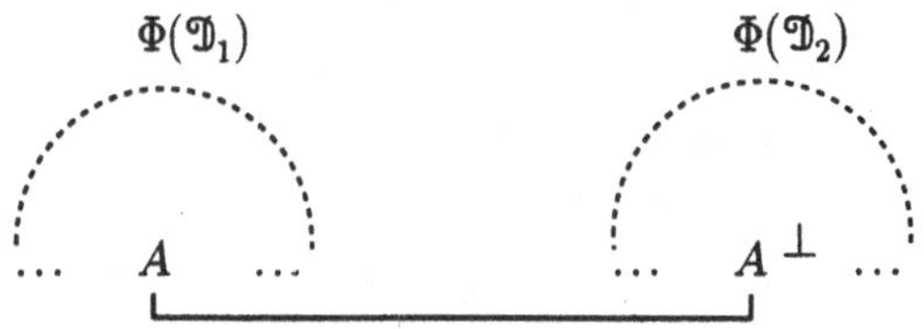

(iii) if $\mathfrak{D} = \dfrac{\begin{matrix}\mathfrak{D}_1 \\ \vdots \\ \vdash \Gamma, A, \Gamma'\end{matrix} \qquad \begin{matrix}\mathfrak{D}_2 \\ \vdots \\ \vdash B, \Delta\end{matrix}}{\vdash \Gamma, A \otimes B, \Delta, \Gamma'} (\otimes 1)$

or $\mathfrak{D} = \dfrac{\begin{matrix}\mathfrak{D}_1 \\ \vdots \\ \vdash \Gamma, A\end{matrix} \qquad \begin{matrix}\mathfrak{D}_2 \\ \vdots \\ \vdash \Delta, B, \Delta'\end{matrix}}{\vdash \Delta, \Gamma, A \otimes B, \Delta'} (\otimes 2)$

then $\Phi(\mathfrak{D}) =$ $\Phi(\mathfrak{D}_1)$ … A … $\Phi(\mathfrak{D}_2)$ … B … $A \otimes B$

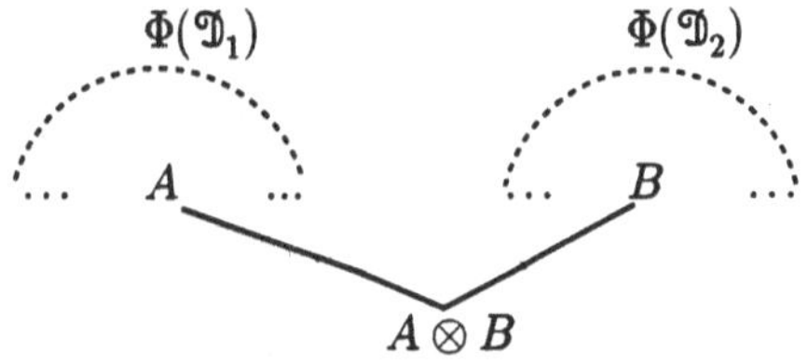

(iv) if $\mathfrak{D} = \dfrac{\begin{matrix}\mathfrak{D}' \\ \vdots \\ \vdash \Gamma, A, B, \Gamma'\end{matrix}}{\vdash \Gamma, A \wp B, \Gamma'} (\wp)$, then $\Phi(\mathfrak{D}) =$ $\Phi(\mathfrak{D}')$ … A B … $A \wp B$

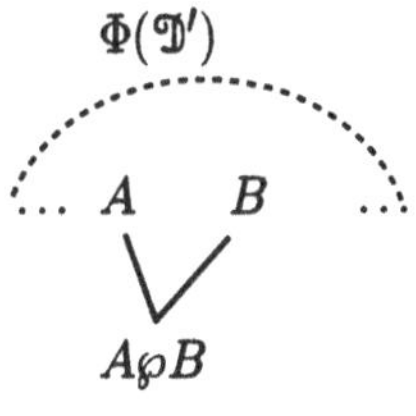

2.5. Proposition.

Let $\mathfrak{D}$ be a proof of the sequent $\vdash \Gamma$ in **MNLL** . Then:

(i) $\mathrm{Tr}(\Phi(\mathfrak{D}))$ is a commutative proof net

(ii) $\Phi(\mathfrak{D})$ is planar

(iii) we can arrange $\Phi(\mathfrak{D})$ in such a way that:

(iii1) there is no crossing of links

(iii2) the conclusions of $\Phi(\mathfrak{D})$, from the left to the right, are exactly the sequence Γ.

Proof .

(i) Simply because $\mathrm{Tr}(\Phi(\mathfrak{D}))$ is the commutative proof structure coming from the proof $\mathrm{Tr}(\mathfrak{D})$ of $\vdash \mathrm{Tr}(\Gamma)$ in **MLL**, and every commutative proof structure coming a proof in **MLL** is a commutative proof net (cf. [Gir87]). - (ii)-(iii) See [Abr91].

2.6. Proposition.

There are noncommutative proof structures π, s.t.

- $\mathrm{Tr}(\pi)$ is a commutative proof net (and moreover $\mathrm{Tr}(\pi) = \pi$)
- π is planar
- π does not come from a proof in **MNLL** (whereas it comes from a proof in **MLL**).

Proof. See the following planar noncommutative proofs structure ψ_1 and ψ_2:

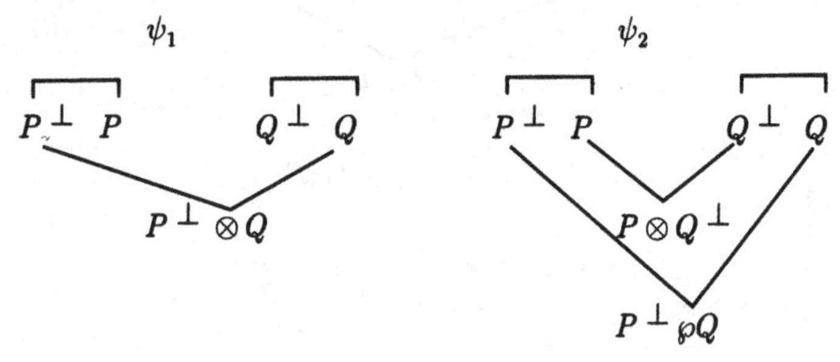

ψ_1 and ψ_2 are commutative proof nets, since they come from the proofs in **MLL**

$$\dfrac{\dfrac{}{\vdash P^\perp, P}(\mathrm{Id}) \qquad \dfrac{}{\vdash Q^\perp, Q}(\mathrm{Id})}{\vdash Q^\perp, P^\perp \otimes Q, P}(\otimes).$$

$$\dfrac{\dfrac{\dfrac{}{\vdash P^\perp, P}(\mathrm{Id}) \qquad \dfrac{}{\vdash Q^\perp, Q}(\mathrm{Id})}{\vdash P^\perp, P \otimes Q^\perp, Q}(\otimes 1) \text{ or } (\otimes 2)}{\vdash P^\perp \wp Q, P \otimes Q^\perp}(\wp)$$

but they do not come from proofs in **MNLL**.

3. Noncommutative proof nets.

3.0. Definition. (Points of π. Special trips in π. Critical vertices of π. $\mathbb{L}(\pi)$)

Let π be a noncommutative proof structure s.t. $\mathrm{Tr}(\pi)$ is a commutative proof net.

(i) T is a *point of* π iff T is $A\downarrow$ ("A as answer") or $\uparrow A$ ("A as question") where A is a vertex of π. If T is a point of π, then $\mathrm{Tr}(T)$ is defined as follows:

$$\mathrm{Tr}(A\downarrow) = \mathrm{Tr}(A)\downarrow\ ,\ \mathrm{Tr}(\uparrow A) = \uparrow\mathrm{Tr}(A)$$

(Remark the elements of *trips* in $\mathrm{Tr}(\pi)$ are $\mathrm{Tr}(T)$ where T is a point of π)

(ii) Let A be a conclusion of π. A sequence of points of π

$$T_1,\ldots,T_n$$

is a *special trip in* π *from* A iff $\mathrm{Tr}(T_1),\ldots,\mathrm{Tr}(T_n)$ is portion of the long trip in $\mathrm{Tr}(\pi)$ from $\uparrow\mathrm{Tr}(A)$ to $\mathrm{Tr}(A)\downarrow$ with the following switching:
- every $\otimes$-link is switched on "R" ("right")
- every $\wp$-link is switched on "L" ("left").

Remark that then each point T of π occurs exactly once in the special trip in π from the conclusion A.

(ii) A is a *critical vertex* of π iff A is a conclusion of π or the second premise of a $\wp$-link in π. - (Remark that A is a *critical vertex* of π iff in the long trip in $\mathrm{Tr}(\pi)$ with all the $\wp$-link switched on "L" ("left") the next step after $\mathrm{Tr}(A)\downarrow$ is just $\uparrow\mathrm{Tr}(A)$).

(iii) $\mathbb{L}(\pi) = [x^C + a\ /\ a \in \mathbb{Z}$ and C is a critical vertex of $\pi]$, where $\mathbb{Z}$ is the set of (positive and negative) integers. Remark that $\mathbb{L}(\pi) \subset \mathbb{Z}[x^{C_1},\ldots,x^{C_k}]$ where $C_1,\ldots,C_k$ are all the critical vertices of π, i.e. $\mathbb{L}(\pi)$ is a set of polynomials on $\mathbb{Z}$.

(iv) We shall use $u,s,t,\ldots$ as ranging over $\mathbb{L}(\pi)$. $u = t$ will mean the identity as defined on polynomials on $\mathbb{Z}$. $u[x^C\ /\ t]$ will mean the the substitution of x^C by t in u.

3.1. Definition. (Assignments for π)

Let π be a noncommutative proof structure s.t. Tr(π) is a commutative proof net. Let A be a conclusion of π.

$\mathcal{L}$ is an *assignment for π from A* iff $\mathcal{L}$ is the partial function from the set of the points of π to $\mathbb{L}(\pi)$, defined as follows by considering the special trip in π from A

$$T_1,\ldots,T_n$$

By induction on $i \leq n$:

- $\mathcal{L}(T_1)=\mathcal{L}(\uparrow A) = x^A$
- Suppose we defined $\mathcal{L}(T_1),\ldots,\mathcal{L}(T_i)$ for $i < n$

(i) if $T_i = B\downarrow$ and B is a critical point of π (so that $T_{i+1}=\uparrow B$), then $\mathcal{L}(T_{i+1}) = x^B$;

(ii) if $T_i = \uparrow B$ and B is the second conclusion of an Ax-link (so that $T_{i+1} = B^{\perp}\downarrow$), then $\mathcal{L}(T_{i+1})=\mathcal{L}(T_i)+1$

(iii) if $T_i = B^{\perp}\downarrow$ and B is the second premise of a Cut-link (so that $T_{i+1} = \uparrow B^{\perp}$), then $\mathcal{L}(T_{i+1})=\mathcal{L}(T_i)-1$

(iv) if $T_i = B\downarrow$ and B first premise of $\wp$-link with C as second premise (so that $T_{i+1} = B\wp C\downarrow$), then

$$\mathcal{L}(T_{i+1})\begin{cases} =\mathcal{L}(C\downarrow), \text{ if } C\downarrow = T_j \text{ with } j<i \text{ and } \mathcal{L}(T_i) = x^C+1 \\ \text{undefined, otherwise} \end{cases}$$

(v) $\mathcal{L}(T_{i+1})=\mathcal{L}(T_i)$, in all the other cases.

- Suppose $\mathcal{L}(T_i)$ is undefined: then $\mathcal{L}(T_{i+1})$ is undefined

An assignment $\mathcal{L}$ for π is *total* iff $\mathcal{L}$ is a total function.

If $\mathcal{L}$ is an assignment for π, and $u,v \in \mathbb{L}(\pi)$, then we write

uA for $\mathcal{L}(\uparrow A) = u$, Av for $\mathcal{L}(A\downarrow) = v$, uAv for $\mathcal{L}(\uparrow A) = u$ and $\mathcal{L}(A\downarrow) = v$

If $\mathcal{L}$ is an assignment for π from a conclusion A, the values of $\mathcal{L}$ may be set as in the following figure, where $\rightarrow$ means the transition from a point of the long trip to the immediately following one.

Conclusion C $\qquad x^C Cs$

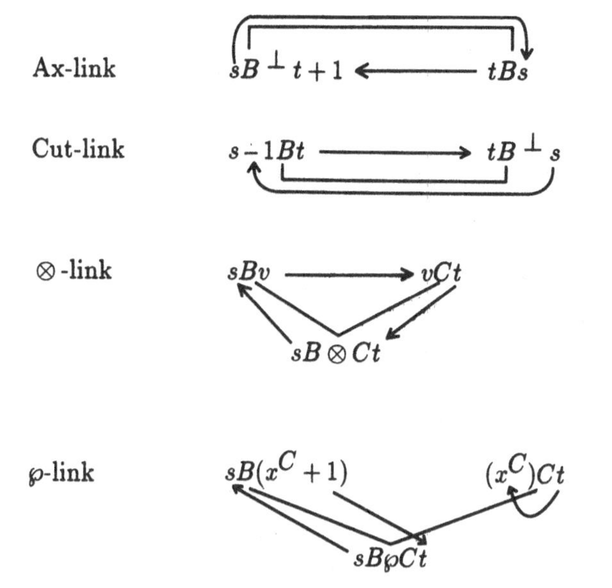

if in the special trip from $\uparrow\mathrm{Tr}(A)$, $\mathrm{Tr}(C)\downarrow$ occurs before $\mathrm{Tr}(B)\downarrow$

3.2. Proposition.

Let π be a noncommutative proof structure, s.t. $\mathrm{Tr}(\pi)$ is a proof net. Let $\mathfrak{L}$ be a total assignment for π. If $\mathfrak{L}'$ is an assignment for π, then $\mathfrak{L}'$ is total and $\mathfrak{L} = \mathfrak{L}'$.

Proof

Suppose $\mathfrak{L}$ be a total assignment for π from a conclusion A. Let $\mathfrak{L}'$ be an assignment for π, from a conclusion B. Consider the special trip in π from B: then (from the property of being a long trip) it looks as follows

$T_1 = \uparrow B, , \ldots, T_k = A\downarrow, T_{k+1} = \uparrow A, T_{k+2} \ldots, T_n, T_{n+1} = B\downarrow$

and necessarily the special trip in π from A is

$T_{k+1} = \uparrow A, T_{k+2} \ldots, T_n, T_{n+1} = B\downarrow, T_1 = \uparrow B, \ldots, T_k = A\downarrow$

It is easy to prove, by induction on $i \leq n+1$ and by using the hypothesis that $\mathfrak{L}$ is total, that $\mathfrak{L}'(T_i)$ is defined and $\mathfrak{L}'(T_i) = \mathfrak{L}(T_i)$.

3.3. Definition. (Goodness. Labeled special trips)

Let π be a noncommutative proof structure, s.t. Tr(π) is a proof net.

(i) π is *good* iff every assignment for π is total. (By the previous proposition, if π is good, then all the assignments for π are equal).

(ii) If π is good, the *labeled special trip in* π is obtained from a special trip in π by replacing each point $A\downarrow$ by $A\mathcal{L}(A\downarrow)$ and each point $\uparrow A$ by $\mathcal{L}(\uparrow A)A$, where $\mathcal{L}$ is the unique assignment for π.

3.4. Remark.

The noncommutative proof structure ψ_1 (shown in 2.3) is good, whereas ψ_2 (shown also in 2.3) is not good. - Thus, the property "π is a noncommutative proof structure, Tr(π) is a commutative proof net and π is good" is not equivalent to the property "π comes from a proof in sequent calculus for **MNLL**".

3.5. Proposition.

Let π be a noncommutative proof structure, Tr(π) a commutative proof net and π good. Let A be a conclusion of π. Let $T_1,\ldots,T_k$ be the special trip from A in π.

For every $i \leq k$, if $\mathcal{L}(T_i) = x^C + a$, then $a = n-(m+p)$ (" the number of the critical pairs between $\uparrow C$ and T_i") where:
- $n =$ the number of the "down-Ax-visits" (i.e. the number of pairs of consecutive points $\uparrow B$, $B^\perp\downarrow$) between $\uparrow C$ and T_i
- $m =$ the number of "$\wp$-answers" (i.e. the number of pairs of consecutive points $B_1\downarrow$, $B_1\wp B_2\downarrow$) between $\uparrow C$ and T_i
- $p =$ the number of the "down-Cut-visits" (i.e. the number of pairs of consecutive points $B^\perp\downarrow$,$\uparrow B$) between $\uparrow C$ and T_i.

Proof

By induction on i . - The case i=1 holds trivially. - Suppose that the result holds for all $j \leq i$, and let $\mathcal{L}(T_{i+1}) = x^C + a$.

Let T_i is $\uparrow B$. If B is the first conclusion of an Ax-link or the conclusion of a $\otimes$-link or $\wp$-link, then $\mathcal{L}(T_i)$ is $x^C + a$, and the result holds by induction hypothesis. If B is a the second conclusion of an Ax-link, then $\mathcal{L}(T_i)$ is $x^C + a - 1$, and the result holds by induction hypothesis (since the number of the "down-Ax-visits" is increased by 1).

Let T_i is $B\downarrow$. If B is the first premise of a cut-link or a premise of a $\otimes$-link, then $\mathcal{L}(T_i)$ is $x^C + a$, and the result holds by induction hypothesis. If B is the second premise of a Cut-link, then $\mathcal{L}(T_i)$ is $x^C + a + 1$, and the result holds by induction hypothesis since the number of the "down-Cut-visits" is increased by 1. If B is the first premise of a $\wp$-link, then $\mathcal{L}(T_i)$ is $x^D + 1$ where D is the second premise of the same $\wp$-link and there is $j < i$ s.t. T_j is $D\downarrow$, T_{j+1} is $\downarrow D$, and $\mathcal{L}(T_j) = x^C + a$; by induction hypothesis, the number of the critical pairs between $\uparrow C$ and T_j is a and the number of the critical pairs between T_{j+1} and T_i is 1, so that now the number of the critical steps between $\uparrow C$ and T_{i+1} is a. If B is a critical point, then the result holds trivially.

3.6. Corollary.

Let π be a noncommutative proof structure, with k conclusions. Suppose $\mathrm{Tr}(\pi)$ is a commutative proof net and π is good. Let $\mathcal{L}$ be the unique assignment for π. Consider for each conclusion D the integer a s.t. $\mathcal{L}(D\downarrow) = x^C + a$, and let m be the sum of these integers. Then $m = k - 1$.

Proof. Follows from the above lemma, and from the fact that $\mathrm{Tr}(\pi)$ is a commutative proof net.

3.7. Definition. (Precedence of conclusions)

Let π be a noncommutative proof structure s.t. $\mathrm{Tr}(\pi)$ is a commutative proof net and π is good. Let $\mathcal{L}$ be the unique total assignment for π.

(i) We define the following binary relation $\prec$ (*precedes*) on the conclusions of π:

$$A \prec B \text{ iff } \mathcal{L}(A\downarrow) = \mathcal{L}(\uparrow B) + 1 = x^B + 1$$

(ii) π *induces the linear order of the conclusions* iff $\prec$ is a chain and every conclusion occurs exactly once in the chain.

Remark that, if π induces the linear order of the conclusions, since π is good, by previous corollary we get: if A is the first element of the chain, and B is the last one, then $\mathcal{L}(B\downarrow) = \mathcal{L}(\uparrow A) = x^A$.

3.8. Definition. (Noncommutative proof nets)

Let π be a noncommutative proof structure, and Γ a sequence of formulas of **MNLL**.

π is *a noncommutative proof net with conclusions* Γ iff
(i) Tr(π) is a commutative proof net
(ii) the conclusions of π are exactly the formulas in Γ
(iii) π is good
(iv) π induces the linear order Γ of the conclusions.

(Remark: (i) is the only property which cannot be checked in an efficient way)

3.9. Lemma.

Let π be a noncommutative proof net with conclusions Γ

(i) If $\Gamma = A$, then the labeled special trip in π looks as follows:

$$\ldots,(x^A)A,\ldots,A(x^A),\ldots$$

i.e. $(x^A)A(x^A)$

(ii) if Γ=A,B, then the labeled special trip in π looks as follows:

$$\ldots,(x^A)A,\ldots,B(x^A),(x^B)B,\ldots,A(x^B+1),\ldots$$

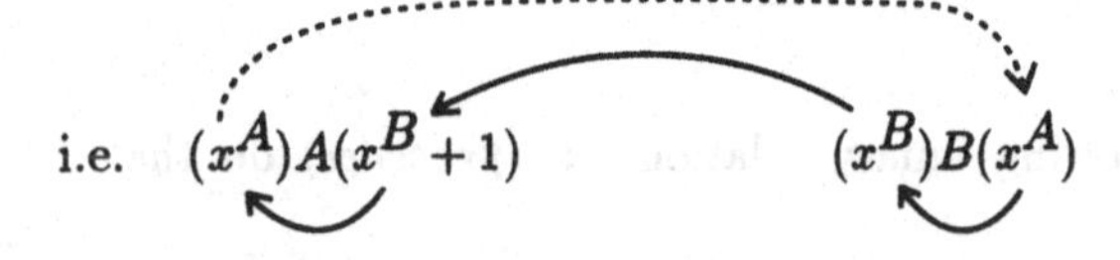

(iii) if $k > 2$, and $\Gamma = A_1,\ldots,A_k$, then the labeled special trip in π looks as follows

$\ldots,(x^{A_1})A_1,\ldots A_k(x^{A_1}),(x^{A_k})A_k,\ldots,A_2(x^{A_3}+1),(x^{A_2})A_2,\ldots,A_1(x^{A_2}+1),$

...

i.e.

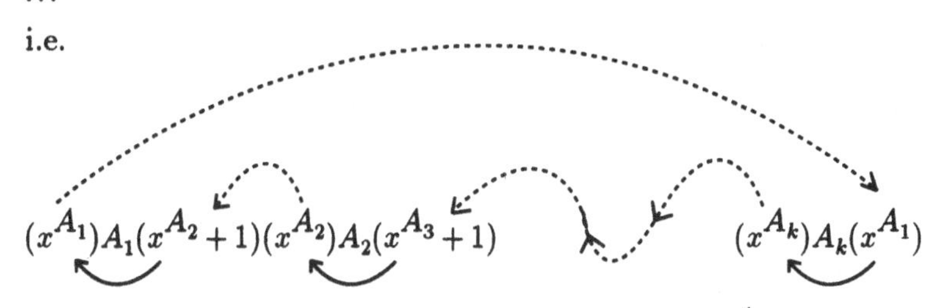

and no conclusion occurs in the portion between $(x^{A_1})A_1,\ldots,A_k(x^{A_1})$ and for every $1 \le i < k$ in the portion between $(x^{A_{i+1}})A_{i+1},\ldots,A_i(x^{A_{i+1}}+1)$

Proof. From above results, and the fact that $\mathrm{Tr}(\pi)$ is a commutative proof net.

4. Equivalence theorem between noncommutative proof nets and noncommutative proof structures coming from proofs in MNLL.

4.0. Theorem

If $\mathfrak{D}$ is a proof of $\vdash\Gamma$ in **MNLL**, then $\Phi(\mathfrak{D})$ is a noncommutative proof net with conclusions Γ.

Proof.

By induction on $\mathfrak{D}$. (In each case, we know already, cf. prop. 2.2, that $\mathrm{Tr}(\Phi(\mathfrak{D}))$ is a commutative proof net and the conclusions of $\Phi(\mathfrak{D})$ are exactly the formulas in Γ).

(i) Let $\mathfrak{D}$ be the axiom, $\vdash A^{\perp},A$. $\Phi(\mathfrak{D})$ is good and induces the linear order $A^{\perp},A$, since there is an unique assignment and the following labeled trip:

$$(x^{A^{\perp}})A^{\perp},A(x^{A^{\perp}}),(x^{A})A,A^{\perp}(x^{A}+1),(x^{A^{\perp}})A^{\perp}$$

i.e. $$(x^{A^\perp})A^\perp(x^A+1) \longleftarrow (x^A)A(x^{A^\perp})$$

(ii) Let $\mathfrak{D}$ be a proof of $\vdash \Gamma,\Delta,\Gamma'$ obtained by (cut,1) from a proof $\mathfrak{D}_1$ of $\vdash \Gamma,A,\Gamma'$ and a proof $\mathfrak{D}_2$ of $\vdash A^\perp,\Delta$. By induction hypothesis $\Phi(\mathfrak{D}_1)$ is a noncommutative proof net with conclusions Γ,A,Γ' and $\Phi(\mathfrak{D}_2)$ is a noncommutative proof net with conclusions $A^\perp,\Delta$.

We prove that $\Phi(\mathfrak{D})$ is good. - Let $\mathcal{L}_1$ be the unique assignment for $\Phi(\mathfrak{D}_1)$, and $\mathcal{L}_2$ be the unique assignment for $\Phi(\mathfrak{D}_2)$; thus, $\mathcal{L}_1(A\downarrow) = x^C + a$ and $\mathcal{L}_2(A^\perp\downarrow) = x^D + b$, where C is a conclusion of $\Phi(\mathfrak{D}_1)$, D is a conclusion of $\Phi(\mathfrak{D}_2)$, $a,b \in \mathbb{Z}$; remark that $C \neq A$ or $D \neq A^\perp$ since **MNLL** is not degenerate. Consider the following function from the points of $\Phi(\mathfrak{D})$ to $\mathbb{L}(\Phi(\mathfrak{D}))$ (each point T of $\Phi(\mathfrak{D})$ is a point of $\Phi(\mathfrak{D}_1)$ or $\Phi(\mathfrak{D}_2)$):
- for every point T of $\Phi(\mathfrak{D}_1)$,

$$\mathcal{L}(T) = \begin{cases} \mathcal{L}_1(T)[x^A \,/\, x^D + b - 1]\,,\ \text{if } D \neq A^\perp \\ \mathcal{L}_1(T)[x^A \,/\, x^C + a + b - 1]\,,\ \text{if } D = A^\perp \text{ and } C \neq A \end{cases}$$

- for every point T of $\Phi(\mathfrak{D}_2)$,

$$\mathcal{L}(T) = \begin{cases} \mathcal{L}_2(T)[x^{A^\perp} \,/\, x^C + a]\,,\ \text{if } C \neq A \\ \mathcal{L}_2(T)[x^{A^\perp} \,/\, x^D + a + b]\,,\ \text{if } D \neq A^\perp \text{ and } C = A \end{cases}$$

It is easy to verify that $\mathcal{L}$ is a total assignment for $\Phi(\mathfrak{D})$, and so $\mathcal{L}$ is the unique assignment for $\Phi(\mathfrak{D})$.

We prove that $\Phi(\mathfrak{D})$ induces the order Γ,Δ,Γ' of the conclusions. Let $\mathcal{L}$ be the unique assignment for π, defined above. If B,B' are two consecutive formulas in Γ (or in Γ'), then by induction hypothesis $\mathcal{L}_1(B\downarrow) = x^{B'} + 1$, and so $\mathcal{L}(B\downarrow) = x^{B'} + 1$ i.e. $B \prec B'$ in $\Phi(\mathfrak{D})$; if B,B' are two consecutive formulas in Δ, then by induction hypothesis $\mathcal{L}_2(B\downarrow) = x^{B'} + 1$, and so $\mathcal{L}(B\downarrow) = x^{B'} + 1$ i.e. $B \prec B'$ in $\Phi(\mathfrak{D})$. Therefore, we have now to state:

1) if B is the last formula in Γ and D the first formula in Δ, then $B \prec D$ in $\Phi(\mathfrak{D})$

2) if B is the last formula in Γ and Δ is empty and C is the first formula in Γ', then $B \prec D$ in $\Phi(\mathfrak{D})$
3) if B is the last formula in Δ and C is the first formula in Γ', then $B \prec D$ in $\Phi(\mathfrak{D})$

Proof of 1). Since $B \prec A$ in $\Phi(\mathfrak{D}_1)$ i.e. $\mathcal{L}_1(B{\downarrow}) = x^A + 1$, and $A^\perp \prec D$ in $\Phi(\mathfrak{D}_2)$, by definition we have that $\mathcal{L}(B{\downarrow}) = x^D + 1$, i.e. $B \prec D$ in $\Phi(\mathfrak{D})$. - Proof of 2). Since $B \prec A$ in $\Phi(\mathfrak{D}_1)$ i.e. $\mathcal{L}_1(B{\downarrow}) = x^A + 1$, and $A^\perp$ is the unique conclusion of $\Phi(\mathfrak{D}_2)$ i.e. $\mathcal{L}_2(A^\perp{\downarrow}) = x^{A^\perp}$, and $A \prec C$ in $\Phi(\mathfrak{D}_1)$ i.e. $\mathcal{L}_1(A{\downarrow}) = x^C + 1$, by definition we have that $\mathcal{L}(B{\downarrow}) = x^C + 1 + 1 - 1 = x^C + 1$, i.e. $B \prec C$ in $\Phi(\mathfrak{D})$. - Proof of 3). Since $\mathcal{L}_2(B{\downarrow}) = x^{A^\perp}$, and $A \prec C$ in $\Phi(\mathfrak{D}_1)$ i.e. $\mathcal{L}_1(A{\downarrow}) = x^C + 1$, by definition we have that $\mathcal{L}(B{\downarrow}) = x^C$, i.e. $B \prec C$ in $\Phi(\mathfrak{D})$.

(iii) Let $\mathfrak{D}$ be a proof of $\vdash \Delta,\Gamma,\Delta'$ obtained by (cut,2) from a proof $\mathfrak{D}_1$ of $\vdash \Gamma,A$ and a proof $\mathfrak{D}_2$ of $\vdash \Delta,A^\perp,\Delta'$. By induction hypothesis $\Phi(\mathfrak{D}_1)$ is a noncommutative proof net with conclusions Γ,A and $\Phi(\mathfrak{D}_2)$ is a noncommutative proof net with conclusions $\Delta,A^\perp,\Delta'$. - The proof that $\Phi(\mathfrak{D})$ is good and induces the order Δ,Γ,Δ' of the conclusions is analogous to the proof of (ii).

(iv) Let $\mathfrak{D}$ be a proof of $\vdash \Gamma,A \otimes B,\Delta,\Gamma'$ obtained by $(\otimes,1)$ from a proof $\mathfrak{D}_1$ of $\vdash \Gamma,A,\Gamma'$ and a proof $\mathfrak{D}_2$ of $\vdash B,\Delta$. By induction hypothesis $\Phi(\mathfrak{D}_1)$ is a noncommutative proof net with conclusions Γ,A,Γ' and $\Phi(\mathfrak{D}_2)$ is a noncommutative proof net with conclusions B,Δ.

We prove that $\Phi(\mathfrak{D})$ is good. - Let $\mathcal{L}_1$ be the unique assignment for $\Phi(\mathfrak{D}_1)$, and $\mathcal{L}_2$ be the unique assignment for $\Phi(\mathfrak{D}_2)$; thus, $\mathcal{L}_1(A{\downarrow}) = x^C + a$ and $\mathcal{L}_2(B{\downarrow}) = x^D + b$, where C is a conclusion of $\Phi(\mathfrak{D}_1)$, D is a conclusion of $\Phi(\mathfrak{D}_2)$, $a,b \in \mathbb{Z}$. Consider the following function from the points of $\Phi(\mathfrak{D})$ to $\mathbb{L}(\Phi(\mathfrak{D}))$ (each point T of $\Phi(\mathfrak{D})$ is a point of $\Phi(\mathfrak{D}_1)$ or a point of $\Phi(\mathfrak{D}_2)$ or $A \otimes B{\downarrow}$ or ${\uparrow}A \otimes B$):
- for every point T of $\Phi(\mathfrak{D}_1)$, $\mathcal{L}(T) = \mathcal{L}_1(T)[x^A / x^{A \otimes B}]$
- for every point T of $\Phi(\mathfrak{D}_2)$,

$$\mathcal{L}(T) = \begin{cases} \mathcal{L}_2(T)[x^B / x^C + a]\ , \text{if } C \neq A \\ \mathcal{L}_2(T)[x^B / x^{A \otimes B} + a]\ , \text{if } C = A \end{cases}$$

- $\mathcal{L}(\uparrow A \otimes B) = x^{A \otimes B}$

- $\mathcal{L}(A \otimes B\downarrow) = \begin{cases} x^D + b, \text{ if } D \neq B \\ x^C + a + b \text{ , if } D = B \end{cases}$

It is easy to verify that $\mathcal{L}$ is a total assignment for $\Phi(\mathfrak{D})$, and so $\mathcal{L}$ is the unique assignment for $\Phi(\mathfrak{D})$.

We prove that $\Phi(\mathfrak{D})$ induces the order $\Gamma, A \otimes B, \Delta, \Gamma'$ of the conclusions. Let $\mathcal{L}$ be the unique assignment for π, defined above. If E and E' are two consecutive formulas in Γ (or in Γ'), then by induction hypothesis $E \prec E'$ in $\Phi(\mathfrak{D}_1)$ i.e. $\mathcal{L}_1(E\downarrow) = x^{E'} + 1$, so that $\mathcal{L}(B\downarrow) = x^{E'} + 1$ i.e. $E \prec E'$ in $\Phi(\mathfrak{D})$; if E and E' are two consecutive formulas in Δ, then by induction hypothesis $E \prec E'$ in $\Phi(\mathfrak{D}_2)$ i.e. $\mathcal{L}_2(E\downarrow) = x^{E'} + 1$, so that $\mathcal{L}(E\downarrow) = x^{E'} + 1$ i.e. $E \prec E'$ in $\Phi(\mathfrak{D})$. Therefore we have to state:

1) If E is the last formula in Γ, then $E \prec A \otimes B$ in $\Phi(\mathfrak{D})$
2) If D is the first formula in Δ, then $A \otimes B \prec E$ in $\Phi(\mathfrak{D})$
3) If C is the first formula in Γ' and Δ is empty, then $A \otimes B \prec C$ in $\Phi(\mathfrak{D})$
4) If E is the last formula in Δ, and C is the first formula in Γ', then $E \prec C$ in $\Phi(\mathfrak{D})$

Proof of 1). Since $E \prec A$ in $\Phi(\mathfrak{D}_1)$ i.e. $\mathcal{L}_1(E\downarrow) = x^A + 1$, by definition we have that $\mathcal{L}(E\downarrow) = x^{A \otimes B} + 1$, i.e. $E \prec A \otimes B$ in $\Phi(\mathfrak{D})$. - Proof of 2) Since $B \prec D$ i.e. $\mathcal{L}_2(B\downarrow) = x^D + 1$, by definition we have that $\mathcal{L}(A \otimes B\downarrow) = x^D + 1$, i.e. $A \otimes B \prec D$ in $\Phi(\mathfrak{D})$. - Proof of 3) Since $A \prec C$ in $\Phi(\mathfrak{D}_1)$ i.e. $\mathcal{L}_1(A\downarrow) = x^C + 1$, and B is the unique conclusion of $\Phi(\mathfrak{D}_2)$ i.e. $\mathcal{L}_2(B\downarrow) = x^B$, by definition we have that $\mathcal{L}(A \otimes B\downarrow) = x^C + 1$, i.e. $A \otimes B \prec C$ in $\Phi(\mathfrak{D})$. - Proof of 4) Since B is the first formula in $\Phi(\mathfrak{D}_2)$ i.e. $\mathcal{L}_2(E\downarrow) = x^B$, and $A \prec C$ in $\Phi(\mathfrak{D}_1)$ i.e. $\mathcal{L}_1(A\downarrow) = x^C + 1$, by definition we get that $\mathcal{L}(E\downarrow) = x^C + 1$, i.e. $E \prec C$ in $\Phi(\mathfrak{D})$.

(v) Let $\mathfrak{D}$ be a proof of $\vdash \Delta, \Gamma, A \otimes B, \Delta'$ obtained by (cut,2) from a proof $\mathfrak{D}_1$ of $\vdash \Gamma, A$ and a proof $\mathfrak{D}_2$ of $\vdash \Delta, B, \Delta'$. By induction hypothesis, $\Phi(\mathfrak{D}_1)$ is a noncommutative proof net with conclusions Γ, A and $\Phi(\mathfrak{D}_2)$ is a noncommutative proof net with conclusions Δ, B, Δ'. - The proof that

$\Phi(\mathfrak{D})$ is good and induces the order $\Delta,\Gamma,A\otimes B,\Delta'$ of the conclusions is analogous to the proof given in (iv).

(vi) Let $\mathfrak{D}$ be a proof of $\vdash\Gamma,A\wp B,\Delta$ obtained by ($\wp$) from a proof $\mathfrak{D}'$ of $\vdash\Gamma,A,B,\Delta$. By induction hypothesis, $\Phi(\mathfrak{D}')$ is a noncommutative proof net with conclusions Γ,A,B,Δ.

We prove that $\Phi(\mathfrak{D})$ is good. - Let $\mathfrak{L}'$ be the unique assignment for $\Phi(\mathfrak{D}')$; thus, $\mathfrak{L}'(A\downarrow)=x^B+1$ (since $A\prec B$ in $\Phi(\mathfrak{D}')$) and $\mathfrak{L}'(B\downarrow)=x^D+b$ where D is a conclusion of $\Phi(\mathfrak{D}')$ and $b\in\mathbb{Z}$. Consider the following function from the points of $\Phi(\mathfrak{D})$ to $\mathbb{L}(\Phi(\mathfrak{D}))$ (each point T of $\Phi(\mathfrak{D})$ is a point of $\Phi(\mathfrak{D}')$ or $A\wp B\downarrow$ or $\uparrow A\wp B$):
- for every point T of $\Phi(\mathfrak{D}')$, $\mathfrak{L}(T)=\mathfrak{L}'(T)[x^A\ /\ x^{A\wp B}]$
- $\mathfrak{L}(\uparrow A\wp B)=x^{A\wp B}$

- $\mathfrak{L}(A\wp B\downarrow)=\begin{cases}x^D+b, \text{ if } D\neq A\\ x^{A\wp B}+b\text{ , if } D=A\end{cases}$

It is easy to verify that $\mathfrak{L}$ is a total assignment for $\Phi(\mathfrak{D})$, and so $\mathfrak{L}$ is the unique assignment for $\Phi(\mathfrak{D})$.

We prove that $\Phi(\mathfrak{D})$ induces the order $\Gamma,A\wp B,\Delta$ of the conclusions. Let ℓ be the unique assignment for π, defined above. If E and E' are two consecutive formulas in Γ (or in Δ), then by induction hypothesis $E\prec E'$ in $\Phi(\mathfrak{D}')$ i.e. $\mathfrak{L}'(E\downarrow)=x^{E'}+1$, and thus $\mathfrak{L}(E\downarrow)=x^{E'}+1$ i.e. $E\prec E'$ in $\Phi(\mathfrak{D})$. Therefore, we have to state:
1) If E is the last formula in Γ, then $E\prec A\wp B$ in $\Phi(\mathfrak{D})$
2) If D is the first formula in Δ, then $A\wp B\prec D$ in $\Phi(\mathfrak{D})$.

Proof of 1). Since $E\prec A$ in $\Phi(\mathfrak{D}')$ i.e. $\mathfrak{L}'(E\downarrow)=x^{A'}+1$, by definition we have that $\mathfrak{L}(E\downarrow)=x^{A\wp B}+1$, i.e. $E\prec A\wp B$ in $\Phi(\mathfrak{D})$. - Proof of 2). Since $B\prec D$ in $\Phi(\mathfrak{D}')$ i.e. $\mathfrak{L}'(B\downarrow)=x^D+1$, by definition we have that $\mathfrak{L}(A\wp B\downarrow)=x^D+1$, i.e. $A\wp B\prec D$ in $\Phi(\mathfrak{D})$.

4.1. Theorem

If π is a noncommutative proof net with conclusions Γ, then there is a proof $\mathfrak{D}$ of the sequent $\vdash\Gamma$ in **MNLL** s.t $\pi=\Phi(\mathfrak{D})$.

Proof.

By induction on the number of links in π.

If π is a noncommutative proof net with only one link, so that π must be of this form:

$$\overline{A^{\perp} \qquad A}$$

so, π comes from the proof $\vdash A^{\perp}, A$ in **M*N*LL**.

Suppose that the theorem holds for every noncommutative proof net with at most k links. Let π be a noncommutative proof net with $k+1$ links and with conclusions Γ. We distinguish two cases (as in the sequentialization theorem for commutative proof nets, cf. [Gir87]):

(i) π has a terminal $\wp$-link

(ii) π has no terminal $\wp$-link.

(i) π has a terminal $\wp$-link whose conclusion is $A\wp B$, and let the order Γ of the conclusions of π be $\Delta_1, A\wp B, \Delta_2$; thus π can be represented as

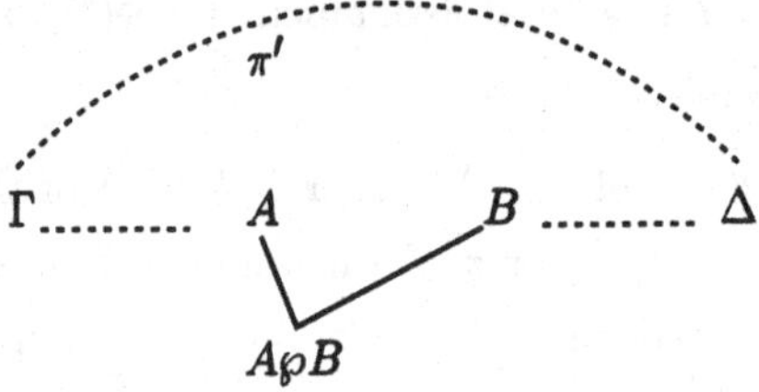

where we know (from [Gir87]) that $\mathrm{Tr}(\pi')$ is a commutative proof net.

We shall prove below that π' is a noncommutative proof net with conclusions Δ_1, A, B, Δ_2. Therefore, by induction hypothesis, $\pi' = \Phi(\mathfrak{D}')$, where $\mathfrak{D}'$ is a proof of $\vdash \Delta_1, A, B, \Delta_2$ in **MNLL**, and then by definition $\pi = \Phi(\mathfrak{D})$ where $\mathfrak{D}$ is the proof of $\vdash \Delta_1, A\wp B, \Delta_2$ obtained by ($\wp$) from $\mathfrak{D}'$:

$$\dfrac{\begin{array}{c}\mathfrak{D}' \\ \vdots \\ \vdash \Delta_1, A, B, \Delta_2\end{array}}{\vdash \Delta_1, A\wp B, \Delta_2}(\wp)$$

In order to prove that π' is a noncommutative proof net with conclusions Δ_1, A, B, Δ_2, we must state: 1) π' is good; 2) π' induces the order Δ_1,A,B,Δ_2 of the conclusions.

Proof of 1). - Take the unique assignment $\mathcal{L}$ for π, and define the following function $\mathcal{L}'$ from the points of π' to $\mathbb{L}(\pi')$: $\mathcal{L}'(T) = \mathcal{L}(T)[x^{A\wp B}/x^A]$; it is easy to prove that $\mathcal{L}'$ is an assignment and is total, so that $\mathcal{L}'$ is the unique assignment for π'.

Proof of 2). - Take the unique assignment $\mathcal{L}$ for π and the unique assignment $\mathcal{L}'$ for π'. - If E and E' are consecutive formulas in Δ_1 (or in Δ_2), then $E \prec E'$ in π i.e. $\mathcal{L}(E\downarrow) = x^{E'} + 1$, so that $\mathcal{L}'(E\downarrow) = x^{E'} + 1$, i.e. $E \prec E'$ in π'; $A \prec B$ in π', since $\mathcal{L}(A\downarrow) = x^B + 1$ and $\mathcal{L}'(A\downarrow) = \mathcal{L}(A\downarrow)$. We have to state:

2.1) If E is the last formula in Δ_1, then $E \prec A$ in π'

2.2) If E is the first formula in Δ_2, then $B \prec E$ in π'.

Proof of 2.1). Since $E \prec A\wp B$ in π i.e. $\mathcal{L}(E\downarrow) = x^{A\wp B} + 1$, by definition $\mathcal{L}(E\downarrow) = x^A + 1$, i.e. $E \prec A$ in π'. - Proof of 2.2). Since $A\wp B \prec E$ in π i.e. $\mathcal{L}(A\wp B\downarrow) = x^E + 1$, and because $\mathcal{L}(B\downarrow) = \mathcal{L}(A\wp B\downarrow)$, by definition we have that $\mathcal{L}'(B\downarrow) = x^E + 1$, i.e. $B \prec E$ in π'.

(iii) π has $k+1$ links, without terminal $\wp$-links and with conclusions Γ. By *splitting theorem* proved in [Gir87], there is at least one *splitting* $\otimes$-link or Cut-link in Tr(π). Suppose the splitting link is a $\otimes$-link with conclusion $A \otimes B$ (the case in which the splitting link is a Cut-link is very analogous to this case and is left to the reader). So, the order Γ of the conclusions of π is Θ_1,$A \otimes B$,Θ_2 and π can be represented as

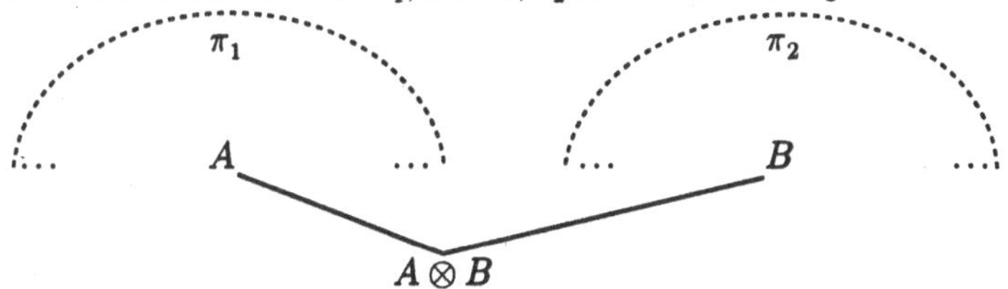

where Tr(π_1) and Tr(π_2) are commutative proof nets and π_1 and π_2 are connected only through the indicated $\otimes$-link.

We shall prove that:

1) In the order Θ_1,$A \otimes B$,Θ_2 of the conclusions induced by π, $\Theta_1 = \Gamma_2,\Gamma_1$ and $\Theta_2 = \Delta_2,\Delta_1$ where Γ_2 and Δ_2 are conclusions of π_2 and Γ_1 and Δ_1 are

conclusions of π_1, and Γ_2 is empty or Δ_1 is empty;
2) π_1 is a noncommutative proof net with conclusions Γ_1,A,Δ_1;
3) π_2 is a noncommutative proof net with conclusions Γ_2,B,Δ_2.
Therefore, since π_1 and π_2 are noncommutative proof nets with conclusions Γ_1,A,Δ_1, resp. Γ_2,B,Δ_2, by induction hypothesis $\pi_1 = \Phi(\mathfrak{D}_1)$ where $\mathfrak{D}_1$ is a proof in **MNLL** of the sequent $\vdash\Gamma_1,A,\Delta_1$ and $\pi_2 = \Phi(\mathfrak{D}_2)$ where $\mathfrak{D}_2$ is a proof in **MNLL** of the sequent $\vdash\Gamma_2,B,\Delta_2$. Moreover, as stated in 1) above, Δ_1 is empty or Γ_2 is empty, so that from $\mathfrak{D}_1$ and $\mathfrak{D}_2$ by applying $(\otimes,2)$ (in the case Δ_1 is empty) or $(\otimes 1)$ (in the case Γ_2 is empty) we get a proof $\mathfrak{D}$ of the sequent $\vdash\Gamma_2,\Gamma_1,A\otimes B,\Delta_2,\Delta_1$, and so by definition $\pi = \Phi(\mathfrak{D})$.

Proof of 1).

The result follows easily from the following statement:

Let C be a conclusion of π and a conclusion of π_1. If $C \prec D$ in π, then D is a conclusion of π_1 or $D = A\otimes B$. If C is the last conclusion of π and D is the first conclusion of π, then D is a conclusion of π_1 or $D = A\otimes B$.

We prove this statement ($\mathfrak{L}$ being the unique assignment for π).

Suppose $C \prec D$ in π. So, $\mathfrak{L}(C\downarrow) = x^D+1$, and (since π_1 is connected with π_2 only through the indicated $\otimes$-link) the labeled special trip in π looks as follows

$$\ldots,(x^{A\otimes B})A\otimes B,(x^{A\otimes B})A,\ldots C(x^D+1),(x^C)C,\ldots At,\ldots$$

If D is not a conclusion of π_1 and is not $A\otimes B$, then D does not occur between $(x^{A\otimes B})A\otimes B$ and At (since D does not belongs to π_1) and so the labeled special trip from the conclusion D is

$$(x^D)D,\ldots,A\otimes Bs,(x^{A\otimes B})A\otimes B,(x^{A\otimes B})A,\ldots C(x^D+1),(x^C)C,\ldots At,\ldots$$

and thus between $(x^D)D$ and $C(x^D+1)$ there is a conclusion, i.e. $A\otimes B$: contradiction with the fact that π is a noncommutative proof net (lemma 3.9).

Suppose C be the last conclusion of π and D the first conclusion of π. Then $\mathfrak{L}(C\downarrow) = x^D$: the argument leading to the fact that D is a conclusion of π_1 or $A \otimes B$ is analogous to the previous one.

Proof of 2)

First, we prove that π_1 is good. Let $\mathcal{L}$ be the unique assignment for π. Define the following function $\mathcal{L}_1$ from the points of π_1 to $\mathbb{L}(\pi_1)$: $\mathcal{L}(T) = \mathcal{L}(T)[x^{A \otimes B}/x^A]$. Then it is easy to prove that $\mathcal{L}_1$ is a total assignment for π_1 (using the fact that the portion of the special trip in π from $\uparrow A$ to $A\downarrow$ is a special trip in π_1).

Now, we prove that π_1 induces the order Γ_1, A, Δ_1 of the conclusions. Let $\mathcal{L}_1$ be the unique assignment for π_1. If E and E' are two consecutive formulas in Γ_1 (or in Δ_1), then $E \prec E'$ in π, and thus $E \prec E'$ in π_1 since $\mathcal{L}(E\downarrow) = \mathcal{L}_1(E\downarrow)$. We have to state:
2.1) If E is the last formula in Γ_1, then $E \prec A$ in π_1
2.2) If E is the first formula in Δ_1, then $A \prec E$ in π_1.

Proof of 2.1). Since $E \prec A \otimes B$, i.e. $\mathcal{L}(E\downarrow) = x^{A \otimes B} + 1$, by definition we have that $\mathcal{L}_1(E\downarrow) = x^A + 1$, i.e. $E \prec A$ in π_1.

Proof of 2.2). Let $\mathcal{L}_1(A\downarrow) = x^C + a$ for some conclusion C of π_1 and some $a \in \mathbb{Z}$. First, suppose Γ_1 be not empty: the first formula F of Γ_1 is the first conclusion of π and the last formula G of Δ_1 is the last conclusion of π, i.e. $\mathcal{L}(G\downarrow) = x^F$, and so $\mathcal{L}_1(G\downarrow) = x^F$; moreover, by above, $D \prec D'$ for every pair of consecutive formulas in Γ_1, A and in Δ_1; so, by property of the labeled special trip, C must be just the first formula E of π_1, and a must be 1 (because $\mathcal{L}_1(D\downarrow) = x^{D'} + 1$ for every pair of consecutive formulas D, D' in Γ_1, A and in Δ_1, and $\mathcal{L}_1(G\downarrow) = x^F$ for G last formula in Δ_1 and F first formula in Γ_1, and then apply corollary 3.6). - Now suppose Γ_1 empty: $A \otimes B$ is the first conclusion of π and the last formula G of Δ_1 is the last conclusion of π, i.e. $\mathcal{L}(G\downarrow) = x^{A \otimes B}$, and so $\mathcal{L}_1(G\downarrow) = x^A$; moreover, by above, $D \prec D'$ for every pair of consecutive formulas in Δ_1; so, by property of the labeled special trip, C must be just the first formula E of π_1, and a must be 1 (because $\mathcal{L}_1(D\downarrow) = x^{D'} + 1$ for every pair of consecutive formulas D, D' in Δ_1, and $\mathcal{L}_1(G\downarrow) = x^A$ for G last formula in Δ_1, and then apply corollary 3.6).

Proof of 3)

First, we prove that π_2 is good. Let $\mathcal{L}$ be the unique assignment for π. Define the following function $\mathcal{L}_2$ from the points of π_2 to $\mathbb{L}(\pi_2)$:

$$\mathcal{L}_2(T) = \begin{cases} \mathcal{L}(T)[x^C/x^B - 1], \text{ if } \Delta_1 \text{ non empty and } A \prec C \text{ in } \pi_1 \\ \mathcal{L}(T)[x^C/x^B], \text{ if } \Delta_1 \text{ empty and } C \text{ is the first formula in } \Gamma_1, A \otimes B \end{cases}$$

Then it is easy to prove that $\mathcal{L}_2$ is a total assignment for π_2 (from the fact that the portion of the special trip in π from $\uparrow B$ to $B\downarrow$ is a special trip in π_2).

Now, we prove that π_2 induces the order Γ_2, B, Δ_2 of the conclusions. Let $\mathcal{L}_2$ be the unique assignment for π_2. If E and E' are two consecutive formulas in Γ_2 (or in Δ_2), then $E \prec E'$ in π, and so $E \prec E'$ in π_2 (since $\mathcal{L}(E\downarrow) = \mathcal{L}_2(E\downarrow)$). So, we have to state:
3.1) if E is the last formula in Γ_2, then $E \prec B$ in π_2
3.2) if E is the first formula in Δ_2, then $B \prec E$ in π_2.

Proof of 3.1). Since E is the last formula in Γ_2, and so Γ_2 is non empty, Δ_1 is empty and A is the last conclusion of π_1; thus, $E \prec C$ in π, where C is the first formula in $\Gamma_1, A \otimes B$, i.e. $\mathcal{L}(E\downarrow) = x^C + 1$, so that $\mathcal{L}_2(E\downarrow) = x^B + 1$ i.e. $E \prec B$ in π_2. - Proof of 3.2). Since $A \otimes B \prec E$ in π, i.e. $\mathcal{L}(A \otimes B\downarrow) = x^E + 1$, and $\mathcal{L}(B\downarrow) = \mathcal{L}(A \otimes B\downarrow)$, by definition we have that $\mathcal{L}_2(B\downarrow) = x^E + 1$ i.e. $B \prec E$ in π_2.

References

[Abr91] V.Michele Abrusci, *Sequent calculus and phase semantics for pure noncommutative propositional linear logic*, JSL, 1991.
[Gir87] Jean-Yves Girard, *Linear Logic*, TCS, 1987.

Volume of multiplicative formulas and provability

François Métayer*

Abstract

In [Mét], we introduced a homological condition of correctness for paired-graphs. As an application of this result, we establish here new conditions of provability in multiplicative linear logic, by defining the volume of a formula. We also extend the complexity results of [Kan] and [LW] to abstract graphs.

1 Introduction

This paper introduces a correspondance between modules with a given border (see sect.2) and subspaces of an euclidian space, in such a way that the correct pasting of two modules M_1 and M_2 along their common border can be expressed by an equation

$$a(X_1, X_2^{\perp}) = \lambda$$

Here X_i denotes the subspace associated with M_i, a is a geometrical invariant of a pair of subspaces, and λ is a constant depending on M_1 and M_2 as separate modules, not on the way they are pasted together.

Since this interpretation stems from the homological correctness criterion we introduced in [Mét], we first recall the main definitions and results of this paper. Let G be a graph. A *pair* of G is a pair of edges having a unique vertex in common. A *paired-graph* (pg) is an ordered pair $(G, \mathcal{P})$ where G is a

*Équipe de Logique, Université Paris VII–Cnrs, 45-55, 5ème étage, 2 place Jussieu 75251 Paris Cedex 05 France. e-mail: metayer@logique.jussieu.fr

graph and $\mathcal{P} = \mathcal{P}(G)$ is a set of mutually disjoint pairs of G. The motivation for considering this notion is that proof-structures in multiplicative linear logic (see [Girl]) can be seen as pg's, where pairs correspond to *par*-links. We denote by $\mathcal{V}(G)$ (resp. $\mathcal{E}(G)$) the set of vertices (resp. edges) of G. We call an edge a *paired-edge* if it belongs to a pair, and a *free* edge otherwise. Now a *morphism* $f : G \longrightarrow G'$ is a map $\mathcal{V}(G) \longrightarrow \mathcal{V}(G')$, $x \mapsto x'$, such that, if uv is a free edge, $u'v'$ is free, or $u' = v'$, and, if $\{uw, vw\}$ is a pair, then $\{u'w', v'w'\}$ is a pair, or $u' = v' = w'$. Abstract proof-nets can be defined in this setting: let G_1, G_2 be pg's, $u \in \mathcal{V}(G_1)$, $v \in \mathcal{V}(G_2)$. We denote by $\mathrm{t}(G_1, G_2, u, v) = G_1 \coprod G_2/u \sim v$ the graph obtained by identifying u and v in the disjoint reunion of G_1 and G_2. Likewise, if G is a pg with distinct vertices u and v, we denote by $\mathrm{p}(G, u, v)$ the graph obtained by adjoining a new vertex w, and a new pair $\{uw, vw\}$ (fig.1).

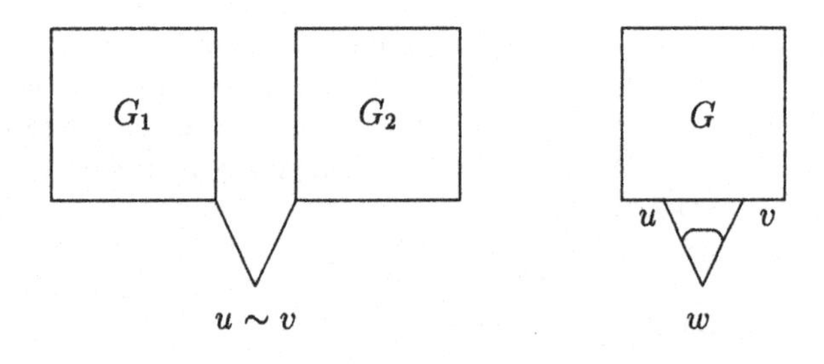

fig.1

The smallest class of pg's containing trees (connected, acyclic graphs, without pairs), and closed by t and p is the class of *proof-nets*. The starting point of [Mét] is the following remark: a paired-graph G, just like an ordinary one, gives rise to a complex of abelian groups:

$$0 \longrightarrow C_1(G) \xrightarrow{\partial} C_0(G) \xrightarrow{\epsilon} \mathbf{Z} \longrightarrow 0$$

where $C_0(G) = \mathbf{Z}[\mathcal{V}(G)]$ and $C_1(G)$ is the subgroup of $\mathbf{Z}[\mathcal{E}(G)]$ generated by the free edges and the elements $e + e^*$ where e runs over paired edges. ∂ is the restriction to $C_1(G)$ of the boundary morphism defined by $\partial(uv) = v - u$, and ϵ is the *augmentation* morphism defined by $\epsilon(u) = 1$ for each vertex u. Notice that edges have to be oriented in order to define ∂. The only requirement is that, in each pair, both edges point towards their common vertex, or both in the opposite direction. Since $\epsilon\partial = 0$ we can define homology groups $H_0(G) = \ker \epsilon / \mathit{im}\, \partial$ and $H_1(G) = \ker \partial$. Morphisms have been defined such that each $f : G \longrightarrow G'$ determines morphisms of groups $f_*^i : H_i(G) \longrightarrow H_i(G')$, for

$i = 1, 2$, making H_i a functor from paired-graphs to abelian groups. As an example, the reader may check that if $\mathcal{V}(G) = \{u, v, w\}$, $\mathcal{E}(G) = \{uw, vw, uv\}$ and $\mathcal{P}(G) = \{\{uw, vw\}\}$, then $H_1(G) = 0$ and $H_0(G) = \mathbf{Z}/2\mathbf{Z}$ (fig.2).

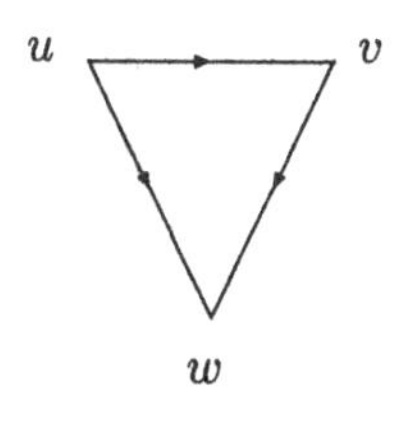

fig.2

The relevance of this homology lies in the fact that it characterizes proofnets among arbitrary paired-graphs, as shown by the main result of [Mét]:

Theorem 1.1 *G is a proofnet if and only if $H_1(G) = 0$ and $card\, H_0(G) = 2^p$, where $p = card\, \mathcal{P}(G)$.*

It should be pointed out that, for *each* graph G, not necessarily a proofnet, satisfying $H_1(G) = 0$ and $H_0(G)$ finite, the cardinal of $H_0(G)$ cannot be greater than $2^{card\, \mathcal{P}(G)}$. Finally, if G is a paired-graph and K a subgraph of G we may define *relative homology groups* $H_i(G, F)$ for $i = 1$, 2, as usual: with the above notations, we denote by $C_i(G, K)$ the factor group $C_i(G)/C_i(K)$. Now ∂ induces $\overline{\partial}$:

$$0 \longrightarrow C_1(G, K) \xrightarrow{\overline{\partial}} C_0(G, K) \longrightarrow 0$$

so that $H_0(G, K) = C_0(G, K)/\, im\, \overline{\partial}$ and $H_1(G, K) = \ker \overline{\partial}$.

2 Modules

Let M, N, and F be paired-graphs, and $F \xrightarrow{m} M$, $F \xrightarrow{n} N$ be injective morphisms. $M *_F N$ is defined, up to isomorphism, by the following pushout diagram:

$$\begin{array}{ccc} F & \xrightarrow{m} & M \\ \downarrow n & & \downarrow i \\ N & \xrightarrow{j} & M *_F N \end{array}$$

We only consider the case where (1) $M \cap N = F$ (as subgraphs of $M *_F N$) and (2) F is a set of vertices. As in [Tro], we say that M and N are *connectable* along F if $M *_F N$ is a proofnet (fig.3). We are asking under which conditions two graphs M and N are connectable.

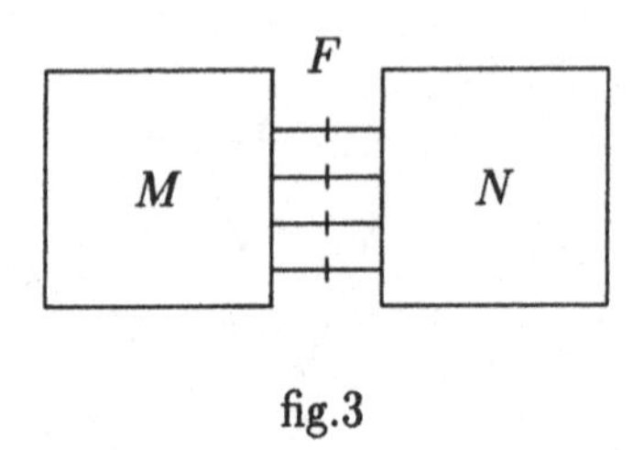

fig.3

In fact, we can restrict our attention to *modules*, in the following sense:

Definition 2.1 *A* module *is an ordered pair (M, F) where M is a graph and F is a subgraph of M consisting of vertices only, such that $H_1(M) = 0$ and $H_0(M, F)$ is a finite group. We call F the* border *of the module.*

It should be noticed that 2.1 is not strictly equivalent with the definition of [Tro] but it is better suited to our homological approach. Now the above injections m and n give rise to morphisms

$$H_0(F) \xrightarrow{m_*} H_0(M)$$

and

$$H_0(F) \xrightarrow{n_*} H_0(N)$$

If F has $k+1$ vertices, we get $H_0(F) \cong \mathbf{Z}^k$ and a Mayer-Vietoris exact sequence

$$0 \longrightarrow \mathbf{Z}^k \xrightarrow{\phi} H_0(M) \oplus H_0(N) \xrightarrow{i_*+j_*} H_0(G) \longrightarrow 0 \quad (1)$$

By using the homological correctness criterion of [Mét], we can show:

Proposition 2.2 *Let (M, F) and (N, F) be two modules, and π_M (resp. π_N) the number of pairs in M (resp. N). They are connectable if and only if*

$$card\, H_0(M,F) \times card\, H_0(N,F) \times [H_0(F) : \ker m_* \oplus \ker n_*] = 2^{\pi_M + \pi_N} \quad (2)$$

Notice that (2) implicitely contains the conditions: $\ker m_* \cap \ker n_* = 0$ and: $rank(\ker m_* \oplus \ker n_*) = rank\, H_0(F)$.

To each multiplicative formula A we associate as usual a binary tree M, which can be seen as a paired-graph where pairs represent the *par* connectives of A. The vertices of M correspond to the subformulas: in particular, vertices of degree 1 are associated to the atomic subformulas, and form a subgraph F. There is a unique vertex of degree 2, which is associated to the whole formula and is the *root* of M. Then (M, F) is a module. Such an M will be called a *graph of formula* (gf). The number of vertices in the border F is the *size* of M and will be denoted by τ_M. Clearly, if A is provable, then $\tau_M = 2k$.

Let M be a gf of size $2k$. It is called *provable* if and only if we can obtain a proofnet by adjoining k disjoint edges having their vertices in F. In other words, if N_k denotes the graph of k disjoint edges, M is provable if and only if there is an injection $n : F \longrightarrow N_k$ such that (M, F) and (N_k, F) are connectable. As a consequence of the finiteness of $[H_0(F) : \ker m_* \oplus \ker n_*]$ in (2), we find again a well-known condition of provability:

Proposition 2.3 *If M is provable, $\tau_M = 2\pi_M$.*

Before we turn to more interesting conditions, we briefly investigate the complexity of the decision problem for gf's.

3 Complexity

We sketch here the proof of the following result:

Theorem 3.1 *The decision problem for graphs of formulas is NP-complete.*

Kanovich has proved NP-completeness for the full propositional fragment of multiplicative linear logic [Kan], and the same is true for the neutral fragment, as shown by Lincoln and Winkler [LW]. We reduce our problem to the latter in two steps.

Step 1: encoding of the neutrals in the fragment $L(a)$, whose formulas are builded with literals a and $a^\perp$ only. We translate neutrals in $L(a)$ as follows:

- $1^\circ = a \wp a^\perp$ and $\perp^\circ = a \otimes a^\perp$.
- For all formulas A and B, $(A \wp B)^\circ = A^\circ \wp B^\circ$ and $(A \otimes B)^\circ = A^\circ \otimes B^\circ$.
- For each sequent $\Gamma = A_1, \ldots, A_p$, $\Gamma^\circ = A_1^\circ, \ldots, A_p^\circ$.

Then

Lemma 3.2 *The translation $A \longrightarrow A^\circ$ is sound and faithful.*

Step 2: encoding of $L(a)$ in (abstract) gf's.

For each formula $A \in L(a)$, we denote its graph by M_A. Let $B = (a \wp a) \wp (a \wp a)$ and $A^* = A[B/a; B^\perp/a^\perp]$. Then we define $T_A = M_{A^*}$. Recall that a formula is *balanced* if it has the same number of occurences of a and $a^\perp$. We can prove

Lemma 3.3 *A balanced formula A is provable in $L(a)$ if and only if the gf T_A is provable.*

Proof. Suppose first that A is provable (hence balanced). Then A^* is provable, as well as its gf, which is precisely T_A.

Suppose conversely that A is a balanced formula such that $T = T_A$ is provable.

If $M = M_A$, $\tau_M = 2k$ hence $\tau_T = 8k$. Let $F^1 = F_{8k}$ be the border of T and $N^1 = N_{4k}$. By hypothesis, there is an injection $F^1 \xrightarrow{n} N^1$ such that $G = T *_{F^1} N^1$ is a proofnet.

Thus $2\pi_T = \tau_T = 8k$ but $\pi_M = \pi_T - 3k = k$ hence $H_0(M) \cong \mathbf{Z}^k$. On the other hand $G = M *_F N$ where F is the border of M and $N = G \setminus M$, which is obtained by pasting together N^1 with k copies of M_B and k copies of $M_B^\perp$ along F^1. If we now chose a switching σ *of* N we know that $G^\sigma = M *_F N^\sigma$ is again a proofnet. By (1) we get

$$rank(H_0(M)) + rank(H_0(N^\sigma)) = rank(H_0(F)) = 2k - 1$$

hence also $rank(H_0(N^\sigma)) = k - 1$. Then N^σ is an ordinary graph having k connected components.

Now the vertices of F^1 are distibuted in sets of four vertices, according to the graphs M_B or $M_B^\perp$ where they belong. We denote these sets by

$$X_i = \{a_{i1}, a_{i2}, a_{i3}, a_{i4}\} \quad for \quad i = 1, \ldots, k$$

$$Y_j = \{a_{j1}^\perp, a_{j2}^\perp, a_{j3}^\perp, a_{j4}^\perp\} \quad for \quad j = 1, \ldots, k$$

where X_i (resp. Y_j) is the border of $M_i \cong M_B$ (resp. $M_j^\perp \cong M_B^\perp$). s_i (resp. $s_j^\perp$) will be the root of M_i (resp. $M_j^\perp$).

Let $X = \bigcup_i X_i$ and $Y = \bigcup_i Y_i$. The edges of N^1 induce a partition of $X \cup Y$ in pairs. Two vertices of Y cannot belong to the same pair, otherwise G has a cycle in contradiction with $H_1(G) = 0$.

By $card(X) = card(Y) = 4k$, all pairs consist of a vertex of X and a vertex of Y. This gives a bijective mapping $\phi : X \longrightarrow Y$. Consider now $\Phi : \{1, \ldots, k\} \longrightarrow \mathcal{P}(\{1, \ldots, k\})$ which associates to each index i the set $\{j / \phi^\bullet(X_i) \cap Y_j \neq \emptyset\}$. Let I be any subset of $\{1, \ldots, k\}$, we verify that

$$card(\bigcup_{i \in I} \Phi(i)) \geq card(I)$$

Indeed, if $C = \bigcup_{i\in I} \Phi(i)$:

$$\phi^\bullet(\bigcup_{i\in I} X_i) \subset \bigcup_{j\in C} Y_j$$

Since X_i and Y_j are pairwise disjoint sets of four elements and ϕ is bijective, the cardinals in the above inclusion are respectively $4 \times card(I)$ and $4 \times card(C)$. This proves the inequality.

Now the wedding lemma (see [Hal, p.48]) applies, giving an injective—hence bijective—ψ from $\{1, \ldots, k\}$ in itself such that, for all i, $\psi(i) \in \Phi(i)$. We can chose in each X_i a vertex x_i such that $\phi(x_i) \in Y_{\psi(i)}$. Take in each M_i the switching connecting s_i to x_i (*in* M_i). It induces a switching σ_0 of N such that for all $i \in \{1, \ldots, k\}$, s_i and $s^\perp_{\psi(i)}$ are in the same connected component of N^{σ_0}. But this graph has k connected components; each one can be replaced by a unique edge $s_i s^\perp_{\psi(i)}$ (fig.4). The result is a proof of A.◇

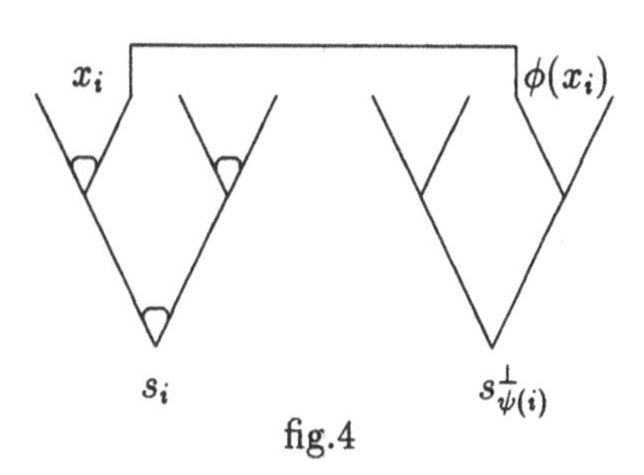

fig.4

We get 3.1 as an immediate consequence. The problem belongs to NP: given (M, F) of size $2k$ and $F \xrightarrow{n} N_k$, the correctness of $M *_F N_k$ is decidable in polynomial time (in k). As regards completeness, the composition of the two translations $A \longrightarrow A^\circ \longrightarrow T_{A^\circ}$ reduces polynomially the decision problem for neutrals to the problem for gf's. But the former is already NP-complete.

4 Volume of modules and formulas

The complexity result we have just proved opposes a simple characterization of provable gf's. It is however possible to improve 2.3, as we shall see. Let (M, F) be a module, with $F = \{s_1, \ldots, s_l\}$ and $F \xrightarrow{m} M$ the inclusion morphism. Recall that $H_0(F) \cong \mathbf{Z}^{l-1}$, and if $rank\, H_0(M) = p$, then $\ker m_* \cong \mathbf{Z}^r$ where $r = l - 1 - p$.

Now $A = (s_1, \ldots, s_l)$ can be seen as the canonical basis of $E = \mathbf{R}^l$ so that $H_0(F)$ becomes a discrete subgroup of the hyperplane:

$$L: \quad x_1 + \cdots + x_l = 0$$

If $e_i = s_i - s_l$, then $\mathcal{B} = (e_i)_{1 \leq i \leq l-1}$ is a basis of L. If we provide E with its canonical (w.r.t A) scalar product $\langle , \rangle$, the *volume* of a family of vectors $v_1, \ldots, v_p$ is well defined: it will be denoted by $\|v_1, \ldots, v_p\|$ (see [Ber]). In particular, if $(m_1, \ldots, m_r)$ is a $\mathbf{Z}$-basis of $\ker m_*$, $\|m_1, \ldots, m_r\|$ only depends on M and F, so that we may define:

Definition 4.1 *The* volume *of M, denoted by $\|M\|$ is the volume of any $\mathbf{Z}$-basis of* $\ker m_*$.

Let M_1 and M_2 be two modules with common border $F = \{s_1, \ldots, s_l\}$ and $F \xrightarrow{m^i} M_i$ the inclusion morphism. Let $p_i = \pi_{M_i}$ and $d_i = card(H_0(M_i, F))$. By 2.2 M_1 and M_2 are connectable if and only if

$$[H_0(F) : \ker m_*^1 \oplus \ker m_*^2] = \frac{2^{p_1+p_2}}{d_1 d_2} \tag{3}$$

If $(m_1^1, \ldots, m_{r_1}^1)$ and $(m_1^2, \ldots, m_{r_2}^2)$ are $\mathbf{Z}$-bases of $\ker m_*^1$ and $\ker m_*^2$ respectively, then $r_1 + r_2 = l - 1$ and the left member of (3) is

$$\left| det_{\mathcal{B}}(m_1^1, \ldots, m_{r_1}^1, m_1^2, \ldots, m_{r_2}^2) \right| = \frac{\|m_1^1, \ldots, m_{r_1}^1, m_1^2, \ldots, m_{r_2}^2\|}{\|\mathcal{B}\|}$$

and we may rewrite (3) as

$$\|m_1^1, \ldots, m_{r_1}^1, m_1^2, \ldots, m_{r_2}^2\| = \frac{2^{p_1+p_2}\sqrt{l}}{d_1 d_2}$$

On the other hand

$$\|m_1^1, \ldots, m_{r_1}^1, m_1^2, \ldots, m_{r_2}^2\| = a \, \|M_1\| \|M_2\|$$

where a only depends on the vector spaces X_1 and X_2 generated in L by $\ker m_*^1$ and $\ker m_*^1$ respectively.

The precise value of a is defined as follows: if Y and Z are subspaces of an euclidian space and if p_Y denotes the orthogonal projector on Y, we may consider $p_Y \circ p_Z$ as an endomorphism u of Y—by restriction to Y. Then $det\, u \geq 0$ and if

$$a(Y, Z) = \sqrt{det\, u}$$

we have

$$a = a(X_1, X_2^{\perp}) = a(X_2, X_1^{\perp})$$

Thus, with the above notations

Theorem 4.2 *The correctness of $M_1 *_F M_2$ is expressed by an equation:*

$$a(X_1, X_2^\perp) = \lambda$$

where λ only depends on M_1 and M_2 separately.

We now apply the previous results to the particular case of formulas. Here $M_1 = M$ is a gf of size $l = 2k$ and $M_2 = N_k = N$. Then $p_1 = \pi_M$—and we may suppose that $p_1 = k$ by 2.3—$p_2 = 0$, $d_1 = 2^h$ and $d_2 = 1$. We also denote $m = m_1$, $n = m_2$, $X_1 = X$ and $X_2 = Y$.

A few calculations show that, with the above notations,

Proposition 4.3 *A gf (M, F) is provable if and only if there is an $n : F \longrightarrow N$ such that*

$$a(X_1, X_2^\perp)\|M\| = (\sqrt{2})^{k-2h+1}\sqrt{k} \tag{4}$$

Any bound $a(X, Y^\perp) \leq A$ independent of n gives a necessary condition of provability for M of the form:

$$\|M\| \geq \frac{(\sqrt{2})^{k-2h+1}\sqrt{k}}{A}$$

Consider for instance

$$A = (((a\wp(a \otimes a)) \otimes a)\wp(a \otimes a))\wp(((a^\perp\wp a^\perp)\wp(a^\perp\wp a^\perp)) \otimes (a^\perp \otimes a^\perp))$$

For $M = M_A$, according to 4.1, $\|M\| = 4\sqrt{33}$. Now formulas of the form $\Phi'\wp\Phi''$ where Φ' (resp. Φ'') has k atoms of type a (resp. $a^\perp$) verify

$$a(X, Y^\perp) \leq \left(\frac{1}{\sqrt{2}}\right)^{k-1}$$

Thus M can be provable only if

$$\|M\| \geq 2^{k-h}\sqrt{k}$$

but here $k = 6$ and $h = 2$, so that $2^{k-h}\sqrt{k} = 16\sqrt{6} > 4\sqrt{33}$. Hence M is not provable, although it satisfies $\tau_M = 2\pi_M$.

Let us examine more generally the problem of finding a proof of a given gf M of size $2k$, if it exists. In the euclidian space L, M defines a subspace X generated by $\ker m_*$. Each partition p of F in pairs determines an injection $n^p : F \longrightarrow N_k$, and consequently a subspace Y^p generated by $\ker n^p$. Now every correct choice (i.e. giving an actual proof) *maximizes* $a(X^\perp, Y^p)$ over all possible p's. Therefore, the decision problem for M amounts to calculate

$$\max_p a(X^\perp, Y^p)$$

and verify that this maximum satisfies (4) in 4.3.

The group S_{2k} of permutations of F can be seen as the subgroup A_{2k-1} (see [Hum, p.5]) of isometries of L generated by the reflections associated to tranpositions of vertices. It acts transitively on the above set of subspaces Y^p. If we chose an arbitrary $Y_0 = Y^p$, we must solve the optimization problem for

$$g \mapsto a(X^\perp, gY_0)$$

over A_{2k-1}. We may then ask if there is any *small* set of generators of A_{2k-1} for which successive approximations provide the right answer, at least for a wide class of formulas.

References

[Ber] M.Berger *Géométrie, vol.2.* (Cedic/Nathan 1977)

[Dan] V.Danos *La Logique Linéaire appliquée à l'étude de certains processus de normalisation.* (Thèse de Doctorat, Université Paris VII, 1990)

[DR] V.Danos & L.Régnier *The structure of multiplicatives.* (Arch. Math.Logic 28, 1990)

[Girl] J.Y.Girard *Linear Logic.* (Theor.Comput.Sci.50, 1987)

[Gir2] J.Y.Girard *Quantifiers in Linear Logic II.* (Prépublications Université Paris VII 19, 1991)

[Hal] P.J.Hall *Combinatorial Theory.* (Wiley, 1986)

[Hum] J.E.Humphreys *Reflection groups and Coxeter groups.* (Cambridge, 1990)

[Kan] M.I.Kanovich *The multiplicative fragment of Linear Logic is NP-complete.* (TR X-91-14 University of Amsterdam, 1991)

[LW] P.Lincoln & T.Winkler *Constant only Multiplicative Linear Logic is NP-complete.* (1992)

[Mét] F.Métayer *Homology of proof-nets.* (Arch.Math.Logic 33, 1994)

[Tro] A.S.Troelstra *Lectures on linear logic.* (CSLI, 1992)

Proof-nets and the Hilbert space

V. Danos & L. Regnier

Abstract

Girard's execution formula (given in [Gir88a]) is a decomposition of usual β-reduction (or cut-elimination) in reversible, local and asynchronous elementary moves. It can easily be presented, when applied to a λ-term or a net, as the sum of maximal paths on the λ-term/net that are not cancelled by the algebra L^* (as was done in [Dan90, Reg92]).

It is then natural to ask for a characterization of those paths, that would be only of geometric nature. We prove here that they are exactly those paths that have residuals in any reduct of the λ-term/net. Remarkably, the proof puts to use for the first time the interpretation of λ-terms/nets as operators on the Hilbert space.

1 Presentation

λ-Calculus is simple but not completely convincing as a real machine-language. Real machine instructions have a fixed run-time; a β-reduction step does not. Some implementations do map β-reductions into sequences of real elementary steps (as in environment machines for example) but they use a global time to achieve this. The "geometry of interaction" (GOI) is an attempt to find a low-level combinatorial code within which β-reduction could be implemented and such that:

- elementary reduction steps are local;
- parallelism shows up and global time disappears;
- some mathematics dealing with syntax is uncovered.

— Goal and organization of this paper.
A *persistent* path is a path on a λ-term which survives the action of any reduction (defined in [Reg92]). A *regular* path is a path which is not cancelled by Girard's algebraic device L^* (defined in [Gir88a]). The paper mainly aims at proving the equivalence of these two definitions and proceeds in the following order: first the geometric viewpoint on reduction is given, yielding a

definition of persistence; second the algebraic viewpoint, yielding a definition of regularity; and last the announced equivalence is proved.

Recently, Girard extended the definition of regularity to accomodate additives ([Gir95]). It would be interesting to investigate whether this matching between persistence and regularity still holds in the full case.

— Context of this paper.
The main result proved here was already known (see [Reg92]), but the method is new and leads to a simpler proof which is the first use of the linear structure of the Hilbert space in GOI. In another paper [ADLR94], together with A. Asperti and C. Laneve, we prove that persistence can be defined in two other ways: *consistency* ([AGL92]) and *legality* ([AL93]). Persistence is a most natural concept and these different ways to define it make it inescapable.

— Paths and execution.
It is fairly obvious that persistent paths describe the reduction. Intuitively they just are all the (anticipations of) cuts/redexes that ever happen during the reductions, and that's why they survive the action of any reduction. Indeed, Asperti and Laneve showed in [AL93] that each redex family (see [Lév78] for a definition) generated along the reduction of a given λ-term M corresponds to a unique persistent path in M. The converse also holds modulo a small technical detail (a balance condition). A similar result relating persistent paths to families of cut in nets was given earlier in [Reg92].

— Nets vs λ-calculus.
We present GOI with nets and not λ-terms. First because there are reduction-preserving embeddings of intuitionistic logic and classical logic into linear logic (LL). The intuitionistic embeddings, e.g., mapping $A \to B$ to $!A \multimap B$ or $!A \multimap !B$, allow to study λ-calculus directly inside nets. In the same way, the classical embeddings, e.g., mapping $A \to B$ to $!?A \multimap ?B$ or $!A \multimap ?!B$ (see [DJS94]), allow to study "classical λ-calculi" (for example, the $\lambda\mu$-calculus of Parigot [Par92]).

A second reason for choosing nets is that they provide a syntax that allows easy description of geometric entities like paths.

Everything here works for LL proof-nets as well as for any known kind of (pure) net(s) obtained by translations of λ-calculus, $\lambda\mu$-calculus. This is why we simply speak of nets.

2 A geometric point of view

In this first section we define nets and their reduction, the action of reduction on paths, say what persistent paths are, then prove there is a finite number of them in case the proof net is strongly normalizing (SN).

2.1 Nets

Nets are oriented graph structures whose nodes are called *links* and whose edges are labeled by formulae of linear logic. When drawing a net we represent edges oriented up-down so that we may speak of moving *downwardly* or *upwardly* in the graph. Possibly, e.g., when encoding pure λ-calculus in linear logic, formulae are quotiented by some duality-preserving equivalence relation.[1] Links are defined together with an arity and a coarity, i.e., a given number of incident edges called the *premises* of the link and a given number of emergent edges called the *conclusions* of the link.

- An **axiom** link has no premise and two conclusions labeled by dual formulae;
- a **cut** link has two premises labeled by dual formulae and no conclusion;
- *multiplicative links* **par** and **times** have two premises and one conclusion labeled by the **par** or the **times** of the premises;
- an **of course** link has one premise and one conclusion labeled by the **of course** of the premise;
- a **why not** link has n premises, for some integer n, all labeled by the same formula and one conclusion labeled by the **why not** of the premises.

Each edge is the conclusion of a unique link and the premise of at most one link. Edges which are not premise of a link are the *conclusions* of the net. Nets are required to fulfill two additional conditions:

Box condition: to each **of course** link **n** in the net R is associated a subnet B of R, called a *box* (represented by a rectangular frame), such that one conclusion of B, called the *principal door (pal)* of B, is the premise of **n**. All the other conclusions, called the *auxiliary doors (pax)* of B are premise of **why not** links. Two boxes are either disjoint or included one in the other. In the latter case, they may share some pax's but have distinct pal's.

Sequentialization condition: a net must be definable inductively in the following way: **axiom** is the basic case, **times** or **cut** of two nets, **par**, **why not** or boxing of one net. Purely geometrical conditions known as *correctness conditions* exist that are equivalent to the inductive one. We shall not discuss that here. The original definition of proof-nets is in [Gir87a]. Also a discussion on various correctness conditions may be found in [Dan90, DR89].

[1] I.e., quotiented in such a way that unsound cuts, e.g., between a **times** and an **of course**, never appear during reductions.

Remark. We use one general **why not** link with any number of premises. Possibly none, in which case we speak of a **weakening** link. There is a one-to-one correspondence between **of course** links and boxes. A premise of a **why not** link may be pax of several nested boxes.

The *height* of a link is the number of boxes in which it is contained and the height $h(\alpha)$ of an edge α is the height of its final link. An *exponential branch* is a premise of a **why not** link and an *exponential tree* is the set of premises of a **why not** link. The *lift* $l(\alpha)$ of an exponential branch α is the difference between the height of its initial link and the height of its final link. Thus it is always positive or null.

2.2 Paths in a net

A path is a sequence of edges or reverted edges, i.e., a path may take an edge from its goal to its source.[2] Let $\mathcal{P}(R)$ be the free category over R, that is, $\mathcal{P}(R)$ is the category of paths on R. We invert the usual convention by taking $\phi_1\phi_2$ to mean "ϕ_2 then ϕ_1", because in the sequel we will weight paths by functions. Thus $\phi_1\phi_2$ is a path if ϕ_2 ends where ϕ_1 begins. The operation of reversing paths (denoted ϕ^*) is a contravariant involution making $\mathcal{P}(R)$ an *involutive category.*

Edges will be denoted by α, β, γ, δ, σ whereas we use ϕ and ψ for general paths. The notation $\alpha^{(*)}$ means α and/or α^* depending on the context.

Let ϕ be a path. We say that ϕ is *non-bouncing* (nB) if it doesn't contain any $\alpha^*\alpha$; that ϕ is *non-twisting* (nT) if it doesn't contain any $\alpha_i^*\alpha_j$, with α_i and α_j distinct premises of a same link; that ϕ is *straight* if it is neither bouncing nor twisting. Note that neither nB nor nT is preserved by composition.

A straight path is a path changing direction only in axioms and cuts, bouncing up and down between the ceiling and the floor; so there is a finite number of them in a normal form, because there is no floor.

Let $\mathcal{P}^+(R)$ be the set of finite multisets of paths in $\mathcal{P}(R)$. We denote by $\sum \phi_i$ the multiset whose elements are the ϕ_i's. In particular we denote by ϕ the multiset whose single element is the path ϕ and respectively by 0 and 1 the empty multiset and the multiset whose elements are the empty paths on all the nodes.[3] If ψ and ϕ are two uncomposable paths in $\mathcal{P}(R)$, we write $\phi\psi = 0$ in $\mathcal{P}^+(R)$ and define composition in $\mathcal{P}^+(R)$ by:

$$\left(\sum \phi_i\right)\left(\sum \psi_j\right) = \sum \phi_i\psi_j.$$

Note that 0 is neutral for +, absorbing for composition and 1 is neutral for composition in $\mathcal{P}^+(R)$. If $\Phi = \sum \phi_i$ is an element of $\mathcal{P}^+(R)$ then Φ^* is $\sum \phi_i^*$. Thus $\mathcal{P}^+(R)$ is an *involutive semiring.*

[2] Think of paths as walks (not drives) in one-way streets.

[3] That is, the sum of the identities of $\mathcal{P}(R)$.

2.3 Reduction of a net

Reductions are sequences of *elementary reduction steps* (*ers*). Apart from the axiom step (a), there is exactly one such ers for each couple of dual links: the exponential one (e), the multiplicative one (m). We will forget the weakening ers here, assuming that all the reductions are strict. This assumption doesn't make much sense in λ-calculus (since restriction to strict reductions may in some cases turn a non SN term into a SN one) but it is innocuous when working with nets. Also, since they are much simpler to handle, we will discard the cases of the axiom and multiplicative steps, only dealing with the exponential step.

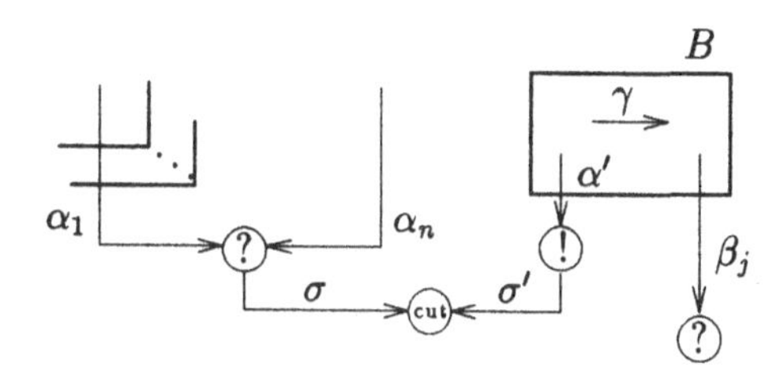

reduces in:

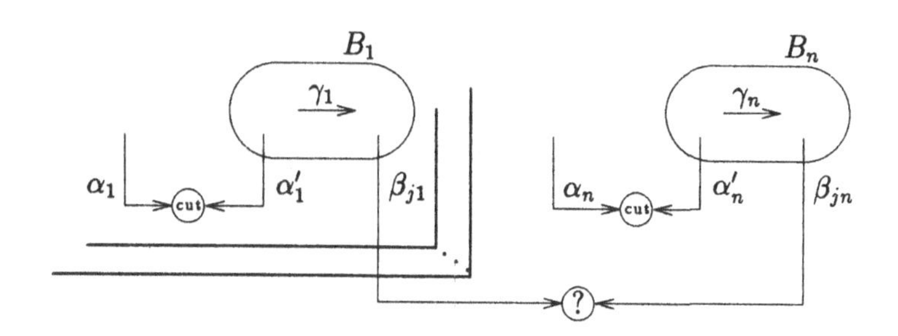

Figure 1: exponential reduction

What happens during the exponential reduction step (see fig. 1 for notations) is that the box B dispatches n copies $B_1, \ldots, B_n$ of itself glued:

- by the (once pal) α' to the top of $\alpha_1, \ldots, \alpha_n$ (thus creating n new cuts);
- and by the (pax's) $\beta_1, \ldots, \beta_m$ to their anchor ?-vertices.

For each i, the ith copy is moved inside the $l(\alpha_i)$ boxes of which α_i is a pax.

Let's classify the edges of R into three groups:

- *consumed* edges: σ and σ';
- *duplicated* edges: γ's inside the box become $\gamma_1, \ldots, \gamma_n$'s, and edges crossing the box, namely α' and β_j's become respectively $\alpha'_1, \ldots, \alpha'_n$ and $\beta_{j1}, \ldots, \beta_{jn}$'s;
- *intact* edges: α_i's and δ's lying elsewhere.

2.4 Reduction on paths

2.4.1 The lifting functor

Let ρ be a reduction on a net R leading to a net $\rho(R)$. The *lifting functor* $\overline{\rho} : \mathcal{P}(\rho(R)) \longrightarrow \mathcal{P}(R)$ is the functor defined by:

- if ρ is a sequence of ers, then $\overline{\rho}$ is the composition of the lifting functors associated to each ers in ρ;
- if ρ is an exponential ers then with the notations of the previous section: $\overline{\rho}(\alpha_i) = \sigma\alpha_i$, $\overline{\rho}(\alpha'_i) = \sigma'\alpha'$, $\overline{\rho}(\beta_{ji}) = \beta_j$, $\overline{\rho}(\gamma_i) = \gamma$ and $\overline{\rho}(\delta) = \delta$ for any intact edge δ;
- in the cases of the a and m ers, the definition is similar.

Because of the exponential ers, $\overline{\rho}$ is not injective (faithful): for instance the α'_i's have the same image. It is not surjective (full) either: for example α_i is not in the image of $\overline{\rho}$, only $\sigma\alpha_i$ is. Some paths even have no extensions in the image of $\overline{\rho}$: typically $\alpha_1^*\alpha_2$ and $\alpha_1^*(\sigma^*\sigma'\alpha')\psi(\sigma^*\sigma'\alpha')^*\alpha_2$ where ψ is any path entirely contained in the box B.

2.4.2 The deformation pseudo-functor

Let ρ be an ers reducing a cut c. A path ψ is *long enough* for ρ if its first and last nodes are neither the cut link c, nor one of the links connected by c. Note that a conclusion-to-conclusion path is long enough for any reduction. Given a nonempty path ϕ in $\mathcal{P}(R)$, we say that ϕ' in $\mathcal{P}(\rho(R))$ is a *residual* of ϕ along ρ if ϕ' is minimal such that ϕ is a subpath of $\overline{\rho}(\phi')$.

In case ϕ is long enough for ρ, if ϕ' is a residual of ϕ then $\overline{\rho}(\phi') = \phi$ so that the set of residuals of ϕ is exactly $\overline{\rho}^{-1}(\phi)$. If ϕ is not long enough for ρ then ϕ is a proper subpath of $\overline{\rho}(\phi')$.

If ϕ is nonempty and has a residual ϕ' along ρ (which is not always the case as the example above show) then ϕ' is nonempty.

We extend by induction on the length of a reduction ρ the definitions of residuals along ρ and paths long enough for ρ. For each reduction ρ, we define the *deformation* action $\rho : \mathcal{P}(R) \longrightarrow \mathcal{P}^+(\rho(R))$ by:

$$\rho(\phi) = \sum \text{residuals}(\phi).$$

The deformation ρ is not a functor. However, if we restrict to long enough paths we have:

$$\rho(\phi) = \sum \{\phi' \in \mathcal{P}(\rho(R)), \quad \overline{\rho}(\phi') = \phi\},$$

and ρ is functorial from long enough paths of $\mathcal{P}(R)$ into $\mathcal{P}^+(\rho(R))$.

Note that because $\overline{\rho}$ is not injective, in general $\rho(\phi)$ is a sum of paths (of equal lengths if ρ is one-step). Also, because $\overline{\rho}$ is not surjective, there are some ϕ such that $\rho(\phi) = 0$. The kernel of ρ, i.e., the set of paths with no residuals, measures the connectivity loss of the net under the deformation ρ.

2.4.3 Computing the deformation of a long enough straight path

Let ρ be an exponential ers, ϕ be a straight path long enough for ρ and assume the notations of fig. 1. Note that ϕ may cross the box B in three different ways (see fig. 2), and since the remaining parts of ϕ are left untouched by ρ, it clearly suffices to compute the action of ρ on *B-crossings* of ϕ, i.e., subpaths of ϕ that cross B. There are three kinds of B-crossings.

palpal (represented by a continuous line in fig. 2): paths starting from an α_i, crossing the cut, entering B by its pal, moving into B, exiting B by its pal, crossing the cut (the other way round) and ending in some α_j^*:

$$\alpha_j^*(\sigma^*\sigma'\alpha')\,\psi\,(\sigma^*\sigma'\alpha')^*\alpha_i$$

where ψ is a path inside B;

palpax (and its dual **paxpal**, represented by a dashed line in fig. 2): paths starting like the preceding ones but this time exiting B by some pax β_j:

$$\beta_j\psi\,(\sigma^*\sigma'\alpha')^*\alpha_i;$$

paxpax (represented by a small dashed line in fig. 2): paths entering and exiting B by some pax's β_i and β_j:

$$\beta_j\psi\beta_i^*$$

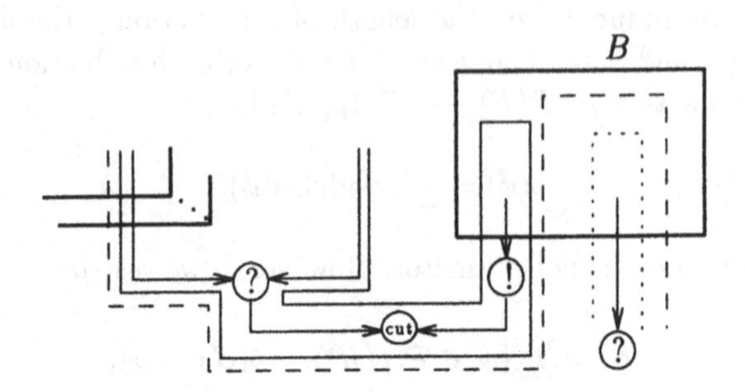

Figure 2: Crossings of a box

It is an easy computation following the definition of ρ to see that our three different kinds of crossings are deformed by ρ in the following way:

$$\begin{aligned}
\alpha_j^*(\sigma^*\sigma'\alpha')\,\psi\,(\sigma^*\sigma'\alpha')^*\alpha_i &\xrightarrow{\rho} \alpha_i^*\alpha_i'\psi_i\alpha_i'^*\alpha_i \quad \text{if } j=i,\ 0 \text{ otherwise}\\
\beta_j\psi\,(\sigma^*\sigma'\alpha')^*\alpha_i &\xrightarrow{\rho} \beta_{ji}\psi_i\alpha_i'^*\alpha_i\\
\beta_j\psi\beta_k^* &\xrightarrow{\rho} \sum_l \beta_{jl}\psi_k\beta_{kl}^*
\end{aligned}$$

In the case of the first type of crossing (palpal) with $i \neq j$ the deformation simply cancels the path. In this case we say that ϕ *exchanges the premises* of the **why not** link. Conversely, it is only if ϕ contains such a crossing that it is cancelled by the deformation. For example, if ρ is a reduction of R such that no residual of ϕ ever crosses a cut reduced by ρ, then $\rho(\phi) \neq 0$.

2.4.4 Persistent paths

Definition 1 *A* persistent *path is any straight path ϕ such that for all reduction ρ, $\rho(\phi) \neq 0$ in $\mathcal{P}^+(\rho(R))$.*

Persistent paths are those that survive all possible deformations. An equivalent definition would be: ϕ in $\mathcal{P}(R)$ is persistent iff for any reduction ρ, there is a ϕ' in $\mathcal{P}(\rho(R))$ such that ϕ is a subpath of $\overline{\rho}(\phi')$.

Note that being straight is preserved by the deformation ρ. And also that any nonempty subpath of a persistent path is still persistent. Thanks to that remark, in the future we will often restrict to long enough paths.

As said before, a typical example of non-persistent path is a palpal crossing exchanging the premises of the **why not** link; and persistence amounts to saying that such a crossing never happen along any reduction.

2.5 Preservation of finiteness of computations

Lemma 2 (Contraction of paths) *Let ρ be an ers, ϕ be a path long enough for ρ and $|\rho(\phi)|$ denote the common length of all terms in the sum $\rho(\phi)$:*

$$|\phi|/2 - 2 \leq |\rho(\phi)| \leq |\phi|$$

under the condition for the left inequation that ϕ is nB and $\rho(\phi) \neq 0$. Furthermore the right inequation is strict iff ϕ crosses the cut reduced by ρ.

Theorem 3 *R is SN iff there is a finite number of persistent paths in $\mathcal{P}(R)$.*

Proof. The left contraction inequation says that reducing a persistent path costs about the logarithm of its length in the fastest case, hence: $|\phi| \leq O(2^r)$ where r is the length of a longest reduction of R.[4]

The right contraction inequation says an ers (through the cut of which the path goes) contracts the path. Hence, following an infinite branch in the reduction space of R, one will get arbitrarily long paths by lifting back cuts to R. But will these lifted cuts be persistent ? It needs the adequacy theorem (regular equivalent to persistent), proved in the last section, to conclude that they are, since they are certainly regular. □

3 An algebraic point of view

In this section we borrow from Girard's original paper ([Gir88a]) the algebraic tool to characterize regular paths, namely L^*. We then build models of this algebra, most notably within $B(H)$ the set of bounded operators on the Hilbert space, and show that any model is as good as L^* in catching regular paths. This property will be at the heart of the proof of the main result in the next and last section.

3.1 The weight

Let R be a net and $T_1, \ldots, T_n$ be an enumeration of the exponential trees of R. We define the equational theory L^*_R attached to R and then the so-called *weight* functor w.

The theory L^*_R is defined by the following generators and relations: L^*_R is a free involutive monoid, i.e., there is a mapping $u \longrightarrow u^*$ in L^*_R such that:

$$(u^*)^* = u \qquad \text{and} \qquad (uv)^* = v^*u^*,$$

[4][DR93] was our reaction against this frightening bound which is reached by the family $(\cdots(I\,I)\cdots I)$ where I denotes the term $\lambda x.x$.

for any u and v. Note that this entails in particular that $1^* = 1$.

There are a 0, a morphism !:

$$!(0) = 0, \quad !(1) = 1, \quad !(u^*) = !(u)^* \quad \text{and} \quad !(u)!(v) = !(uv);$$

two *multiplicative generators* p and q satisfying:

$$p^*p = q^*q = 1, \tag{1}$$
$$p^*q = 0; \tag{2}$$

and for each exponential tree $T_i = (\beta_{ij})_j$ in R, a family $(b_{ij})_j$ of *exponential generators* satisfying the zapping (or annihilation, or orthogonality) relations:

$$b^*_{ij}\, b_{ik} = \delta_{jk}; \tag{3}$$

and the swapping (or commutation) relations:

$$!(u)\, b_{ij} = b_{ij}\, !^{e_{ij}}(u) \tag{4}$$

where u is any element and $e_{ij} = l(\beta_{ij})$ is the lift of β_{ij}. We say that the generator b_{ij} has lift $l(b_{ij}) = e_{ij}$.

In the sequel, unless explicitly stated, we shall write $u = v$ for "u and v are provably equal in the equational theory L^*_R".

The *weight* functor w from $\mathcal{P}(R)$ to L^*_R is defined on edges by: if α is the left (resp. right) premise of a multiplicative link then $w(\alpha) = !^{h(\alpha)}(p)$ (resp. $!^{h(\alpha)}(q)$); if β_{ij} is an exponential branch then $w(\beta_{ij}) = !^{h(\beta_{ij})}(b_{ij})$; if α is any other edge then $w(\alpha) = 1$.

Remarks.

- The equations for p and q are the same as the zapping equations of a binary exponential tree;
- commutations are linear equations (a remark of use only if one finds a model of L^*_R that allows linear combinations);
- annihilations are self-dual; by dualizing the commutations equations one finds: $b^*_{ij}\, !(u) = !^{e_{ij}}(u)\, b^*_{ij}$;
- the equations have a natural orientation: from lhs to rhs. W.r.t. that orientation, starred generators are inclined to travel eastwards and unstarred ones westwards;
- w is a local valuation on each exponential tree of R, i.e., there are no relations between exponential generators of two different exponential trees. On the other hand, the same p and q are used to weight all multiplicative links.

Definition 4 (regular paths) *A path ϕ in $\mathcal{P}(R)$ is* regular *iff* $w(\phi) \neq 0$.

Note that twisting paths are not regular; neither are the palpal crossings with i distinct from j as the reader can check.

We denote by S_R the quotient category $\mathcal{P}(R)/w$ of *weighted paths*. By analogy with $\mathcal{P}^+(R)$, we denote S_R^+ the set of finite multisets of weighted paths in which all the 0-paths are identified to the empty multiset, and composition is defined by distributivity over sums (multiset unions). Therefore S_R^+ is in turn an involutive semiring.

3.2 Execution formula

To a net R, we associate a self-dual element of S_R^+ by the so-called *execution formula*:

$$\mathsf{EX}(R) = \sum_{\phi \in \mathcal{P}_c(R)} w(\phi),$$

where $\mathcal{P}_c(R)$ denotes the set of conclusion-to-conclusion straight paths of $\mathcal{P}(R)$. We'll see at the end of the next section (second part of the adequacy theorem), that it contains the same information as the set of persistent conclusion-to-conclusion paths, and hence may be thought of as a computation of R.

This is another presentation of the formula given in [Gir87b, Gir88a].

3.3 Small models of L_R^*

We will now present the *small models* of the equational theory L_R^*. This construction proves by the way that $0 \neq 1$ in L_R^*.

The basic structure. Let $\mathbf{N}$ be the set of natural numbers and PI($\mathbf{N}$) be the inverse monoid of partial injective mappings from $\mathbf{N}$ to $\mathbf{N}$. Product is (partial) composition, * is the inversion, 0 is the nowhere defined mapping and 1 is the full identity on $\mathbf{N}$.

The shift morphism. Fix a bijection $\lceil .,. \rceil : \mathbf{N} \times \mathbf{N} \longrightarrow \mathbf{N}$ and, given u and v in PI($\mathbf{N}$), define $u \otimes v$ in PI($\mathbf{N}$) by:

$$(u \otimes v)(\lceil x, y \rceil) = \lceil u(x), v(y) \rceil.$$

Then $!(u) = 1 \otimes u$ is a morphism on PI($\mathbf{N}$), which we call the *shift morphism*. Define $\lceil x_1, \ldots, x_n \rceil$ as an abbreviation for $\lceil x_1, \ldots \lceil x_{n-1}, x_n \rceil \ldots \rceil$.

Modeling the swapping relations. Let $\tau : \mathbf{N}^n \longrightarrow \mathbf{N}$ be a partial injection (where in the case $n = 0$, τ is a constant in $\mathbf{N}$); then $u_\tau(\lceil x_1, \ldots, x_{n+1} \rceil) = \lceil \tau(x_1, \ldots, x_n), x_{n+1} \rceil$ satisfies the swapping relation with lift n ($n \geq 0$). Any solution to that relation, $\lceil ., . \rceil$ and $!(.)$ being given, has to be of that form; in particular, no solution with lift zero is to be found here with a full range.

Modeling the zapping relations. If τ has full domain, then $u_\tau^* u_\tau = 1$; if τ and τ' have disjoint ranges in $\mathbf{N}$, then $u_\tau^* u_{\tau'} = 0$. This is enough to interpret the zapping relations.

3.4 How to make them bigger

Take H to be the Hilbert space with a countable basis $X = (e_n)$. The *big model* of L_R^* is the set of bounded operators on H, denoted by $B(H)$. The interest of this new model is that we are going to use our remark about the linearity of the commutation relation by superposing "small" solutions; another interest may be understood by reading [Gir88b] or [MR91].

The basic structure. $B(H)$ is clearly an involutive monoid (w.r.t. composition and adjunction) wherein there is a zero and a one. It even is a C^*-algebra.

The shift morphism. Take $\mu : H \otimes H \longrightarrow H$ to be the isomorphism defined by $\mu(e_i \otimes e_j) = e_{\lceil i,j \rceil}$, where $H \otimes H$ means the completion of the usual vector space tensor product w.r.t. the inner product defined by $\langle h_1 \otimes h_1', h_2 \otimes h_2' \rangle = \langle h_1, h_2 \rangle . \langle h_1', h_2' \rangle$.

For each U in $B(H)$ put $!(U) = \mu(1 \otimes U)\mu^{-1}$. Then $!(.)$ is a C^*-algebra isomorphism mapping $B(H)$ isometrically into itself.

The only point which is not obvious is $\|!(U)\| = \|U\|$. Take h in H of norm 1, then $\|(1 \otimes U)(h \otimes h)\| = \|U(h)\|$ so $\|!(U)\| = \|1 \otimes U\| \geq \|U\|$.

There is another way to see this. Take D as defined in the second paragraph below with (d_i) such that D is an isometry. We have $!(U)D = DU$, so $\|!(U)\| \geq \|!(U)D\| = \|DU\| = \|U\|$.

Conversely, for any h_1, h_2 in $B(H)$, $\|(1 \otimes U)(h_1 \otimes h_2)\| = \|h_1\|.\|U(h_2)\| \leq \|U\|.\|h_1\|.\|h_2\|$ which yields the converse inequality.

We have:

$$\langle !(U)(e_{\lceil i_1, i_2 \rceil}), e_{\lceil j_1, j_2 \rceil} \rangle = \delta_{i_1 j_1} \langle U(e_{i_2}), e_{j_2} \rangle \qquad (5)$$

Hence the shift morphism preserves being a partial permutation of the basis (see next paragraph) and being a partial isometry.

Upgrading partial injections to partial isometries. For any u in PI($\mathbb{N}$) there is a partial isometry U in $B(H)$ defined by:

$$U(e_n) = e_{u(n)} \text{ if } u(n) \text{ is defined, 0 otherwise.}$$

Such U's are partial permutations of the basis X. One can easily check that zapping and swapping equations are preserved by upgrading.

Modeling the 0-lift swapping equation. The following lemma is proved by calculation.

Lemma 5 *The solutions D in $B(H)$ of the 0-lift swapping equation $!(U)D = DU$ have the form: $D = \sum_i d_i D_i$ where $D_i(e_n) = e_{\lceil i,n \rceil}$ (upgrading of the 0-lift injection $n \longrightarrow \lceil i,n \rceil$) and $\sum_i |d_i|^2 = d^2 < \infty$. Moreover $\|D\| = d$ (so such a D is indeed continuous).*

Note that:

$$D^*D = \sum_{ij} \bar{d}_i d_j D_i^* D_j = d^2,$$

so if $d = 1$, then D^*D is a self-adjoint idempotent (a projector) and D an isometry. Its final projector satisfies:

$$DD^*(e_{\lceil n_1,n_2 \rceil}) = \sum_{ij} d_i \bar{d}_j D_i D_j^*(e_{\lceil n_1,n_2 \rceil}) = \sum_i d_i \bar{d}_{n_1} D_i(e_{n_2}) = \bar{d}_{n_1} \sum_i d_i e_{\lceil i,n_2 \rceil}$$

Hence:

$$\langle DD^*(e_{\lceil n_1,n_2 \rceil}), e_{\lceil n_1,n_2 \rceil} \rangle = \|DD^*(e_{\lceil n_1,n_2 \rceil})\|^2 = |d_{n_1}|^2, \qquad (6)$$

therefore, if we assume that $d_n \neq 0$ for all n, then $DD^*(e_n) \neq 0$ for all n, which is a key property of our interpretation.

X-positivity. An operator U on H is said to be X-*positive* if for all i, j, we have $\langle U(e_i), e_j \rangle \geq 0$. In other words U sends H^+ (all h such that $\langle h, e_i \rangle \geq 0$ for all i) into itself. We denote the set of X-positive bounded operators on H by $B(H^+)$.

Lemma 6 (Everybody can be taken X-positive) *The algebra $B(H^+)$ is stable by product, sum, adjunction, shift, multiplication by a positive real and contains all upgraded partial injections as well as D-like operators with $d_i > 0$ for all i.*

The interpretations of L_R^* in $B(H)$ that we will consider in the next section will be in fact in $B(H^+)$.

The non-interception argument. This is a preliminary step for the adequacy proof, see next section.

Define $h_1 \geq h_2$ iff $h_1 - h_2$ is in H^+.

Lemma 7 *V is X-positive iff it is increasing w.r.t. this partial ordering.*

Proof. By linearity of V we have: $\langle V(h), e_i\rangle = \sum_j \langle h, e_j\rangle\langle V(e_j), e_i\rangle$, so if $h_1 \geq h_2$, since V is X-positive, for all i, j we have that $\langle h_1, e_j\rangle\langle V(e_j), e_i\rangle$ is greater than $\langle h_2, e_j\rangle\langle V(e_j), e_i\rangle$. Hence $\langle V(h_1), e_i\rangle$ is greater than $\langle V(h_2), e_i\rangle$. This computation shows that V, if not zero, is even strictly increasing. Conversely, $e_i \geq 0$, hence if V is increasing, for all j, $\langle V(e_i), e_j\rangle \geq 0$. □

Lemma 8 (No-interception) *Take U_1, U_2 and P in $B(H^+)$, P being a projector. Suppose further that for all n, $P(e_n) \neq 0$, then:*

$$U_1 P U_2 = 0 \quad \Rightarrow \quad U_1 U_2 = 0.$$

Proof. Suppose $U_1U_2 \neq 0$. There exist a vector of the basis, say f_2, such that $U_1U_2(f_2) \neq 0$. Now: $U_1U_2(f_2) = \sum_i \langle U_2(f_2), e_i\rangle U_1(e_i)$, so there also exist another vector of the basis, say f_1, such that $\langle U_2(f_2), f_1\rangle \neq 0$ and $U_1(f_1) \neq 0$

Put $\langle U_2(f_2), f_1\rangle = \lambda$ and $\langle P(f_1), f_1\rangle = \lambda'$. Since U_2 and P are X-positive, λ and λ' are non-negative reals; by construction of f_2, $\lambda > 0$; by the further assumption on P, $P(f_1) \neq 0$. But P being a projector, $\langle P(f_1), f_1\rangle = \|P(f_1)\|^2$, so $\lambda' > 0$. Now, f_1 and f_2 being basis vectors, $U_2(f_2) \geq \lambda f_1$; since P is X-positive, by the preceding lemma, P is increasing w.r.t. $\geq$, so $PU_2(f_2) \geq \lambda P(f_1) \geq \lambda\lambda' f_1$. Finally, U_1 being also X-positive (by the preceding lemma again), we get $U_1PU_2(f_2) \geq \lambda\lambda' U_1(f_1) > 0$. □

3.5 The ab^*-like theorems

We consider the rewriting system associated to L^*_R by orienting the equations of section 3.1 from left to right. An ab^*-form (or stable form) is a term of the equational theory L^*_R of the form: $m := a!(m)b^* \mid 1$ where a and b are flat (no !(.)'s), positive (no *'s) and possibly empty words on the generators. Such a form is normal w.r.t. the oriented equations, although other normal forms exist: consider for instance $b^*_{00}\, b_{10}$.

Theorem 9 (ab^* version 1) *For all straight ϕ, either there is a unique form ab^* into which $w(\phi)$ rewrites, or $w(\phi)$ rewrites to zero by means of the oriented equations.*

Proof. (Sketchy, see [Reg92] for a complete proof) By the right "contraction inequation", we can sort out an inductive proof on the length of ϕ using *special* exponential steps ρ w.r.t. ϕ. Special means that ϕ contains no cut edge below the pax's of the door associated to ρ and crosses the cut associated to ρ.

Given ϕ, such a ρ always exists; this follows from the correctness condition on nets. Now in the case ρ is special relative to ϕ, paxpax crossings can't occur, so $\rho(\phi)$ is a path (not a sum of paths) the weight of which rewrites to ab^* iff the weight of ϕ does as an easy calculation shows. □

In fact the theorem is true for any path, not necessarily straight.

A first easy consequence of ab^* is the endogamy of exponential trees: during the rewriting of $w(\phi)$ no two generators of different exponential trees are ever in zapping position. This is why we don't need any relation between b_{ij} and $b_{i'j'}$ when $i \neq i'$ in the definition of L^*_R. An immediate but interesting corollary is that any straight path that never uses more than one branch per exponential tree is regular.

A second consequence of ab^*, stating a kind of no-backtrack property, is:

Lemma 10 (Stretching) *Any regular straight path that has straight extensions, has at least one regular proper straight extension.*

Another more significant consequence, which needs the extension of ab^* to bouncing paths to be proved is:

Theorem 11 (ab^* version 2) *The category of weighted paths S_R is an involutive category which is:*
(i) inversive : it satisfies $(x^*)^* = x$, $xx^*x = x$ and $xx^*yy^* = yy^*xx^*$ for all weighted paths x and y;
(ii) decidable in linear time.

Finally, the computation of regularity doesn't depend on the model of L^*_R. This lemma will be at the heart of the proof of the adequacy theorem in the next section.

Lemma 12 *Let $\mathcal{L}$ be a non trivial model of L^*_R and ϕ be a path in $\mathcal{P}(R)$. Then $w(\phi) = 0$ iff $\mathcal{L} \models w(\phi) = 0$.*

Proof. The only if part is just the definition of model. For the if part, suppose $w(\phi) \neq 0$ in L^*_R. By ab^* version 1, there exist some positive words a and b such that $w(\phi) = ab^*$. Thus $a^*w(\phi)b = 1$ in L^*_R, therefore also in $\mathcal{L}$. Since $\mathcal{L}$ is non trivial we deduce that $\mathcal{L} \models w(\phi) \neq 0$. □

Conversely, lemma 12 implies ab^* version 1. Take a counterexample to ab^*, it must be of the form x^*x' where x and x' are either generators belonging to different exponential trees, either one of them is an exponential generator whereas the other is multiplicative, or one of them is a multiplicative whereas the other has the form $!(y)$; these are the only cases in which no oriented equation of L^*_R applies. A look at the small models shows that one can choose x and x' orthogonal (i.e., such that $x^*x' = 0$), hence by the lemma they are already zero in L^*_R.

This last version of ab^* explains why the "context semantics" of Lamping and [AGL92] can be made to compute persistent paths as well as the two original models of Girard in [Gir88a].

4 Where they meet

Now we are all set to prove what we were aiming at. We first state the fundamental lemma.

Lemma 13 (Preservation of kernels) *Let ϕ be in $\mathcal{P}(R)$, and ρ be a reduction of R:*

$$w(\phi) = 0 \qquad \textit{iff for all } \phi' \in \rho(\phi), \quad w(\phi') = 0.$$

The if part is called "hardlemma", and the converse "easylemma".

Proof. First a word on tactics. By the semantic formulation of the ab^* property (lemma 12), it is enough to prove the preservation in some non-trivial model $\mathcal{L}$ of both L^*_R and $\mathsf{L}^*_{\rho(R)}$. Ideally we would like to build $\mathcal{L}$ together with an interpretation of the weight functors such that $w(\phi) = w(\rho(\phi))$ is satisfied in $\mathcal{L}$. That's almost what we are going to achieve (that is except in one limit case, where ρ is such that all α_i have lift zero, which calls for a weaker result).

Before continuing, let us fix a small gap in what we have just said. We want to compare $w(\phi)$ and $w(\rho(\phi))$, in some model to be built. However, $\rho(\phi)$ being, in general, a multiset of paths, the expression $w(\rho(\phi))$ has no sense. For this reason, during the proof we shall add a formal sum in the usual way to the definition of L^*_R, and define $w(\sum_i \psi_i)$ to be $\sum_i w(\psi_i)$ for any family of paths ψ_i. We interpret this sum in the big model $B(H)$ by the sum of operators. In the small models we define a partial sum by: $f+g$ is defined only if $fg^* = f^*g = 0$ (i.e., f and g have disjoint domains and disjoint codomains), in which case $f+g(x)$ is $f(x)$ if x is in the domain of f, $g(x)$ is x is in the domain of g. Then $f+g$ is obviously a partial injection and composition is distributive over partial sum. Note that the upgrading from small to big models maps this partial sum on the sum of operators. It is enough to define only a partial sum since as we will see, we only consider sums of weights $\sum_i u_i$ such that $u_i u_j^* = u_i^* u_j = 0$ in L^*_R.

After these preliminaries let us go in the proof. First note that we obviously may restrict to a one-step reduction. We will consider only the case of an exponential ers ρ. By the stretching property, we may without loss of generality assume that ϕ is long enough w.r.t. ρ. Remark also that we may as well assume that ϕ neither twists nor contains palpal crossings with i distinct from j, for if it does the lemma boils down to $0 = 0$ which holds indeed.

We use the notations of the definition of the exponential ers in section 2.3. If δ is an intact edge of R we assume it has the same weight in R and $\rho(R)$.

Let h be the height of the box B and $l_i = l(\alpha_i)$ be the lift of α_i. If γ is an edge lying inside B, its weight is $!^{h+1}(u)$ where u is some generator of L^*_R. By definition, γ has n residuals $(\gamma_i)_{1\leq i\leq n}$ in $\rho(R)$ and the weight of γ_i is $!^{h+l_i}(u_i)$ where u_i satisfy the same zapping and swapping equations as u. Therefore we may interpret u and u_i by the same object in $\mathcal{L}$ and that's what we do. We call this property of the interpretation the *box relation*. With the box relation, if ψ is a path contained in the box B and ψ_i is one of its residual in $\rho(R)$, then we have $w(\psi) = !^{h+1}(u)$ and $w(\psi_i) = !^{h+l_i}(u_i)$ where u and u_i are elements of L^*_R and $\mathsf{L}^*_{\rho(R)}$ such that $\mathcal{L} \models u = u_i$.

Let's call the weights of α_i, β_j and the residuals β_{ji} of the latter, respectively x_i, y_j and z_{ji}. Thus x_i is an exponential generator of lift $l(x_i) = l_i$.

Assuming the notations of section 2.4.3 we now inspect each way ϕ crosses the box B.

(a) For the palpal crossing the (weight of the) lhs is:

$$w(\alpha_i^*(\sigma^*\sigma'\alpha')\,\psi\,(\sigma^*\sigma'\alpha')^*\alpha_i) = !^h(x_i^*)!^{h+1}(u)!^h(x_i) = !^{h+l_i}(u)$$

while the rhs is:

$$w(\alpha_i^*\alpha_i'\psi_i\alpha_i'^*\alpha_i) = !^{h+l_i}(u_i)$$

so that the equality holds in $\mathcal{L}$ by the box relation.

(b) For the palpax crossing let h_j denote the height of β_j (which by the way is the same as that of β_{ji}); thus h_j is lower or equal than h. The lhs is:

$$w(\beta_j\psi\,(\sigma^*\sigma'\alpha')^*\alpha_i) = !^{h_j}(y_j)!^{h+1}(u)!^h(x_i) = !^{h_j}(y_j)!^h(x_i)!^{h+l_i}(u),$$

while the rhs is:

$$w(\beta_{ji}\psi_i\alpha_i'^*\alpha_i) = !^{h_j}(z_{ji})!^{h+l_i}(u_i).$$

We ask that $\mathcal{L}$ satisfies the so-called *coagulation relation:*

$$z_{ji} = y_j!^{h-h_j}(x_i). \tag{7}$$

An easy computation shows that this definition of (the interpretation of) z_{ji} is consistent w.r.t. the zapping and swapping relations, because $l(z_{ji}) = l_i + l(y_j) - 1$. With the box and the coagulation relations we get identity of weights of palpax crossings in $\mathcal{L}$.

(c) For the paxpax crossing, the lhs is:

$$w(\beta_j\psi\beta_k^*) = !^{h_j}(y_j)!^{h+1}(u)!^{h_k}(y_k^*).$$

Now by the coagulation relation above, the rhs in $\mathcal{L}$ is:

$$\begin{aligned}\sum_i w(\beta_{ji}\psi_i\beta^*_{ki}) &= \sum_i !^{h_j}(z_{ji})!^{h+l_i}(u_i)!^{h_k}(z^*_{ki}) \\ &= \sum_i !^{h_j}(y_j)!^h(x_i)!^{h+l_i}(u_i)!^h(x^*_i)!^{h_k}(y^*_k) \\ &= !^{h_j}(y_j)\left(\sum_i !^h(x_ix^*_i)\right)!^{h+1}(u_i)!^{h_k}(y^*_k).\end{aligned}$$

Let $P = \sum_i !^h(x_ix^*_i)$; since $\mathcal{L}$ satisfies the box and coagulation relations we have:

$$\mathcal{L} \models w(\rho(\phi)) = v_1P \dots v_pPv_{p+1}$$

where the v_i's are such that:

$$\mathcal{L} \models w(\phi) = v_1 \dots v_{p+1}.$$

Since the x_i's satisfy the zapping equations together, P is easily seen to be an idempotent. Assume $\mathcal{L}$ is a small model and suppose one of the residuals of ϕ in $\rho(R)$ is regular. Being the sum of the weights of the residuals of ϕ, $w(\rho(\phi))$ is non null, thus by lemma 12, non null in $\mathcal{L}$. But idempotents in $\mathcal{L}$ are identities on subsets of N so that $w(\phi)$ is in turn non null in $\mathcal{L}$, thus in L^*_R and this ends the proof of easylemma.

To get hardlemma we should ask that $\mathcal{L}$ satisfy the so called *partition relation*:

$$\sum_i !^h(x_ix^*_i) = 1. \tag{8}$$

Now set $\mathcal{L} = B(H)$ and $x_i = U_{\tau_i}$ in $B(H)$. Then the partition relation above is equivalent to: $H = \bigoplus_i \mathrm{Rg}(U_{\tau_i})$. But this in turn only means that the ranges of the τ_i are a partition of N, which is possible except in the degenerate case where all x_i have zero lift.

Therefore, to end the proof of hardlemma, it only remains to consider this degenerate case.

A side-remark, before concluding: up to this point we only had to use small models, yet there is no hope of the proof going through in that context in degenerate cases. This because any 0-lift injection's range has the form $\lceil i_0, X\rceil$ for some i_0 in X. In fact it's even worse, for adding $dd^* = 1$ to the theory for some 0-lifted d yields a proof of $0 = 1$.

Suppose, to begin with, that $h = 0$ and that $n = 1$ (where n is the number of premises of the **why not** link being cut). We interpret x_1 by $D = \sum_i d_iD_i$ with $d_i > 0$ for all i, and z_{j1} according to the coagulation relation. Since everything lies in $B(H^+)$ and DD^* satisfies the further assumption of the no-interception lemma, we may apply the latter to (the interpretations of) $w(\rho(\phi))$ and $w(\phi)$. But this is just hardlemma.

To deal with situations where $h > 0$, observe that by instantiating equation (5) one get:

$$\langle !(U)(e_{\lceil i_1,i_2 \rceil}), e_{\lceil i_1,i_2 \rceil}\rangle = \langle U(e_{i_2}), e_{i_2}\rangle,$$

which when iterated yields:

$$\langle !^h(U)(e_{\lceil i_1,\ldots,i_{h+1} \rceil}), e_{\lceil i_1,\ldots,i_{h+1} \rceil}\rangle = \langle U(e_{i_{h+1}}), e_{i_{h+1}}\rangle,$$

so $!^h(P)$ is going to satisfy the further assumption of the no-interception lemma if P does. To deal with the case where there is more than one x_i, say n, just cut D in n pieces of which the final projectors are orthogonal. □

Now for a vague but intuitive remark: as announced in the tactics, the argument goes through, though the partition relation is not satisfied by D, for $DD^* \neq 1$. This is no accident that no model equates fully $w(\rho(\phi))$ to $w(\phi)$, and is due to the fact that the L^*-way to compute is less duplicating than the usual.

4.1 The adequacy theorems

Theorem 14 (Adequacy, part 1) *Let ϕ be straight in $\mathcal{P}(R)$. Then ϕ is persistent iff it is regular (that is $w(\phi) \neq 0$ in L^*_R).*

Proof. (a) Suppose ϕ is not persistent. Then by definition there is a ρ such that $\rho(\phi) = 0$ in $\mathcal{P}^+(\rho(R))$. Hence $w(\rho(\phi)) = 0$ and by hardlemma $w(\phi) = 0$.

(b) Suppose conversely that ϕ is persistent. There are two cases.

(b1) Suppose further that ϕ is an *elementary path*, i.e., ϕ crosses no cut. Being straight, ϕ is an upward-then-downward path, for to switch from traveling downward to traveling upward needs either a rebounce, a twist or a cut. Therefore $w(\phi)$ is not zero, since it is already in stable form.

(b2) Suppose on the contrary that ϕ crosses a cut c and let ρ be the one-step reduction associated to c. At least one of the terms of $\rho(\phi)$ is persistent and shorter than ϕ. Hence by induction hypothesis $w(\rho(\phi)) \neq 0$, and by easylemma $w(\phi) \neq 0$. □

The adequacy theorem has an immediate corollary inducing a simplification of persistence.

Corollary 15 *A straight path ϕ is persistent iff there is reduction ρ such that an element of $\rho(\phi)$ is an elementary path.*

Proof. Suppose there is a reduction ρ such that an element of $\rho(\phi)$ is an elementary path. Such a path has a non null weight as proved in case b1 above, therefore by the preservation lemma 13, ϕ has a non null weight which, by adequacy, is equivalent to ϕ is persistent.

Conversely, the argument sketched in the proof of the ab^* theorem (theorem 9) builds a special reduction ρ of the net such that at each step of ρ, ϕ has exactly one residual the size of which strictly decreases along ρ. Therefore at some point $\rho(\phi)$ must be an elementary path and we are done. □

Theorem 16 (Adequacy, part 2) *L^*_R separates persistent paths, i.e., if ϕ_1 and ϕ_2 are persistent paths in R with same source and goal and not subpath one of the other, then $w(\phi_1\phi_2^*) = 0$ in L^*_R.*

Proof. These two paths have a first bifurcation point, i.e., there are paths ϕ, ϕ'_1 and ϕ'_2 and two distinct premises α_1 and α_2 of a link such that $\phi_i = \phi'_i\alpha^*_i\phi$, $i = 1, 2$. Let w_i, x_i and w be respectively the weights of ϕ'_i, α_i and ϕ. By ab^* version 3, we may apply the equations of inversive category to weights of paths so that we get:

$$w(\phi_1)w(\phi_2)^* = w_1x^*_1ww^*x_2w^*_2 = w_1x^*_1ww^*x_1x^*_1x_2w^*_2 = 0.$$

□

Remark. Note that L^*_R does not separate regular bouncing paths because $w(\phi\phi^*\phi) = w(\phi)$.

This last result says that not only L^*_R catches all persistent paths, but also, in some sense, that there is as much information in $\mathsf{EX}(R)$ as in the set of persistent paths.

References

[ADLR94] Andrea Asperti, Vincent Danos, Cosimo Laneve, and Laurent Regnier. Paths in the lambda-calculus. In *Proceedings of the Ninth Annual Symposium on Logic in Computer Science*, Paris, 1994. IEEE Computer Society Press.

[AGL92] Martin Abadi, Georges Gonthier, and Jean-Jacques Lévy. Linear logic without boxes. In *Proceedings of the Seventh Annual Symposium on Logic in Computer Science*, San Diego, 1992. IEEE Computer Society Press.

[AL93] Andrea Asperti and Cosimo Laneve. Paths, computations and labels in lambda-calculus. In *Proceedings of the fifth International Conference on Rewriting Techniques and Applications*, volume 690 of *Lecture Notes in Computer Science*, pages 152–167. Springer-Verlag, 1993.

[Dan90] Vincent Danos. *Une Application de la Logique Linéaire à l'Étude des Processus de Normalisation (Principalement du λ-Calcul).* Thèse de doctorat, Université Paris 7, 1990.

[DJS94] Vincent Danos, Jean-Baptiste Joinet, and Harold Schellinx. LKQ and LKT: Sequent calculi for second order logic based upon dual linear decompositions of classical implication. In *this volume*, 1994.

[DR89] Vincent Danos and Laurent Regnier. The structure of multiplicatives. *Archives for Mathematical Logic*, 28:181–203, 1989.

[DR93] Vincent Danos and Laurent Regnier. Local and asynchronous beta-reduction. In *Proceedings of the Eight Annual Symposium on Logic in Computer Science*, Montreal, 1993. IEEE Computer Society Press.

[Gir87a] Jean-Yves Girard. Linear logic. *Theoretical Computer Science*, 50(1):1–102, 1987.

[Gir87b] Jean-Yves Girard. Multiplicatives. In Lolli, editor, *Logic and Computer Science : New Trends and Applications*, pages 11–34, Torino, 1987. Università di Torino. Rendiconti del seminario matematico dell'università e politecnico di Torino, special issue 1987.

[Gir88a] Jean-Yves Girard. Geometry of interaction I: interpretation of system F. In Ferro, Bonotto, Valantini, and Zanardo, editors, *Proceedings of the Logic Colloquium 88*, pages 221–260, Padova, Italy, 1988. North Holland.

[Gir88b] Jean-Yves Girard. Geometry of interaction II: Deadlock free algorithm. In Martin-Löf & Mints, editor, *Proceedings of COLOG'88*, volume 417 of *Lecture Notes in Computer Science*, pages 76–93. Springer-Verlag, 1988.

[Gir95] Jean-Yves Girard. Geometry of interaction III: Accomodating the additives. In *this volume*, 1995.

[Lév78] Jean-Jacques Lévy. *Réductions Correctes et Optimales dans le Lambda-Calcul.* Thèse de doctorat, Université Paris 7, 1978.

[MR91] Pasquale Malacaria and Laurent Regnier. Some results on the interpretation of λ-calculus in operator algebra. In *Proceedings of the Sixth Annual Symposium on Logic in Computer Science*, Amsterdam, 1991. IEEE Computer Society Press.

[Par92] Michel Parigot. $\lambda\mu$-calculus: an algorithmic interpretation of classical natural deduction. In *Proceedings LPAR'92*, volume 624 of *Lecture Notes in Computer Science*, pages 190–201. Springer-Verlag, 1992.

[Reg92] Laurent Regnier. *Lambda-Calcul et Réseaux.* Thèse de doctorat, Université Paris 7, 1992.

Geometry of Interaction III : Accommodating the Additives

Jean-Yves Girard
Laboratoire de Mathématiques Discrètes
UPR 9016 – **CNRS**
163, Avenue de Luminy, Case 930
13288 MARSEILLE Cedex 09

girard@lmd.univ-mrs.fr

Abstract

The paper expounds geometry of interaction, for the first time in the full case, i.e. for all connectives of linear logic, including additives and constants. The interpretation is done within a C^*-algebra which is induced by the rule of resolution of logic programming, and therefore the execution formula can be presented as a simple logic programming loop. Part of the data is public (shared channels) but part of it can be viewed as private dialect (defined up to isomorphism) that cannot be shared during interaction, thus illustrating the theme of communication without understanding. One can prove a nilpotency (i.e. termination) theorem for this semantics, and also its soundness w.r.t. a slight modification of familiar sequent calculus in the case of exponential-free conclusions.

1 Introduction

1.1 Towards a monist duality

Geometry of interaction is a new form of semantics. In order to understand what is achieved, one has to discuss the more traditional forms of semantics.

1.1.1 Classical model theory

The oldest view about logic is that of an external observer : there is a pre-existing reality (mathematical, let us say) that we try to understand (e.g. by

proving theorems). This form of *dualism* is backed by the so-called completeness theorem of Gödel (1930), which says that a formula is provable iff it is true in all models (i.e. in all realizations). There is strong heterogeneity in the duality world/observer (or model/proof) proposed by model-theory, since the latter is extremely finite whereas the former is infinite. Hilbert's attempt at reducing the gap between the two actors failed because of the renowned incompleteness theorems, also due to Gödel (1931), whose basic meaning is that infinity cannot be eliminated.
A paradoxical situation arose because Gentzen proved in 1934 a *cut-elimination* theorem for a formulation of logic of his own, *sequent calculus*, yielding the lineaments of an elimination of infinity in the style proposed by Hilbert. The paradox was mainly ideological, since the applications that Gentzen gave of his method to Peano arithmetic (in 1936 and later), make use of very strong forms of infinity... and therefore achieved no elimination of infinity, but for true believers.
The cut-elimination procedure introduced by Gentzen was a pure syntactical rewriting technique for proofs, enabling one to eliminate infinitary notions from finitary theorems, but whose termination could not be proved without even stronger infinitary techniques... Such a hybrid animal had a difficult life, and in particular could not find his status within the narrow duality models/proofs.

1.1.2 The semantics of proofs

Of course classical model-theory does not refuse the observer as a minor part of the reality (the same is true for classical physics) : it makes the assumption of an objective reality where notions like *true, false...* make sense. A formula A is true or false in the world, and if I prove A it is also true that I prove A. Brouwer, by introducing *intuitionism*, radically changed the classical paradigm, by excluding the external reality and focusing on the interaction of proofs (the first consequence was taxonomical : the creation of *classical* logic, the new name for ordinary simple-minded logic).
Instead of explaining a proof Π of $A \Rightarrow B$ as *the justification that whenever A is true then B is also true*, intuitionism takes the viewpoint of the function which, given as input a proof Σ of A, yields a proof $\Pi(\Sigma)$ of B as output ; the basic example is the identity function, which maps a proof of A to itself, and which can be seen as a proof of the basic tautology $A \Rightarrow A$. This functional viewpoint yields the so-called *semantics of proofs.*
This change of viewpoint should not be too easily styled subjectivistic, even if this was the ideology of Brouwer. This must be seen as a critics of simple-minded realism, analogous to relativity theory, which considers time as the quantified result (and no longer the cause) of interaction. A pure subjectivis-

tic reading is so-called *realizability* which interprets the functions of semantics of proofs as purely syntactical operations, taking the code (e.g. the Gödel-number) of a proof of A to the code of a proof of B. But this interpretation is slightly regressive, since the functions involved in the semantics of proofs have a very high degree of *naturality*, which is conspicuous in the following extraordinary fact : whether we take Gentzen's sequent calculus or neighboring paradigms (lambda-calculi, natural deduction), equipped with the rewriting used for eliminating infinity (cut-elimination or β-conversion or normalization), the functions which are induced by proofs are exactly the canonical morphisms of a *cartesian closed category*, a beautiful kind of category (even if one had to wait until 1969 to find the first truly interesting example, *Scott domains*).

It must be observed that intuitionism does not formally contradict the duality models/proofs. For instance there is still a notion of model with a completeness theorem for intuitionistic logic. But the problem is not that one cannot import classical model-theory in the intuitionistic world : it is that it becomes inefficient. Take for instance the notion of *consistency* : in classical logic, in conformity with Hilbert's view, the consistency of A produces a model of A, i.e. an object ; but intuitionistic consistency only produces the mockery of an object, a so-called *Kripke model.* The notion of a consistent intuitionistic theory is therefore as ridiculous as the idea of fixing a tire with a horseshoe, nay feeding a horse with leadfree gasoline.

The central point of intuitionism is indeed *constructivity* (which can take the more ideological dress of *constructivism*, that we shall not discuss). Like classical logic (and like any reliable bank), one should be able *on request*, to exhibit something ; no longer a model, but some explicit information[1]. Typically, a proof of a disjunction $A \vee B$, could on request be replaced with a proof of A or with a proof of B. From this viewpoint, classical logic which allows the principle $A \vee \neg A$ without being able to tell which one holds, is "inconsistent" (in the common sense, as in the famous saying by king François 1er : *women are inconsistent*)[2].

Technically speaking, this exhibition is made possible by the fact that the cut-free proofs (or the normal terms in λ-calculus) that are the outputs of cut-elimination are proved by very restricted means : typically, in the case of intuitionistic disjunction, the only cut-free way to get $A \vee B$ is to get it from A or to get it from B.

But this explanation would be incomplete, if one were not stressing its "on request" aspect : in real life, a proof of $A \vee B$ is *never* a proof of A or a proof

[1]In classical logic the model is constructed from the absence of a proof, whereas constructivity tries to extract information from the presence of a proof

[2]Souvent femme varie/Bien fol est qui s'y fie

of B (only a masochist would state $A \vee B$ when he knows A). In other terms, a proof of $A \vee B$ is a proof of A or B, but we don't know which one ; only on request, we can be more explicit, and tell you which side holds. But this requires a painful work (typically cut-elimination) to replace everywhere the implicit by the explicit.
Here we have to be very precise, *mathematics cannot be explicit* : as soon as one departs from trivial elementary facts like $2 + 2 = 4$, mathematics (and all forms of reasoning) become abstract, i.e. *implicit*. This implicit character is conspicuous by the use of *lemmas*, which are combined by transitivity, by means of the rule of *Modus Ponens* : from A and $A \Rightarrow B$ deduce B, rewritten under the form of the CUT-rule of Gentzen (we adopt here the formalism of linear logic, see below) :

$$\frac{\vdash \Gamma, A \quad \vdash A^{\perp}, \Delta}{\vdash \Gamma, \Delta} CUT$$

Without the CUT-rule, it is practically impossible to make any deduction (typically, if I know that $27 \cdot 37 = 999$, I cannot use the lemma "commutativity of multiplication" to infer $37 \cdot 27 = 999$). Hence the truly paradoxical result of Gentzen is that this essential rule can be eliminated (for pure logic, i.e. predicate calculi), i.e. that mathematics can (within some limits) be explicated... not by a human being, but surely by a machine.
These observations are the starting point for applications to computer science : the choice L/R between the two premises of a disjunction can be used to represent a boolean datum, a proof of an implication can represent a computable function, the CUT-rule can take care of application of a function to an argument and cut-elimination, suitably implemented, can handle evaluation. The existence of a categorical model basically asserts the robustness of this approach (e.g. independence of any implementation). This is the basis for *functional programming*, more precisely *typed λ-calculus* an experimental language popular among theoreticians.

1.1.3 The monist duality

The intuitionistic world replaces the relation *model of* $\neg A$*/proof of* A of classical logic with the relation *proof of* $A \Rightarrow B$*/proof of* A. This is not a duality : to get a duality, one at least needs an involutive negation, and such a thing does not exist in the intuitionistic world. Linear logic ([G86]) is based on a refined analysis of the categorical semantics of intuitionistic proofs (replacement of Scott domains by *coherent spaces*), and individualizes new logical connectives. The basic point is to remove the rule of contraction of Gentzen (which amounts to saying that from A one can deduce $A \wedge A$) and also the rule of weakening (which allows fake hypotheses), in which case the basic symmetries missing in intuitionistic logic are restored ; these symmetries are expressed by the involutive connective of *linear negation* $A^\perp$. It is now possible to think of *monist* duality, *proof of* $A^\perp$/*proof of* A based on the CUT-rule, with the essential difficulty that A and $A^\perp$ cannot be simultaneously provable (because in the CUT-rule, Γ and Δ never happen to both empty). In the classical case, the duality is of a very strange kind : a proof of A and a model of $\neg A$ cannot simultaneously exist, in particular classical duality cannot account for the difference between two proofs of A, since there is model of $\neg A$ on which to compare them. If we want duality to account for the distinction between proofs of the same formula, then the two partners should not be exclusive one of another... The construction of a satisfactory monist duality is therefore delicate (we have to replace *proof of* A with something slightly more liberal) and has been attacked by various methods, from categorical semantics to game semantics, without yet any definite answer. *Geometry of interaction* is one of the major approaches to this question.

1.2 Linear Logic

Coherent semantics ([G86]) is built in analogy to linear algebra. The basic constructions of linear algebra can be mimicked by coherent semantics, and yield the basic linear connectives :

- There is an involutive duality (analogous to vector space duality), which induces *linear negation* $A^\perp$;

- There are operations analogous to the tensor product of vector spaces, and which yield the so-called *multiplicative* connectives $\otimes$, $⅋$, $\multimap$; $⅋$, which is the disjunction *par*, is the dual of $\otimes$, which is the conjunction *times* (the tensors here are strongly non self-dual) ; the *linear implication* $\multimap$ can be defined by $A \multimap B = A^\perp ⅋ B$. To the multiplicative universe should be attached the dual constants 1 and $\perp$.

- There are operations analogous to the direct sum of vector spaces, and which yield the so-called *additive* connectives, $\&, \oplus$; $\&$, which is the conjunction *with* is the dual of $\oplus$, which is the disjunction *plus* ; here too the sums are strongly non-involutive. To the additive universe should be attached the dual constants $\top$ and 0.

- There are operations analogous to symmetric tensor algebras, and which yield the so-called *exponential* connectives $!, ?$. ! (*of course*) is the dual of ? (*why not*).

The main categorical properties of these connectives are expressed by a certain number of canonical isomorphisms :

- Involutivity of negation (which allow in practice to ignore double negation symbols) ;
- Commutativity of $\otimes, \parr, \&, \oplus$;
- Associativity of $\otimes, \parr, \&, \oplus$;
- Distributivity of $\otimes$ over $\oplus$ (and dually of $\parr$ over $\&$) ;
- Exponentiation isomorphisms $!(A\&B) \simeq !A\otimes !B$ (and dually $?(A \oplus B) \simeq ?A \oplus ?B$) ;
- Neutrality of the constants 1 w.r.t. $\otimes$, $\perp$ w.r.t. $\parr$, $\top$ w.r.t. $\&$, 0 w.r.t. $\oplus$, together with $!\top \simeq 1$ and $?0 \simeq \perp$.

1.3 Linear sequent calculus

Linear logic is organised into a sequent calculus, (in which we can in particular derive the canonical proofs of the isomorphisms just mentioned). Its standard syntax is one-sided, with defined negation (and implication). Sequents are of the form $\vdash A_1, \ldots, A_n$, where $A_1, \ldots, A_n$ is a sequence of formulas. The rules are organised in three groups :

1.3.1 Identity/negation

$$\frac{}{\vdash A, A^{\perp}}ID \qquad \frac{\vdash \Gamma, A \quad \vdash A^{\perp}, \Delta}{\vdash \Gamma, \Delta}CUT$$

This group asserts that A is A, the only absolute evidence of logic ; this fact is expressed by two rules. In the one-sided calculus, the identity can be seen as the definition of negation.

1.3.2 Structure

$$\frac{\vdash \Gamma, A, B, \Delta}{\vdash \Gamma, B, A, \Delta}X$$

a rule that can be avoided if one considers sequents as multisets instead of sequences. The two other traditional structural rules

$$\frac{\vdash \Gamma}{\vdash A, \Gamma}W \qquad \frac{\vdash A, A, \Gamma}{\vdash A, \Gamma}C$$

also called *weakening* and *contraction* are not allowed in linear logic. These rules are essential in classical and (when written in two-sided style) in intuitionistic logics.

1.3.3 Logic

$$\frac{\vdash \Gamma, A \quad \vdash \Delta, B}{\vdash \Gamma, \Delta, A \otimes B}\otimes \qquad \frac{\vdash A, B, \Gamma}{\vdash A \mathbin{⅋} B, \Gamma}⅋ \qquad \frac{}{\vdash 1}1 \qquad \frac{\vdash \Gamma}{\vdash \bot, \Gamma}\bot$$

The rule for $\otimes$ concatenates (adds up) the contexts ; if the context is a price to pay (by destruction) to get a formula, a natural price for $A \otimes B$ is the sum of price for A and of a price for B.

$$\frac{\vdash \Gamma, A \quad \vdash \Gamma, B}{\vdash \Gamma, A \& B}\& \qquad \frac{\vdash A, \Gamma}{\vdash A \oplus B, \Gamma}\oplus_1 \qquad \frac{\vdash B, \Gamma}{\vdash A \oplus B, \Gamma}\oplus_2 \qquad \frac{}{\vdash \top, \Gamma}\top$$

The rule for & assumes that the two contexts, i.e. prices, are the same ; in other terms a possible price for $A \& B$ is a price for which one can get any of A and B, which means that if both of them are simulataneously available (as expected from a conjunction) for this price, only one of them (up to our choice) will eventually be bought.

$$\frac{\vdash ?\Gamma, A}{\vdash ?\Gamma, !A}! \qquad \frac{\vdash A, \Gamma}{\vdash ?A, \Gamma}d? \qquad \frac{\vdash \Gamma}{\vdash ?A, \Gamma}w? \qquad \frac{\vdash ?A, ?A, \Gamma}{\vdash ?A, \Gamma}c?$$

These rules are called *promotion, dereliction, weakening, contraction.* The interpretation in terms of prices (or resources) is no longer very convincing for exponentials (especially because of the contraction rule) : this is because the exponential group is a "classical" group, which enables one to instill traditional classical or intuitionistic features inside a calculus which otherwise (in spite of the novelty of its connectives) would not be expressive enough. Indeed exponentials allow weakening and contraction, but no longer as structural (i.e. universal) rules, but as controlled logical rules : indeed the role of exponentials is precisely to control the use of these two rules.

The main connective of linear logic is linear implication (which does not appear in the official right-handed syntax) and which is extremely different from the familiar intuitionistic implication : from A and $A \multimap B$ one can still derive B... but we have lost A in the process. This strong difference with preexistent logical systems is due to the disappearance of the contraction rule which enabled one to produce two copies of A from one. This fact is responsible for the resource-sensitive character of linear logic. This is also expressed as a

form of *causality* : $A \multimap B$ is the action "from A get B" which involves, like in physics, a reaction : the destruction of the cause. Now observe that among the rules for exponentials, reappears the contraction rule, but now limited to the holders of the front symbol ?. This is why the interpretation of intuitionistic implication as $!A \multimap B$ is possible. $!A$ has the meaning of A *ad libitum*, i.e. allows us from a single use of A to its unlimited reuse. One can also say, in analogy with quantum mechanics, that linear logic is about microscopic facts and that the exponentials ensure the transition with the macroscopic world... but such tantalizing analogies should be taken with care ; in particular quantum non-determinism has not yet find a definite analogue in terms of linear logic.
The idea of causality is that of a reciprocal annihilation of A and $A^{\perp}$ by the CUT -rule ; this goes very well with our idea of a monist (or homogeneous) duality. (Notice the change w.r.t. intuitionistic logic, where there inputs and outputs ; linear logic says that these roles are exchanged by linear negation, and what is called input or output depends on the user). But how does this annihilation work ?

1.4 Proof-nets

Although the basic formalisms for explication are basically equivalent, it makes in practice a lot of difference to work with natural deduction (or typed λ-calculus) instead of sequent calculus. If sequent calculus remains the best possible formulation of logic, its cut-elimination procedure spends too much time on bureaucratic details, typically permutation of rules. Natural deduction has the immense advantage (in the absence of disjunction and existence) of being insensitive to the order of rules, and therefore is much more efficient than sequent calculus. The idea was therefore to build a kind of natural deduction for linear logic, the main difficulty being that the tree form of natural deduction (so successful in the absence of disjunction and existence) could no longer be exploited in the presence of an involutive negation, for which the distinction input/output or hypothesis/conclusion no longer makes sense.
Proof-nets are this "linear natural deduction", and their discovery has been a long process : first limited to multiplicatives (without the constants), they have been extended to quantifiers, and more recently to additives [G94]. The main difficulty came because of the graph-like structure of such unfamiliar proofs : given a graph $\mathcal{G}$ which pretends to be a proof-net (a so-called *proof-structure*), can we decide whether this claim is grounded ? This is the problem of correctness criterions : the solutions found are of the form $\mathcal{G}$ *successfully passes a certain set of tests*, see [L93] in this volume. Soon after the introduction of multiplicative proof-nets in [G86], the paper [G86A] introduced a major dualist idea, namely that the tests to be passed by a proof of A are like

virtual proofs of its negation $A^\perp$; "virtual" means that these tests (which in general do not represent proofs at all) are "dense" in a set containing all proofs of $A^\perp$ if such objects were to exist.
Proofs and tests are therefore homogeneous in nature : both can be seen as finite "wirings" of the atoms of formulas, and cut-elimination basically amounted to connect the wires and to "follow the current". Mathematically such wires are permutations and cut-elimination can be written as a pure operation between permutations, the only role of logic being to guarantee that a certain expression makes sense, i.e. converges. It "only" remained to extend this paradigm to full logic...

1.5 Geometry of interaction

The philosophy is the following : elimination of infinity (or better : *explication*) is not the implementation of a clever, but artificial algorithm. On the contrary, there is an intrinsic dynamics of interaction (expressing the physical reciprocal annihilation of two antagonist actors), and logic is here only as a comment on this "physical" phenomenon. The comment is about :

- Specifications of the antagonistic actors in the interaction (in analogy with the shape of plugs in electricity which are not responsible for the passage of the current, but which limits in principle the plugging of an *acceptor of 220V* to a *giver of 220V)* ; the specification is therefore a formula.

- The building of a step-by-step justification of the specification ; this justification is what we usually see as a syntactical proof.

The problem was to find the right kind of mathematical objects. The multiplicative case was handled by means of finite permutations. A permutation of n can be represented by an isometry of the finite-dimensional Hilbert space $\mathcal{C}^n$[3]. The need to represent the contraction rule for the connective "?" (and whose dynamics is duplication) leads one to replace finite dimension by "continuous" dimension. In other terms this means that the general case will be handled by means of certain operators (technically : partial isometries) of the standard separable Hilbert space $\mathcal{H}$. Indeed we are interpreting proofs by objects of a C^*-algebra (acting on the Hilbert space, but consisting only of very specific operators). This C^*-algebraic aspect is not technically essential, but it was the constant leading intuition for the whole program, and without a backbone made of solid mathematics (presumably not yet used in a significant way) nothing would have existed.

[3]The appendix contains the basic definitions relevant to this subsection

Cut-free proofs are interpreted by such operators, the meaning being that of a static linear input/output machinery. Typically the linear logic axiom $\vdash A, A^\perp$ is represented by an extension cord, whose abstract definition is to have two extremities complementarily labelled or shaped (here $A, A^\perp$) and whose physical action is to transfer the input in A as an output to $A^\perp$ and vice versa : such an operator can be written as a 2×2 anti-diagonal matrix. In general all logical rules are interpreted by isomorphisms of C^*-algebras, and in case of binary rules, summations. Typically, in [G88] the interpretation was based on the $*$-isomorphisms induced by an isometry of $\mathcal{H} \oplus \mathcal{H}$ into $\mathcal{H}$ and $\mathcal{H} \otimes \mathcal{H}$ onto $\mathcal{H}$, see subsections 5.8 and 5.9 in appendix, see also [DR93], this volume.
To achieve a dynamical effect, we must give inputs to our operators ; this is achieved in presence of CUT : the general paradigm of interpretation is that of a pair (U, σ) (with $\sigma = 0$ in the cut-free case). The partial isometry σ is hermitian (i.e. it is a partial symmetry) and expresses a feedback. This is our way of saying that CUT is a physical plugging. Take the example of a cut between $\vdash \Gamma, A$ and $\vdash A^\perp, \Delta$: the two operators V and W enjoying these specifications are indeed square matrices (respectively labelled by the indices Γ, A and $A^\perp, \Delta$). We can join the sets of indices into $\Gamma, A, A^\perp, \Delta$, and form U (which corresponds to a "disjoint sum" of V and W). Now we can introduce the partial symmetry σ which exchanges the two opposite labels and is zero everywhere else. From the pair (U, σ) it is possible to express the feedback, namely that the inputs coming through A and $A^\perp$ are equal to the outputs coming out of the same channels, but flipped by σ. This amounts to writing a linear equation (whose parameters are the inputs labelled by the remaining free plugs Γ, Δ). For instance assume that U maps a direct sum $\mathcal{H} \oplus \mathcal{H}$ to itself, and that σ is a feedback on the first coordinate (i.e. σ^2 is the projection on the first coordinate) ; then the problem is, given $x \in \mathcal{H}$, to find $y, a \in \mathcal{H}$ such that :

$$U(\sigma(a) \oplus x) = a \oplus y$$

A sufficient condition for a solution is the invertibility of $1 - \sigma U$, in which case the *execution formula*

$$RES(U, \sigma) = (1 - \sigma^2)U(1 - \sigma U)^{-1}(1 - \sigma^2)$$

expresses the solution, i.e. the input/output dependency of the remaining plugs (i.e. y as a function of x), see subsection 5.7 in appendix.
A stronger existence condition is the *nilpotency* of σU, i.e. that some power $(\sigma U)^n$ is zero, a condition which is experimentally true for all pairs (U, σ) arising from logical proofs[4]. In this case the central part of the execution

[4]For non-logical systems like *pure λ-calculus* there is still a notion of *weak nilpotency*, see [G88A, MR91]

formula can be written as a finite sum :

$$U(1-\sigma U)^{-1} = U + U\sigma U + U\sigma U\sigma U + \ldots$$

the length of the sum being equal to the *order of nilpotency* n of σU. Observe that n can be extremely big : the basic way to get a high order of nilpotency n is to make n cuts with identity axioms, which is not very surprising ! But the logical rules, interpreted by $*$-isomorphisms will alter the pattern, since if $*$-isomorphisms cannot affect invariants like orders of nilpotency etc., they surely can dramatically affect the apparent size (i.e. the description) of objects : think for instance that the sum $U_1 + \ldots + U_{100}$ of 100 isometric copies of U can easily be recovered from a single operator of the form $U \otimes 1$ which has a more compact definition (this is the basic idea behind the interpretation of !). In fact this change of size is so dramatic that the order of nilpotency can hardly be predicted from the pair (U, σ), and that the nilpotency of all pairs $\Pi^\bullet\sigma$ coming from proofs in standard logical systems cannot be proved within usual mathematics : this nilpotency is related to the termination of cut-elimination, and therefore implies the consistency of various systems (in consequence, by incompleteness it cannot be proved without a heavy use of infinitary notions). The *execution formula* does not quite correspond to syntactical cut-elimination, but it is not too far ; in particular for proofs of sufficiently simple formulas, the correspondence is exact and that's enough. The interest of this approach for computer science was later confirmed by its applications to optimal reduction in λ-calculus due to Gonthier ([GAL92]).

1.6 The case of additives

The original paper [G88] only dealt with multiplicatives, exponentials and quantifiers of any order, which was enough to modelize extant typed λ-calculi. However some essential elements were missing, typically the treatment of additive connectives. The situation remained the same for several years, until a satisfactory extension of *proof-nets* to the additive case was found [G94]. The main novelty consists in assigning boolean *weights* to the links of a proof-nets. Since geometry of interaction uses a C^*-algebraic formulation and boolean algebras are basically commutative C^*-algebras, there is no major obstacle to this extension.

The main difficulty arises in the interpretation of the &-rule, typically in presence of a context. For instance, if U and V are operators on $\mathcal{H} \oplus \mathcal{H}$ corresponding to proofs of $\vdash C, A$ and $\vdash C, B$, we must "merge" U and V in order to represent the result of the application of the &-rule. If we see U and V as 2×2-matrices (U_{ij}) and (V_{ij}), this merge is performed in quite different ways, depending on the index, typically :

- U_{22} and V_{22} (which correspond to the formulas A and B) can be merged into a single operator W_{22} (corresponding to the formula $A\&B$), by means of an isometry φ of $\mathcal{H} \oplus \mathcal{H}$ into $\mathcal{H}$, as in the multiplicative case. The isometry φ is used to relate $A\&B$ to its components A and B : it is public knowledge that the coefficient W_{22} of *any* proof of $\vdash C, A\&B$ comes from coefficients U_{22} and V_{22} by means of φ.

- But the case of U_{11} and V_{11} is more delicate : these coefficients correspond to the common context C. A plain summation (i.e. $W_{11} = U_{11} + V_{11}$) would definitely be too brutal, erasing the fact that U_{11} is related to A and not to B. Therefore U_{11} and V_{11} must also be merged by means of an isometry φ' of $\mathcal{H} \oplus \mathcal{H}$ into $\mathcal{H}$. The problem arises from the fact that the formula C is used to label U_{11} and V_{11} and W_{11}, and so that nothing in C indicates that we have to look for a decomposition along φ' ; worse, the main connective of C (e.g. if C begins with a "&") suggests another decomposition, along say φ. These two decompositions should be simultaneously possible.

- The trick is to introduce an auxiliary space : in [G88] a cut-free proof Π of a n-ary sequent was interpreted by a $n \times n$ matrix with entries in a C^*-algebra Λ^*. This operator could be seen as acting on a n-ary direct sum $\mathcal{H} \oplus ... \oplus \mathcal{H}$. Now, our operators will still act on a n-ary direct sum, but of n spaces $\mathcal{H} \otimes \mathcal{H}'$ where $\mathcal{H}'$ is a Hilbert space, seen as a space of *auxiliary messages*[5]. The conflict between the φ and the φ' decomposition which occurs in the case of W_{11} is solved by taking for φ and φ' the isometries of $(\mathcal{H} \otimes \mathcal{H}') \oplus (\mathcal{H} \otimes \mathcal{H}')$ into $\mathcal{H} \otimes \mathcal{H}'$ respectively induced by an isometry of $\mathcal{H} \oplus \mathcal{H}$ into $\mathcal{H}$ and $\mathcal{H}' \oplus \mathcal{H}'$ into $\mathcal{H}'$. The space $\mathcal{H}'$ is therefore specifically used to handle additive merges.

- But this is not enough, since the same C can serve several times as a context to an additive rule : think of a sequent $\vdash C, A\&B, A'\&B'$ which involves the merge of four coefficients : the two possible orders of performance of the &-rules induce two alternative merges of the four coefficients. No commutation trick (like the distinction between $\mathcal{H}$ and $\mathcal{H}'$) can help us any longer. But it is easy to see that the two solutions proposed are isomorphic, i.e. that they are equal up to a change of the auxiliary messages, i.e. up to an isometry of $\mathcal{H}'$. This introduces the most important idea, the real novelty of this paper : everything dealing with $\mathcal{H}'$ is up to isomorphism... This means that everything related to $\mathcal{H}'$ is *private*, and we therefore speak of a *dialect*. The dialect

[5] Of course $\mathcal{H}'$ is isomorphic to h, but we prefer to keep different names here, since both spaces are used in a strongly different spirit

(which behaves to some extent like bound variables in traditional syntax) is up to isomorphism, what we express by introducing an equivalence relation : being *variants*. All constructions are compatible with that equivalence[6]. When we interpret a conjunctive rule, $\otimes$ or &, a common dialect has to be produced, but we should beware of accidental matchings of the respective dialects, exactly like in usual syntax, we have to prevent accidental collision of bound variables. The only possibilities are to make generic operations, tensorization of dialects or summation. On the other hand $\mathcal{H}$ represents (shared) *channels* of communication. The paradigm of interaction is therefore *communication without understanding through shared channels.*

- As in [G88] the *execution* formula does not quite correspond to syntactical normalisation ; in other terms certain operations (typically erasings) are not performed. But the soundness of the execution formula w.r.t. cut-elimination still holds under a reasonable restriction on the proven formula. Notice that the syntax has to be adapted to prove this result (introduction of the $\flat$-*calculus*).

1.7 Closing the system

The interpretation given here, although perhaps not definitive, omits no logical connective. But there is still an essential lingering problem : the problem of "closing the system". This means being able to build a duality between proofs of A and proofs of $A^{\perp}$. We already noticed that the notion of proof has to be liberalized to something wider (like the tests for $A^{\perp}$ in the case of multiplicative proof-nets) but homogeneous to proofs. This is for instance what we do in our definition of *weak types*, see definition 12. But the problem is that, among the elements of the weak type $A^{\bullet}$ interpreting A there is no way to distinguish those objects which are interpretations of proofs. Surely the notion of *orthogonality* introduced in definition 11 and central in the notion of a weak type, is too lax. What is missing is that $U \perp V$ is completely symmetrical in the partners, and that there is no way to tell the wheats (real proofs) from the tares (tests). This could be fixed by defining (but how ?) an output $< U, V >$ of the interaction, sufficiently antisymmetric so that a distinguished value -say 1- for $< U, V >$ would exclude the same value for $< V, U >$. It would of course remain to prove completeness, namely that if $< U, V >= 1$ for all $V \in A^{\perp\bullet}$, then U is the interpretation of a proof of A, a widely open program...

[6]But one, namely the promotion rule for ! which strongly resists to the proof-net spirit : in [G86], the promotion rule is accommodated with a "box", i.e. treated as a global entity ; this globality is perhaps inherent to this rule and technically expressed by a failure of the variance principle

This is where geometry of interaction merges with *game semantics* : the weak orthogonality between U, V means that U, V can be seen as strategies for the two players inside the same game, and our output $< U, V >$ is the result of the game, won by U when the result is 1. Therefore the problem of closing the system is closely related to the search for a complete game semantics for linear logic... (the connection between linear logic and game semantics was initiated by Blass, see [B92]).

1.8 Notes on the style

Geometry of interaction is most naturally handled by means of C^*-algebras ; this yields surely more elegant proofs, but it obscures the concrete interpretation. So we prefer to follow a down to earth description of the interpretation. An unexpected feature will help us : the C^*-algebras used can in fact be interpreted in terms of *logic programming*, since the basic operators are very elementary PROLOG programs, and composition is resolution ! This ultimate explanation of the dynamics of logic in terms of (some theoretical) logic programming cannot be without consequence : we could for instance try to use linear logic as a typing (i.e. specification) discipline for such logic programs... this promising connection is therefore a strong reason to remain concrete. In appendix we give some hints as to the C^*-algebraic presentation, enough to understand the various allusions to the Hilbert space that are scattered in the main text.
Although our main inspiration was our (yet unwritten) work on additive proof-nets, proof-nets will not at all appear below : first this would make too many simultaneous novelties, second certain details ($\flat$-proof-nets, simultaneous treatment of additives and quantifiers) have not yet been fixed.
Finally, we shall give many examples, especially in the section devoted to exponentials ; it must not be inferred from the elementary character of these examples that the global construction is trivial : like in λ-calculus, all atomic steps are elementary, but the combination of few such steps easily becomes... explosive. Concerning the style, we are deeply indebted to the referee who read the first version of this work in details and suggested many modifications that should increase the legibility of the paper.

2 The algebra of resolution

2.1 Resolution

By a *term language* T, we mean the set of all terms t that can be obtained from variables $x_1, \ldots, x_n$ by means of a finite stock of function letters ; we assume that T contains at least one constant (0-ary function letter), so that the set of ground (i.e. closed) terms is non-empty. By a *language* L, we mean the set of all atoms $pt_1 \ldots t_n$, where p is a n-ary predicate letter varying through a finite set of such symbols, given with their arity. An important case is when the n predicates of L are of the same arity m, in which case we shall use the notation $T^m \cdot n$ for L.

Remark 1
The basic need of geometry of interaction is to have two constants g and d and a binary function $\odot$. The predicate letters correspond to the indices of matrices, i.e. a $n \times n$ matrix makes use of $p_1, \ldots, p_n$ (in the sequel we shall rather use formulas as indices : $p_A, p_B, \ldots$, which supposes that they are pairwise distinct, a standard bureaucratic problem, solved by replacing formulas by occurrences etc.). These predicates will be binary, although in the presence of a binary function, the arity of the predicates hardly matters.

Definition 1 (Unification)
Let L be a language ; we say that two expressions (terms or atoms) e and e' are unifiable *when there is a substitution (called a* unifier *of e and e') θ of terms for the variables occurring in e and e' such that $e\theta = e'\theta$.*

Remark 2
In such a case, there is a *most general unifier* (mgu), i.e. a unifier θ of e and e' such that any other unifier θ' can can be written as the composition $\theta\theta''$ of θ with a substitution θ''. Most general unifiers are unique up to renaming of variables.

Definition 2 (Clauses)
Let L be a language ; by a (rudimentary) clause *in L, we mean a sequent $P \mapsto Q$, where P and Q are atoms of L* with the same variables.

Remark 3
Clauses are considered up to the renaming of their variables : we do not distinguish between $p(x) \mapsto q(x)$ and $p(y) \mapsto q(y)$ (but we distinguish between $p(x \odot y) \mapsto q(x \odot y)$ and $p(y \odot x) \mapsto q(x \odot y)$; in other words the variables of clauses are bound, and other notations like $\forall x_1 \ldots x_n P \Rightarrow Q$ could be considered.

Definition 3 (Resolution)
If $P \mapsto Q$ and $P' \mapsto Q'$ are clauses, then we can assume w.l.o.g. that they have no variable in common and

- *either Q and P' are not unifiable, in which case we say that* resolution *of the clauses $P \mapsto Q$ and $P' \mapsto Q'$ (in this order)* fails

- *or Q and P' have a mgu θ, in which case resolution succeeds, with the* resolvant
  $P\theta \mapsto Q'\theta$ *; in that case we introduce the notation*

$$(P \mapsto Q) \cdot (P' \mapsto Q') = P\theta \mapsto Q'\theta.$$

Remark 4
Let us introduce the formal clause 0, and extend resolution by setting $(P \mapsto Q) \cdot (P' \mapsto Q') = 0$ when unification fails ; then it turns out that resolution is associative. All the clauses $P \mapsto P$ are idempotent. Finally the operation defined on clauses by $(P \mapsto Q)^* = Q \mapsto P$ is an anti-involution : $(R \cdot R')^* = R'^* \cdot R^*$.

Remark 5
In case L has a single (let us say : binary) predicate symbol p, the clause $I = p(x,y) \mapsto p(x,y)$ is neutral w.r.t. resolution : if x, y are not free in r, s, t, u then $p(x,y)$ unifies with $p(r,s)$ by means of the mgu $\theta(x) = r, \theta(y) = s$, hence $I(p(r,s) \mapsto p(t,u)) = p(x,y)\theta \mapsto p(t,u)\theta = p(r,s) \mapsto p(t,u)\ldots$

2.2 The algebra $\lambda^*(L)$

Definition 4 (The algebra)
Let $\lambda^*(L)$ be the set of all (finite) formal linear combinations

$$\sum \alpha_i \cdot (P_i \mapsto Q_i),$$

with the scalar α_i in C ; $\lambda^*(L)$ is obviously equipped with

- *A structure of complex vector space.*
- *A structure of complex algebra, the multiplication being extended by bilinearity from resolution :*

$$(\sum \alpha_i \cdot (P_i \mapsto Q_i))(\sum \beta_j \cdot (R_j \mapsto S_j)) = \sum \alpha_i\beta_j \cdot (P_i \mapsto Q_i))(R_j \mapsto S_j)$$

- *An identity : for instance the identity of $T^2 \cdot n$ is $\sum p_i(x,y) \mapsto p_i(x,y)$ (x, y, distinct variables) : easy consequence of remark 5.*
- *An (anti-)involution defined by $(\sum \alpha_i \cdot (P_i \mapsto Q_i))^* = \sum \overline{\alpha_i} \cdot (Q_i \mapsto P_i)$.*

In other terms $\lambda^*(L)$ bears all the features of a C^*-algebra, (see definition 26 in appendix) but for the norm features. Although the uses of this fact are quite limited[7], it is of interest to make a C^*-algebra out of $\lambda^*(L)$. This is done in appendix, see subsection 5.6.

[7]The only known utilization of the Hilbert space in geometry of interaction is due to Danos & Regnier [DR93].

2.3 The execution formula

Definition 5 (Loops)
Let us fix the language L. We adapt the main definitions and notions coming from [G88].

- *A* message *is any finite sum $\sum(P_i \mapsto P_i)$, with the P_i and P_j not unifiable for $i \neq j$. A message is therefore a projection (see subsection 5.5) of $\Lambda^*(L)$, and messages commute with each other. We abbreviate the message $P_i \mapsto P_i$ in P_i.*

- *A* wiring *is any finite sum $\sum(P_i \mapsto Q_i)$, with the P_i and P_j not unifiable for $i \neq j$ and the Q_i and Q_j not unifiable for $i \neq j$; each of the clauses $(P_i \mapsto Q_i)$ is called a* wire. *A finite wiring is therefore a partial isometry (see subsection 5.5) of $\Lambda^*(L)$. (In [G88], wirings were called "observables".) If w and w' are wirings then ww' is a wiring.*

- *If m is a message and w is a wiring, there is a (non-unique) message m', such that $m'w = wm$: take $m' = wmw^*$ (PROOF: $wmw^*w = ww^*wm = wm$. $\square$) ; in other terms wirings* propagate *messages.*

- *A* loop *(U, σ) is a pair of wirings such that σ is hermitian, i.e. $\sigma = \sigma^*$ (in particular σ^2 is a projection and $\sigma^3 = \sigma$).*

- *A loop* converges *when σU is nilpotent, i.e. when $(\sigma U)^n = 0$ for some n. The* execution *of the loop is the element*

$$EX(U, \sigma) = U(1 - \sigma U)^{-1} = U \sum_{k=0}^{n-1} (\sigma U)^k$$

and the result *of the execution is defined as*

$$RES(U, \sigma) = (1 - \sigma^2)U(1 - \sigma U)^{-1}(1 - \sigma^2).$$

The output is a wiring, whereas the execution is not a wiring. Observe that $U = V(1 + \sigma V)^{-1}$, with $V = EX(U, \sigma)$.

Remark 6
Wirings correspond to very specific PROLOG programs. (Here we do not refer to the actual language PROLOG, but rather to the idea of resolution, independently of any implementation). Each *wire* $P_i \mapsto Q_i$ in $\sum(P_i \mapsto Q_i)$ can be seen as a clause in a program : the program consists in the clauses $P_i \mapsto Q_i$. Such programs are very peculiar :

- The *tail* (i.e. the body) of the clause consists of a single literal,

- The same variables occur in the head and in the tail,
- The heads are pairwise non-unifiable,
- The tails are pairwise non-unifiable.

But one has to be (slightly) subtler to relate execution in our sense to the execution of a logic program. In order to interpret a loop, we systematically duplicate all predicate symbols $p, q, r, \ldots$ into pairs (p^-, p^+) (p^- for inputs, p^+ for outputs) ; therefore an atom P receives two interpretations P^- and P^+. The pair (U, σ) yields the following program, consisting in the clauses

- Clauses $P^- \mapsto Q^+$ for each wire $P \mapsto Q$ in U,
- Clauses $P^+ \mapsto Q^-$ for each wire $P \mapsto Q$ in σ.

The execution of (U, σ) consists in finding all clauses $P^\epsilon \mapsto Q^{\epsilon'}$ which are consequences of this program by means of resolution. The result of the execution consists in retaining only those clauses $P^- \mapsto Q^+$ that cannot be unified (by prefixing and/or suffixing) with a clause coming from σ.

Exercise 1
By suitable modifications of the language L, express the execution formula as a logic programming problem such that

- *L has only three unary predicates $\mathrm{e}, \mathrm{c}, s$*
- *(U, σ) is interpreted by only four kinds of clauses*
  - *clauses* $\mathrm{e}(t) \mapsto \mathrm{c}(u)$,
  - *clauses* $\mathrm{e}(t) \mapsto s(u)$,
  - *clauses* $\mathrm{c}(t) \mapsto \mathrm{c}(u)$,
  - *clauses* $\mathrm{c}(t) \mapsto s(u)$,

  in such a way that the result of the execution corresponds to the consequences of the program (by means of resolution) of the form $\mathrm{e}(t) \mapsto \mathrm{c}(u)$.

3 The interpretation of MALL

3.1 Laminated wirings

Strictly speaking, the interpretation can be made within a fixed algebra $\lambda^*(L)$, for instance with one constant, one binary function and one unary predicate... but this not very user-friendly. In practice we shall need two constants g and

d and one function letter $\odot$, together with a variable number of predicates to reflect the structure of *sequents*, which have a variable number of formulas. Also these predicates will be chosen *binary*, for reasons that will be explained below. In the sequel T will be the term language $\{g, d, \odot\}$, and $T^2 \cdot n$ will be the language built from the terms of T by means of n binary predicates, $p_1, \ldots, p_n$. We shall use the following notational convention for terms : $t_1 \ldots t_k$ is short for $t_1 \odot (\ldots \odot (t_{k-1} \odot t_k) \ldots)$, so that tuv is the same as $t(uv)$.

Definition 6 (Variants)
Let U and V be wirings in $\lambda^*(T^2 \cdot n)$; U and V are said to be variants *if there exist* wirings *W and W' of $\lambda^*(T^2 \cdot n)$ such that :*

- *$U = W^*VW$ and $V = W'^*UW'$*
- *W (resp. W') is a sum of wirings of the form $\sum_{i=1}^{n} p_i(x, t) \mapsto p_i(x, u)$ with x not free in t, u ; this means that W and W', as operators, are of the form $Id \otimes Z$.*

The notion of being variants is clearly an equivalence relation, noted $\sim$.

Proposition 1
If $U \sim V$ and σ is a sum of wires of the form $p(t, x) \mapsto q(u, x)$ (x not occurring in t, u), then :

- *$U\sigma$ is nilpotent iff $V\sigma$ is nilpotent ;*
- *In this case, $EX(U, \sigma) \sim EX(V, \sigma)$ and $RES(U, \sigma) \sim RES(V, \sigma)$.*

PROOF: immediate. □

Definition 7 (Changing dialects, see remark 7 below)
Let U be a wiring in $\lambda^*(T^2 \cdot n)$;

- *$\otimes_1(U)$ is defined as follows : every wire $p(t, u) \mapsto q(t', u)$ of U is replaced with the wire $p(t, uy) \mapsto q(t', uy)$, where y is a fresh variable.*
- *$\otimes_2(U)$ is defined as follows : every wire $p(t, u) \mapsto q(t', u)$ of U is replaced with the wire $p(t, xu) \mapsto q(t', xu)$, where x is a fresh variable.*
- *$\&_1(U)$ is defined as follows : every wire $p(t, u) \mapsto q(t', u)$ of U is replaced with the wire $p(t, ug) \mapsto q(t', ug)$.*
- *$\&_2(U)$ is defined as follows : every wire $p(t, u) \mapsto q(t', u)$ of U is replaced with the wire $p(t, ud \mapsto q(t', ud)$.*

Proposition 2

- $\otimes_1(U) \sim \otimes_2(U)$
- $\&_1(U) \sim \&_2(U)$
- $\otimes_1(U) \sim \otimes_1(\otimes_1(U))$

PROOF: let us verify the third fact : let $W = \sum_{i=1}^{n} p_i(x,(yz)z) \mapsto p_i(x,yz)$ and
$W' = \sum_{i=1}^{n} p_i(x,yz_1z_2) \mapsto p_i(x,(yz_1)z_2)$; then $\otimes_1(U) = W^* \otimes_1(\otimes_1(U))W$ and $\otimes_1(\otimes_1(U)) = W'^* \otimes_1(U)W'$. □

Definition 8 (Laminated wirings)
A wiring U in $\lambda^*(T^2 \cdot n)$ is said to be laminated *if it is a sum of wires the form $p(t,u) \mapsto q(t',u)$ and if $U \sim \otimes_1(U)$.*

Remark 7
A proof Π of a n-ary sequent using m cuts will be interpreted by a pair $(\Pi^\bullet, \sigma)$ of laminated wirings in $\lambda^*(T^2 \cdot (2m+n))$. We can see $\Pi^\bullet$ as an operator on the Hilbert space
$(\mathcal{H} \otimes \mathcal{H}) \oplus \ldots \oplus (\mathcal{H} \otimes \mathcal{H})$, a n-ary direct sum of spaces $\mathcal{H} \otimes \mathcal{H}$, which can also be seen as
$(\mathcal{H} \oplus \ldots \oplus \mathcal{H}) \otimes \mathcal{H}$. Now, in $(\mathcal{H} \oplus \ldots \oplus \mathcal{H}) \otimes \mathcal{H}$ the two components are treated in a very different spirit :

- The first component $(\mathcal{H} \oplus \ldots \oplus \mathcal{H})$ is seen as a space of *shared* messages, the *channels* ; typically the decomposition of $(\mathcal{H} \oplus \ldots \oplus \mathcal{H})$ into its summands is part of this public knowledge.

- The second component $\mathcal{H}$ is seen as a space of *private* messages : a private *dialect* useful only for $\Pi^\bullet$ but that cannot be communicated to the environment ; the dialect is a typical additive creation, coming from the need of avoiding overlap in the treatment of the &-rule.

- This distinction between these two components is a consequence of the *lamination* of $\Pi^\bullet$: by definition 8 every wire of $\Pi^\bullet$ is of the form $p(t,u) \mapsto q(t',u)$.

In fact the privacy of the dialect is expressed by the fact that all constructions of geometry of interaction indeed deal with equivalence classes modulo $\sim$. In the case of binary rules, this fact limits the possible ways of merging dialects. If we want to combine U and V, we shall try to replace them by *variants* U' and V' whose respective dialects are in certain relation. We basically can use the constructions of definition 7, and these two possibilities lead to the interpretations of the $\otimes$- and &-rules, that we now sketch : let us assume that U and V are laminated wirings interpreting cut-free proofs of $\vdash \Gamma, A$ and $\vdash \Delta, B$; then

- The first operation is to create a common dialect ; this is done by :
  - in the case of a $\otimes$ -rule, tensorization of the respective dialects : let $U' = \otimes_1(U)$, $V' = \otimes_2(V)$.
  - in the case of a & -rule, summation of the respective dialects : $U' = \&_1(U)$, $V' = \&_2(V)$.
- The next operation is to glue together A and B ; this is done by merging (in $U' + V'$) the predicates p_A and p_B respectively associated to A and B into a new predicate p_C (where C is $A \otimes B$ or $A \& B$, depending on the rule), as follows :
  - every occurrence $p_A(t, u)$ of p_A in U' is replaced with an occurrence of $p_C(tg, u)$,
  - every occurrence $p_B(t, u)$ of p_B in V' is replaced with an occurrence of $p_C(td, u)$.

  the result W of this operation is the desired interpretation.

To end with our informal description of the interpretation, let us consider the case of a CUT-rule : let us assume that U and V interpret cut-free proofs of $\vdash \Gamma, A$ and $\vdash \Delta, A^\perp$; then

- We define $U' = \otimes_1(U)$, $V' = \otimes_2(V)$ as in the case of a $\otimes$-rule, and we sum them ;
- We introduce σ as the sum $(p_A(x, y) \mapsto p_{A^\perp}(x, y)) + (p_{A^\perp}(x, y) \mapsto p_A(x, y))$. The pair $(U' + V', \sigma)$ is the desired interpretation.

Remark 8

The rules for $\otimes$ and & share the last step, which consists in merging two predicates p_A and p_B by means of renaming of the channels ; the choice of the constants g and d to do so is arbitrary, but once it has been set, one cannot touch it. The main difference between a $\otimes$-rule and a &-rule is the way in which the respective dialects are merged

- In the case of the $\otimes$, this is splendid ignorance (the dialects now commute with each other) : this is achieved by replacing u by uy and v by xv ;

- Whereas in the case of the &, the dialects are just disjoint (i.e. incompatible). This is achieved by replacing u by ug and v by vd.

Observe that we could as well choose isomorphic solutions for the first step, for instance :

- Exchange (in the case of $\otimes$) $\otimes_1$ with $\otimes_2$.
- Replace (in the case of &) $\&_1$ with $\&_2$.

This flexibility contrasts with the (relative) rigidity of the choice involved when merging predicates. This is because channels are public whereas dialects are private.

3.2 Interpretation of proofs of MALL

We first interpret the fragment **MALL** of multiplicative-additive linear logic.

Definition 9 (Pattern)
The general pattern is as follows :

- *We are given a proof* Π *of a sequent* $\vdash \Gamma$, *containing cuts on formulas* Δ *; in* Δ *formulas are coupled in pairs* $(B, B^{\perp})$ *; in fact we use the formulas of* Δ *as labels for the cuts performed rather than actual formulas. The expression* "Let Π be a proof of $\vdash [\Delta], \Gamma$" *is short for* "Let Π be a proof of $\vdash \Gamma$, containing cuts on the formulas Δ".
- *We introduce the algebra* $\lambda^*(\Delta, \Gamma)$ *:*
  - *the function letters are the constants* d *and* g *and the binary* $\odot$ *;*
  - *the predicate letters are all binary :* p_A *for all formula* A *in* Δ *and* Γ. *We assume that these formulas are pairwise distinct (if not, they can be distinguished by adding indices).*
  - *an important particular case is* $\Delta = \emptyset$, *in which case we use the notation* $\lambda^*(\Gamma)$ *; if* Γ *were empty as well (a case that will never occur) then we would need to define* $\lambda^*(\emptyset)$, *that can conveniently be taken as the field* C *of scalars.*
- *We define* $\sigma_{\Delta;\Gamma}$ *in the algebra* $\lambda^*(\Delta, \Gamma)$ *by*

  $$\sigma_{\Delta;\Gamma} = \sum_{B \in \Delta} p_B(x, y) \mapsto p_{B^{\perp}}(x, y)$$

  x, y *are distinct variables, the sum is taken over all cut formulas ; observe that* $\sigma = \sigma^*$, *and that* σ^2 *is a projection. Therefore* $1 - \sigma^2$ *is a projection too, and*

  $$\forall u \in \lambda^*(\Delta, \Gamma), \quad (1 - \sigma^2)u(1 - \sigma^2) \in \lambda^*(\Gamma)$$

in other terms prefixing and suffixing with $1 - \sigma^2$ acts like a restriction, see subsection 5.7 in appendix.

- *We interpret Π by a pair $(\Pi^\bullet, \sigma_{\Delta;\Gamma})$ in the algebra $\lambda^*(\Delta, \Gamma)$: the definition of $\Pi^\bullet$ is by induction.*

Definition 10 ($\Pi^\bullet$)
Let Π be a proof of $\vdash [\Delta], \Gamma$; we associate to Π its interpretation $(\Pi^\bullet, \sigma_{\Delta;\Gamma})$, where $\Pi^\bullet$ is defined by induction :

Case 10.1 *If Π consists in the axiom $\vdash A, A^\perp$, then $\Pi^\bullet = \sigma_{A,A^\perp}$; i.e.*

$$\Pi^\bullet = (p_A(x,y) \mapsto p_{A^\perp}(x,y)) + (p_{A^\perp}(x,y) \mapsto p_A(x,y))$$

Case 10.2 *If Π is obtained from Π_1 and Π_2 by means of a CUT-rule with cut-formulas $A, A^\perp$, then*

- *let us replace in $\Pi_1^\bullet$ all atoms $p_C(t,u)$ by $p_C(t,uy)$ (y fresh variable) ; the result is called U_1 ;*
- *let us replace in $\Pi_2^\bullet$ all atoms $p_C(t,v)$ by $p_C(t,xv)$ (x fresh variable) ; the result is called U_2 ; then*

$$\Pi^\bullet = U_1 + U_2.$$

Of course the essential feature of this step is the modification of the component σ : if Π_1 and Π_2 respectively prove $\vdash [\Delta_1], \Gamma_1, A$ and $\vdash [\Delta_2], \Gamma_2, A^\perp$, then Π is a proof of
$\vdash [\Delta_1, \Delta_2, A, A^\perp], \Gamma_1, \Gamma_2$.

Case 10.3 *If Π is obtained from Π_1 by means of an exchange rule, then*

$$\Pi^\bullet = \Pi_1^\bullet.$$

Case 10.4 *If Π is obtained from Π_1 and Π_2 by means of a $\otimes$-rule introducing $A \otimes B$, then*

- *let us replace in $\Pi_1^\bullet$ all atoms $p_C(t,u)$ by $p_C(t,uy)$ (y fresh variable) when C is distinct from A, and all atoms $p_A(t,u)$ by $p_{A\otimes B}(tg,uy)$; the result is called U_1 ;*
- *let us replace in $\Pi_2^\bullet$ all atoms $p_C(t,v)$ by $p_C(t,xv)$ (x fresh variable) when C is distinct from B, and all atoms $p_B(t,v)$ by $p_{A\otimes B}(td,xv)$; the result is called U_2 ; then*

$$\Pi^\bullet = U_1 + U_2.$$

Case 10.5 *If Π is obtained from Π_1 by means of a $⅋$-rule, then let us replace in $\Pi_1^\bullet$ all atoms $p_A(t,u)$ by $p_{A⅋B}(tg,u)$ and all atoms $p_B(t,v)$ by $p_{A⅋B}(td,v)$; the result is by definition $\Pi^\bullet$.*

Case 10.6 *If Π is obtained from Π_1 and Π_2 by means of a &-rule introducing $A\&B$, then*

- *let us replace in $\Pi_1^\bullet$ all atoms $p_C(t,u)$ by $p_C(t,ug)$ when C is distinct from A, and all atoms $p_A(t,u)$ by $p_{A\&B}(tg,ug)$; the result is called U_1 ;*
- *let us replace in $\Pi_2^\bullet$ all atoms $p_C(t,v)$ by $p_C(t,vd)$ when C is distinct from B, and all atoms $p_B(t,v)$ by $p_{A\&B}(td,vd)$; the result is called U_2 ; then*

$$\Pi^\bullet = U_1 + U_2.$$

Case 10.7

- *If Π is obtained from Π_1 by means of a $\oplus_1$-rule, then let us replace in $\Pi_1^\bullet$ all atoms $p_A(t,u)$ by $p_{A\oplus B}(tg,u)$; the result is by definition $\Pi^\bullet$.*
- *If Π is obtained from Π_1 by means of a $\oplus_2$-rule, then let us replace in $\Pi_1^\bullet$ all atoms $p_B(t,u)$ by $p_{A\oplus B}(td,u)$; the result is by definition $\Pi^\bullet$.*

Case 10.8 *If Π is obtained from Π' by means of $\forall$-rule (first or second order), then $\Pi^\bullet = \Pi'^\bullet$.*

Case 10.9 *If Π is obtained from Π' by means of $\exists$-rule (first or second order), then $\Pi^\bullet = \Pi'^\bullet$.*

Remark 9

Definition 10 uses several times rather arbitrary constructions to merge dialects (the cases 10.4 and 10.6, but also the constant g in the case 10.1) . In fact in all these cases, we might have chosen variants, without any difference. This is because of proposition 1 will allows one to replace $\Pi^\bullet$ with a variant.

Example 1

Cut-elimination replaces the cut

$$\dfrac{\vdash \Gamma, A \quad \dfrac{}{\vdash B^\perp, B}\,Id}{\vdash \Gamma, B}\,CUT$$

(with A and B different occurrences of the same formula) by the original proof Π of $\vdash \Gamma, A$. The original proof is interpreted as $(\otimes_1(\Pi^\bullet) + \otimes_2(\sigma_{B^\perp,B\,;\Gamma,A}), \sigma_{A,B^\perp\,;\Gamma,B})$. Nilpotency is easily shown ($(U\sigma)^2 = 0$), and the result of execution is $(\otimes_1(\Pi^\bullet)$ in which p_A has been replaced with p_B, i.e. it is a variant of the interpretation of the modified proof. This shows, in this basic, but essential example, that execution corresponds to cut-elimination, up to variance.

3.3 The nilpotency theorem

Our first goal is to prove that $EX(\Pi^\bullet, \sigma)$ $(= \Pi^\bullet(\sigma\Pi^\bullet)^-1$, see definition 5) makes sense, i.e., $\Pi^\bullet\sigma$ is *nilpotent*. This will be done by an adaptation of the results from [G88].

Theorem 1 (Transitivity of cut)
In the algebra $\lambda^*(\Delta, \Delta', \Gamma)$ let us define σ, τ as respectively $\sigma_{\Delta;\Delta',\Gamma}$ and $\sigma_{\Delta';\Delta,\Gamma}$ (so that $\sigma_{\Delta,\Delta';\Gamma} = \sigma + \tau$) ; let U be a wiring

- *If $U\sigma$ is nilpotent, $RES(U, \sigma)$ $(= (1-\sigma^2)EX(U,\sigma)(1-\sigma^2))$ is a wiring.*
- *$(\sigma + \tau)U$ is nilpotent iff σU and $\tau \cdot RES(U, \sigma)$ are nilpotent ;*
- *In that case $RES(U, \sigma + \tau) = RES(RES(U, \sigma), \tau)$.*

PROOF: see [G88], lemmas 4 and 5, where this result was called *associativity of cut.* □

Theorem 2 (Nilpotency)
$\sigma\Pi^\bullet$ is nilpotent.

PROOF: the proof is adapted from [G88]. (We neglect all features related to quantifiers, which are quite the same.) We first define :

Definition 11 (Weak orthogonality)
Let U and V be laminated wirings in $\lambda^*(T^2)$ (an algebra with a single binary predicate p; U and V are said to be weakly orthogonal *if $\otimes_1(U) \cdot \otimes_2(V)$ is nilpotent. In that case we introduce the notation $U \perp V$.*

The definition makes sense because of the following :

Proposition 3

- *If $U \perp V$, then $V \perp U$.*
- *If $U \sim U'$ and $V \sim V'$ and $U \perp V$, then $U' \perp V'$.*

Definition 12 (Weak types)
Given a subset X of $\lambda^*(T^2)$, we define $X^\perp = \{U \; ; \; \forall V \in X, \quad U \perp V\}$.
A weak type *is any set X of laminated wirings such that $X = X^{\perp\perp}$. The connectives of linear logic can be interpreted as operations on weak types :*

Case 12.1 *If X and Y are weak types, then we define the weak type $X \otimes Y$ as $Z^{\perp\perp}$, where Z consists of all $U' + V'$, where U' and V' are constructed from $U \in X$ and $V \in Y$ as follows :*

- *Replace in U all atoms $p(t,u)$ by $p(tg,uy)$.*
- *Replace in U all atoms $p(t,v)$ by $p(td,xv)$.*

Case 12.2 *If X and Y are weak types, then we define the weak type $X \parr Y$ as $(X^{\perp} \otimes Y^{\perp})^{\perp}$.*

Case 12.3 *If X and Y are weak types, then we define the weak type $X \oplus Y$ as $Z^{\perp\perp}$, where Z consists of : all U' and all V', where U' and V' are constructed from $U \in X$ and $V \in Y$ as follows :*

- *Replace in U all atoms $p(t,u)$ by $p(tg,u)$.*
- *Replace in U all atoms $p(t,v)$ by $p(td,v)$.*

Case 12.4 *If X and Y are weak types, then we define the weak type $X \& Y$ as $(X^{\perp} \oplus Y^{\perp})^{\perp}$.*

Remark 10
The interpretation by weak types is very degenerated ; in particular, it easy to check that & and $\oplus$ are not separated by this rough interpretation.

Definition 13 (Weak types associated to formulas and sequents)
Let us associate an arbitrary weak type $\alpha^{\bullet}$ to each atom α of a language in linear logic ; then each formula A built from such atoms by means of linear negation, and the binary connectives $\otimes$, $\parr$, &, $\oplus$ immediately gets interpreted by a weak type $A^{\bullet}$. (This also extends to quantifiers of any order, provided one works on weak type parameters as in definition 3 of [G88].) The definition of $A^{\bullet}$ immediately induces a definition of $\Gamma^{\bullet}$, when $\vdash \Gamma$ is a sequent : assume that $\Gamma = A_1, \ldots, A_n$, and let U be a laminated wiring in $\lambda^*(\Gamma)$;

- *Given laminated wirings $V_1, \ldots, V_n$ in $A_1^{\perp\bullet}, \ldots, A_n^{\perp\bullet}$ respectively, we can rename the predicates so that $V_i \in \lambda^*(A_i)$, and introduce $V_1' + \ldots + V_n'$ as the result of applying $n-1$ cases 10.4 to the V_i ; the object depends in which order the rules are performed, but different choices yield* variants.
- *Then $U \in \Gamma^{\bullet}$ iff for all $V_1, \ldots, V_n$ in $A_1^{\perp\bullet}, \ldots, A_n^{\perp\bullet}$ respectively, $U \perp V_1' + \ldots + V_n'$.*

of course the definition has two particular cases of interest :

- *The particular case $n = 1$ yields $(\vdash A)^{\bullet} = A^{\bullet}$;*
- *The particular case $n = 0$ yields $\vdash {}^{\bullet} = \mathcal{C}$, see definition 9.*

Lemma 2.1 *If U, V are variants and $U \in A^\bullet$ (resp. $U \in \Gamma^\bullet$), then $V \in A^\bullet$ (resp. $V \in \Gamma^\bullet$).*

PROOF: immediate □

Lemma 2.2 *$U \in (\Gamma, A)^\bullet$ iff for all $V \in A^{\perp\bullet}$ the "cut" $(U_1 + V_1, \sigma_{A,A^\perp;\Gamma})$ (defined as in case 10.2) is such that : $(U_1 + V_1)\sigma_{A,A^\perp;\Gamma}$ is nilpotent and the result $RES(U_1 + V_1, \sigma_{A,A^\perp;\Gamma})$ is in $\Gamma^\bullet$.*

PROOF: easy, see lemma 7 of [G88]. □
The nilpotency theorem follows from the more precise :

Theorem 3
If Π is a proof of $\vdash [\Delta], \Gamma$, then $\Pi^\bullet\sigma_{\Delta;\Gamma}$ is nilpotent and $RES(\Pi^\bullet, \sigma_{\Delta;\Gamma}) \in \Gamma^\bullet$.

PROOF: this theorem is proved by induction on Π ; there is a general pattern that can be followed for all cases. We do this in detail for the first case (the case of a &-rule, which is the truly new case), but we shall only treat simplified versions of the other cases. The pattern basically reduces the general case to the cut-free case (i.e. when the premise(s) is (are) cut-free), and then the cut-free case to a context-free case. To do this in very rigorous way, it would be natural to enlarge the set of rules with an additional one : for each $V \in \vdash \Gamma^\bullet$ add the axiom Γ whose proof Π_V is interpreted as $\Pi_V^\bullet = V$ (in the spirit of model-theory, where new constants for the elements of the model are introduced).

Case 3.1 • We first treat a very limited case : the last rule of Π is a &-rule between proofs Π_1 of $\vdash A_1$ and Π_2 of $\vdash A_2$, (so that $\Delta = \emptyset$ and $\Gamma = A_1 \& A_2$). We have to show that for all U' obtained from $U \in A_1^{\perp\bullet}$ (resp. all V' obtained from $V \in A_2^{\perp\bullet}$) by means of definition 12.3, then $\Pi^\bullet \perp U'$ (resp. $\Pi^\bullet \perp V'$). But this immediately reduces to $\Pi_1^\bullet \perp U$ (resp. $\Pi_2^\bullet \perp V$).

- Then we treat the case where the premises are cut-free : the last rule of Π is a &-rule between proofs Π_1 of $\vdash \Phi, A_1$ and Π_2 of $\vdash \Phi, A_2$, (so that $\Delta = \emptyset$ and $\Gamma = \Phi, A_1 \& A_2$). We can argue by induction on the cardinality of Φ :

  - the case $\Phi = \emptyset$ has already been treated ;
  - if $\Phi = B, \Psi$, then by lemma 2.2, $\Pi^\bullet \in (\Phi, A_1 \& A_2)^\bullet$ (resp. $\Pi_1^\bullet \in (\Phi, A_1)^\bullet$, $\Pi_2^\bullet \in (\Phi, A_2)^\bullet$) iff for all $V \in B^{\perp\bullet}$ $\Pi^\bullet\sigma_{B,B^\perp;\Psi,A_1\&A_2}$ is nilpotent and $RES(\Pi^\bullet, \sigma_{B,B^\perp;\Psi,A_1\&A_2}) \in (\Psi, A_1\&A_2)^\bullet$ (resp. $\Pi_1^\bullet\sigma_{B,B^\perp;\Psi,A_1}$ is nilpotent and $RES(\Pi_1^\bullet, \sigma_{B,B^\perp;\Psi,A_1}) \in (\Psi, A_1)^\bullet$, $\Pi_2^\bullet\sigma_{B,B^\perp;\Psi,A_2}$ is nilpotent and $RES(\Pi_2^\bullet, \sigma_{B,B^\perp;\Psi,A_2}) \in (\Psi, A_2)^\bullet$). The result follows from the induction hypothesis on the size of Φ and the fact that

* $\Pi^\bullet \sigma_{B,B^\perp;\Psi,A_1\& A_2}$ is nilpotent iff $\Pi_1^\bullet \sigma_{B,B^\perp;\Psi,A_1}$ and $\Pi_2^\bullet \sigma_{B,B^\perp;\Psi,A_2}$ are nilpotent ;
* if we apply definition 10.6 to $RES(\Pi_1^\bullet, \sigma_{B,B^\perp;\Psi,A_1})$ and $RES(\Pi_2^\bullet, \sigma_{B,B^\perp;\Psi,A_2})$, then we get a variant of $RES(\Pi^\bullet, \sigma_{B,B^\perp;\Psi,A_1\& A_2})$.

- It remains to treat the general case : the last rule of Π is a &-rule between proofs Π_1 of $\vdash [\Delta_1], \Phi, A_1$ and Π_2 of $\vdash [\Delta_2], \Phi, A_2$, (so that $\Delta = \Delta_1, \Delta_2$ and $\Gamma = \Phi, A_1 \& A_2$). From the induction hypothesis, $\Pi_i^\bullet \sigma_{\Delta_i;\Phi,A_i}$ is nilpotent and $RES(\Pi_i^\bullet, \sigma_{\Delta_i;\Phi,A_i}) \in (\Phi, A_i)^\bullet$ for $i = 1, 2$. Let $u_i = RES(\Pi_i^\bullet, \sigma_{\Delta_i;\Phi,A_i})$; then it is easily seen that $\Pi^\bullet \sigma_{\Delta;\Phi,A\&B}$ is nilpotent and that by definition 10.6 applied to u_1 and u_2, we get $RES(\Pi^\bullet, \sigma_{\Delta;\Phi,A\&B}) \in (\Phi, A\&B)^\bullet$.

Case 3.2 If Π is the axiom $\vdash A, A^\perp$; for reasons of labelling, we prefer to rename the first A as B; then

$$\Pi^\bullet = (p_B(x,y) \mapsto p_{A^\perp}(x,y)) + (p_{A^\perp}(x,y) \mapsto p_B(x,y))$$

select $U = \sum p_A(t_i, v_i) \mapsto p_A(u_i, v_i)$ in $A^\bullet$; then these two wirings are respectively modified into

$$\otimes_1(\Pi^\bullet) = (p_B(x,yz) \mapsto p_{A^\perp}(x,yz)) + (p_{A^\perp}(x,yz) \mapsto p_B(x,yz))$$

and $\otimes_2(U) = \sum p_A(t_i, y'v_i) \mapsto p_A(u_i, y'v_i)$. The nilpotency of $(\otimes_1(\Pi^\bullet) + \otimes_2(U))\sigma_{A,A^\perp;B}$ is rather immediate, and the result is easily seen to be $\sum p_B(t_i, yv_i) \mapsto p_B(u_i, yv_i)$, a variant of U ; the result follows from lemmas 2.2 and 2.1.

Case 3.3 The last rule of Π is a CUT-rule between proofs Π_1 of $\vdash A$ and Π_2 of $\vdash A^\perp$. (This extremely simplified case can only occur because we have extended the set of all proofs by adding a lot of new axioms ; also remark that the reduction of the general case to this case makes a heavy use of transitivity of cut). Due to the peculiar definition of $\vdash^\bullet$, it suffices to show that $\Pi^\bullet \sigma_{A,A^\perp;}$ is nilpotent ; but this immediately reduces to the orthogonality of $\Pi_1^\bullet$ and $\Pi_2^\bullet$, which is precisely the induction hypothesis.

Case 3.4 The last rule of Π is a $\oplus_1$-rule applied to a proof Π_1 of $\vdash A$. This simplified case follows immediately from definition 12.3. The case of a $\oplus_2$-rule is similar.

Case 3.5 The last rule of Π is a $\otimes$-rule between proofs Π_1 of $\vdash A$ and Π_2 of $\vdash B$. This simplified case follows from definition 12.1.

Case 3.6 The last rule of Π is a $\parr$-rule applied to a proof Π_1 of $\vdash A, B$. By definition 12.2 we must show that for all $U \in A^{\perp\bullet}$ and $V \in B^{\perp\bullet}$, then (with U' and V' as in definition 12.3) $\Pi^\bullet \perp U' + V'$. But this is easily reduced to $\Pi_1^\bullet \perp \otimes_1(U) + \otimes_2(V)$, i.e. the induction hypothesis.

Case 3.7 The remaining cases are left to the reader. □

Remark 11
Although the result proved (nilpotency) is rather weak, it is "proof-theoretically strong" : for instance, in the presence of exponentials (next subsection) and second order quantification, it cannot be proved inside second-order arithmetic : this is because the orders of nilpotency are related to the length of normalisation steps (see [DR93] for instance).

3.4 $\flat$-sequent calculus

Our next goal is to prove that the result $RES(\Pi^\bullet, \sigma)$ of the execution is $\Pi_1^\bullet$, where Π_1 is obtained from Π by cut-elimination. Unfortunately

- We can only expect a *variant* of $\Pi_1^\bullet$;
- Worse, the result $RES(\Pi^\bullet, \sigma)$ contains "unerased" information, what we called a *beard* in [G88]. This second feature makes the precise statement of a theorem quite delicate.

Since of the two partners -geometry of interaction and cut-elimination-, the former (although very recent) is the most natural, we shall therefore (slightly) modify sequent calculus : the result is called $\flat$-sequent calculus. The reason for this terminology is that the calculus contains unessential rules corresponding to the erasings that geometry of interaction, in its lazy spirit, does not perform. We introduce a new symbol, $\flat$, that we call *flat* ; this symbol is treated differently from the other formulas of linear logic : $\flat$ cannot be combined ($\flat \parr \flat$, $\flat^\perp$ are not allowed) ; this leaves only the possibility of using $\flat$ inside sequents, like in $\vdash \flat$, $\vdash \flat, A$, $\vdash \flat, A, \flat$ etc. The specific rules of $\flat$ are the following :

$$\frac{}{\vdash \flat, \Gamma}\flat \qquad \frac{\vdash \flat, \flat, \Gamma}{\vdash \flat, \Gamma}c\flat \qquad \frac{\vdash \flat, \Gamma \quad \vdash \Gamma}{\vdash \Gamma}s\flat$$

- The axiom scheme $\vdash \flat, \Gamma$ for any Γ.
- The contraction rule : from $\vdash \flat, \flat, \Gamma$ deduce $\vdash \flat, \Gamma$.
- The summation rule : from $\vdash \flat, \Gamma$ and $\vdash \Gamma$ deduce $\vdash \Gamma$.

These rules are basically minor structural rules.

Remark 12

- $\flat$ can be faithfully interpreted as the constant $\top$ of linear logic : the axiom is the usual axiom for $\top$ and the rules are derivable... but this is not the point.

- Among the derivable rules for $\flat$:

  - a form of inflation : from $\vdash \flat, \Gamma$ derive $\vdash \flat, \Gamma, \Delta$
  - another form of summation : from $\vdash \flat, \Gamma$ and $\vdash \flat, \Gamma$ deduce $\vdash \flat, \Gamma$.

- Contraction and inflation imply that we do not count the multiplicity of $\flat$. This means that in reality, we are working with two kinds of sequents : usual ones (without $\flat$) and *flat* ones (with at least one $\flat$). Of course we get a much simpler presentation with only one kind of sequents, and this explains our choice.

Definition 14 (Geometry of interaction of $\flat$)
A proof Π of $\vdash [\Delta], \Gamma$ is interpreted as a laminated wire in $\lambda^*(\Delta, \Gamma - \flat)$, where $\Gamma - \flat$ is obtained from Γ by the removal of all $\flat$ (observe that since $\flat$ cannot be negated, it does not occur in Δ).

- *An axiom $\vdash \flat, \Gamma$ is interpreted by $0 \in \lambda^*(\Gamma - \flat)$.*

- *Contraction is interpreted identically.*

- *If Π follows from Π_1 and Π_2 by means of a summation : then $\Pi^\bullet = \&_1(\Pi_1^\bullet) + \&_2(\Pi_2^\bullet)$.*

Remark 13
The basic idea behind the symbol "$\flat$" is that a proof of a flat sequent $\vdash \flat, \Gamma$ comes from a "real" proof in which some essential part has been erased. Flat proofs interact with real ones by means of summation : summation typically occurs in the case of cut-elimination on an additive formula. For instance assume that we perform a cut between a proof of $\vdash A \oplus B$ (obtained from a proof Π of A by a $\oplus_1$-rule) and a proof of of a sequent $\vdash A^\perp \& B^\perp, \Gamma$ coming from proofs Π_1 of $\vdash A^\perp, \Gamma$ and Π_2 of $\vdash B^\perp, \Gamma$. The traditional way of eliminating the cut is to replace it with the proof Π' obtained by a cut between Π and Π_1, but this does not quite correspond to geometry of interaction. This is why in the $\flat$-calculus, our cut is replaced by the *summation* of Π' and the proof of $\vdash \flat, \Gamma$ obtained by a cut between the axiom $\vdash \flat, B$ and Π_2.

It is not surprising that the $\flat$-calculus enjoys cut-elimination... In fact we are rather interested in the details of the cut-elimination procedure that we now sketch :

Definition 15 (Cut-elimination procedure)

A cut

$$\dfrac{\dfrac{\cdots}{\vdash \Gamma, A}R \quad \dfrac{\cdots}{\vdash \Delta, A^{\perp}}S}{\vdash \Gamma, \Delta}CUT$$

is replaced as follows :

- *If the rule R is not an introduction of A or an axiom, then R is performed after the cut, for instance :*
  - *If R is a summation rule, with premises $\vdash \Gamma, A$ and $\vdash \Gamma, A, \flat$, then our cut is replaced with :*

$$\dfrac{\dfrac{\vdash \Gamma, A \quad \vdash \Delta, A^{\perp}}{\vdash \Gamma, \Delta}CUT \quad \dfrac{\vdash \Gamma, A, \flat \quad \vdash \Delta, A^{\perp}}{\vdash \Gamma, \Delta, \flat}CUT}{\vdash \Gamma, \Delta}s\flat$$

- *If the rule S is not an introduction of $A^{\perp}$ or an axiom, then S is performed after the cut, as above.*
- *If both R and S are introductions of the cut-formulas or axioms, then there are several cases, including :*
  - *when R is the identity axiom $\vdash A, A^{\perp}$, the proof is replaced with* $\dfrac{}{\vdash \Delta, A^{\perp}}S$ *;*
  - *when R is the axiom $\vdash \Phi, A, \flat$, the solution depends on S, typically :*
    * *if S is the axiom $\vdash \Psi, A^{\perp}, \flat$, then we replace the proof with the axiom $\vdash \Phi, \Psi, \flat, \flat$;*
    * *if S is a $\otimes$-rule, with premises $\vdash \Psi, B^{\perp}$ and $\vdash \Xi, C^{\perp}$, then our proof is replaced with two cuts between the axiom $\vdash \Phi, B, C, \flat$ and the premises of the $\otimes$-rule ;*
    * *if S is a $⅋$-rule, with premise $\vdash \Delta, B^{\perp}, C^{\perp}$, then our proof is replaced with two cuts between the axioms $\vdash \Phi, B, \flat$ and $\vdash C, \flat$ and the premise of the $⅋$-rule ; this yields a proof of $\vdash \Phi, \Delta, \flat, \flat$ that we contract into $\vdash \Phi, \Delta, \flat$;*

* *if S is a &-rule, with premises $\vdash \Delta, B^{\perp}$ and $\vdash \Delta, C^{\perp}$, then we cut these premises respectively with the axioms $\vdash \Phi, B^{\perp}, \flat$ and $\vdash \Phi, C^{\perp}, \flat$, yielding two proofs of $\vdash \Phi, \Delta, \flat$ to which we can apply summation, as seen in remark 12 ;*
* *if S is a $\oplus_1$-rule, with premise $\vdash \Delta, B^{\perp}$, then our proof is replaced with a a cut between this premise and the axiom $\vdash \Phi, B, \flat$...*

- *when R and S are both logical rules, then we get several cases, including :*
  * *if R is a $⅋$-rule, with premise $\vdash \Gamma, B, C$ and S is a $\otimes$-rule, with premises $\vdash \Psi, B^{\perp}$ and $\vdash \Xi, C^{\perp}$, then we replace our proof with*

$$\dfrac{\dfrac{\vdash \Gamma, B, C \quad \vdash \Psi, B^{\perp}}{\vdash \Gamma, \Psi, C}\,CUT \qquad \vdash \Xi, C^{\perp}}{\vdash \Gamma, \Psi, \Xi}\,CUT$$

  * *if R is a $\oplus_1$-rule, with premise $\vdash \Gamma, B$ and S is a &-rule, with premises $\vdash \Delta, B^{\perp}$ and $\vdash \Delta, C^{\perp}$, then we replace our proof with*

$$\dfrac{\dfrac{\vdash \Gamma, B \quad \vdash \Delta, B^{\perp}}{\vdash \Gamma, \Delta}\,CUT \qquad \dfrac{\dfrac{}{\vdash \Gamma, C, \flat}\,\flat \quad \vdash \Delta, C^{\perp}}{\vdash \Gamma, \Delta, \flat}\,CUT}{\vdash \Gamma, \Delta}\,s\flat$$

Proposition 4
Cut-elimination holds for the $\flat$-calculus, using the procedure sketched in definition 15.

PROOF: boring and straightforward ; however notice that the treatment of a cut on an additive formula makes a significant use of flat sequents. □

Theorem 4 (Soundness)
If a cut-free proof Π of $\vdash \Gamma$ is obtained from a proof Σ of $\vdash [\Delta], \Gamma$ by means of the transformations sketched in definition 15, then $\Pi^{\bullet}$ is a variant of $RES(\Sigma^{\bullet}, \sigma_{\Delta;\Gamma})$.

Lemma 4.1 *If a proof Π of $\vdash [\Delta'], \Gamma$ is obtained from a proof Σ of $\vdash [\Delta], \Gamma$ by one step of definition 15, then $RES(\Pi^{\bullet}, \sigma_{\Delta';\Gamma}) \sim RES(\Sigma^{\bullet}, \sigma_{\Delta;\Gamma})$.*

PROOF: we treat only a few distinguished cases :

- Assume that Σ is obtained from Σ_1, Σ_2 respectively proving $\vdash [\Delta], \Phi, A$, and $\vdash A^\perp, A$, (an identity axiom that we note $\vdash A^\perp, B$) via a cut on A. Then $\Sigma^\bullet$ is of the form $\otimes_1(\Sigma_1^\bullet) + \otimes_2(\Sigma_2^\bullet)$. Transitivity of cut yields $RES(\Sigma^\bullet, \sigma_{\Delta,A,A^\perp;\Phi,B}) = RES(RES(\Sigma^\bullet, \sigma_{\Delta;A,A^\perp,\Phi,B}), \sigma_{A,A^\perp;\Delta,\Phi,B})$. The result follows from the fact that $RES(\otimes_1(U) + \otimes_2(\Sigma_2^\bullet), \sigma_{A,A^\perp;\Delta,\Phi,B})$ is a variant of U (with $U = RES(\Sigma_1, \sigma_{\Delta;\Phi,A})$).

- Assume that Σ is obtained from $\Sigma_1, \Sigma_2, \Sigma_3$, respectively proving $\vdash [\Delta_1], \Phi, B, C$, $\vdash [\Delta_2], \Psi, B^\perp$ and $\vdash [\Delta_3], \Xi, C^\perp$, by a $\parr$-rule and a $\otimes$-rule followed by a cut, then the construction of $\Sigma^\bullet$ involves two steps :

  - the formation of the sum $U = \otimes_1(\Sigma_1^\bullet) + \otimes_2(\otimes_1(\Sigma_2^\bullet) + \otimes_2(\Sigma_3^\bullet))$; this sum is a variant of $\Pi^\bullet = \otimes_1(\otimes_1(\Sigma_1^\bullet) + \otimes_2(\Sigma_2^\bullet)) + \otimes_2(\Sigma_3^\bullet)$
  - the merge in U of p_B and p_C into p_A and of $p_{B^\perp}$ and $p_{C^\perp}$ into $p_{A^\perp}$.

  If we define $\Lambda = \Delta_1, \Delta_2, \Delta_3$, then it is immediate that $RES(\Sigma^\bullet, \sigma_{\Lambda,A,A^\perp;\Gamma}) = RES(U, \sigma_{\Lambda,B,B^\perp,C,C^\perp;\Gamma})$. But $\Lambda, B, B^\perp, C, C^\perp$ is Δ' so $RES(U, \sigma_{\Lambda,B,B^\perp,C,C^\perp;\Gamma})$ is a variant of $RES(\Pi^\bullet, \sigma_{\Delta';\Gamma})$.

- Assume that Σ is obtained from $\Sigma_1, \Sigma_2, \Sigma_3$, respectively proving $\vdash [\Delta_1], \Phi, B, C$, $\vdash [\Delta_2], \Psi, B^\perp$ and $\vdash [\Delta_3], \Psi, C^\perp$, by a $\oplus_1$-rule and a &-rule followed by a cut, then the construction of $\Sigma^\bullet$ involves two steps :

  - the formation of the sum $U = \otimes_1(\Sigma_1^\bullet) + \otimes_2(\&_1(\Sigma_2^\bullet) + \&_2(\Sigma_3^\bullet))$; this sum is a variant of $\Pi^\bullet = \&_1(\otimes_1(\Sigma_1^\bullet) + \otimes_2(\Sigma_2^\bullet)) + \&_2(\otimes_2(\Sigma_3^\bullet))$
  - the merge in U of p_B and p_C into p_A and of $p_{B^\perp}$ and $p_{C^\perp}$ into $p_{A^\perp}$.

  If we define $\Lambda = \Delta_1, \Delta_2, \Delta_3$, then it is immediate that $RES(\Sigma^\bullet, \sigma_{\Lambda,A,A^\perp;\Gamma}) = RES(U, \sigma_{\Lambda,B,B^\perp,C,C^\perp;\Gamma})$. But $\Lambda, B, B^\perp, C, C^\perp$ is Δ' so $RES(U, \sigma_{\Lambda,B,B^\perp,C,C^\perp;\Gamma})$ is a variant of $RES(\Pi^\bullet, \sigma_{\Delta';\Gamma})$.

- All the petty cuts between an axiom for $\flat$ and a logical rule (or an axiom for $\flat$) are easily treated : this is because the axiom for flat is interpreted by 0.

- The endless list of commutations of the CUT-rule is easily handled : such steps introduce variants. □

PROOF: the proof of theorem 4 is immediate from the lemma. □
Whether or not we have totally achieved our task, surely execution corresponds exactly to normalisation, but in a slightly exotic sequent calculus. This calculus contains more cut-free proofs than usual, and therefore its use might be problematic. But the question is solved by the following proposition :

Proposition 5
In the $\flat$-calculus, the booleans remain standard : there is a sequent Γ such that the set $\Gamma^\bullet$ of all $\Pi^\bullet$ where Π varies through cut-free proofs of Γ has exactly two elements.

PROOF: we have to give a precise meaning to the proposition.

- We can decide to represent booleans by a sequent $\vdash \alpha^\perp, \alpha \oplus \alpha$, where α is a given atomic formula.
- This formula has only two cut-free proofs in the usual sequent calculus (Identity axiom and a $\oplus_1$-rule, identity axiom and a $\oplus_1$-rule) ; these two proofs can be taken as the two booleans. Geometry of interaction obviously interprets them differently.
- In the $\flat$-calculus, the summation rule offers more possibilities of cut-free proofs ; however it is easily proved that the flat summands must have interpretation 0. □

Therefore the execution formula is consistent with usual syntactic manipulations : this means that for any boolean question that we can ask of the output of some cut-elimination, both methods will yield the same answer. See [G88] for a discussion.

3.5 The neutral elements

The case of the multiplicative units is delicate ; the only "natural" choice for interpreting the rule of introduction of $\perp$ is the trivial one : if a proof Π of $\vdash \Gamma, \perp$ comes from a proof $\vdash \Pi_1$ of Γ by the $\perp$-rule, let $\Pi^\bullet = \Pi_1^\bullet$. This choice has the following consequence : given two proofs Π_1 and Π_2 of $\vdash \Gamma$ (this sequent is written below as $\vdash \Gamma_i$ to distinguish the two proofs), the proofs

$$\dfrac{\dfrac{\vdash \Gamma_1}{\vdash \Gamma, \perp}\perp \quad \dfrac{\vdash \Gamma_2}{\vdash \Gamma, \perp}\perp}{\vdash \Gamma, \perp \& \perp}\& \qquad \text{and} \qquad \dfrac{\dfrac{\vdash \Gamma_2}{\vdash \Gamma, \perp}\perp \quad \dfrac{\vdash \Gamma_1}{\vdash \Gamma, \perp}\perp}{\vdash \Gamma, \perp \& \perp}\&$$

are interpreted by variants, i.e. are not distinguished. In fact proposition 5 would become false. In terms of $\flat$-calculus, a new problematic case arises, namely a cut

$$\dfrac{\dfrac{}{\vdash 1, \flat}\flat \quad \dfrac{\vdash \Gamma}{\vdash \perp, \Gamma}\perp}{\vdash \Gamma, \flat}CUT$$

for which the only natural cut-elimination would be to add a rule of weakening on $\flat$ (from $\vdash \Gamma$, deduce $\vdash \Gamma, \flat$) which contradicts our idea that a flat proof is a proof in which something is actually missing : this rule would acknowledge ordinary proofs as flat ones, thus destroying all our construction. In other terms, the "natural" geometry of interaction for $\perp$ leads to accept the rule : from $\vdash \Gamma$ and $\vdash \Gamma$ deduce $\vdash \Gamma$, which is a strong form of summation. With such a rule, booleans would no longer be standard (any formal sum of booleans would be accepted).
Instead, we modify the $\perp$-rule into : from $\vdash A, \Gamma$ deduce $\vdash \perp, A, \Gamma$, in which one of the formulas in the premise is distinguished. Strictly speaking, there is no need to distinguish A, if we stick to the viewpoint of sequents as sequences (instead of multisets). With this modified rule, the problem of a cut

$$\dfrac{\dfrac{}{\vdash 1, \flat}\flat \quad \dfrac{\vdash A, \Gamma}{\vdash \perp, A, \Gamma}\perp}{\vdash A, \Gamma, \flat}CUT$$

disappears ; for instance the cut can be replaced with a cut between $\vdash A, \Gamma$ and the flat axiom $\vdash A^{\perp}, A, \flat$ ...

Definition 16
The rules for the multiplicative units are interpreted as follows :

- *If Π is the axiom $\vdash 1$, then $\Pi^{\bullet}$ is the message $p_1(x, y)$;*
- *If the proof Π of $\vdash [\Delta], \perp, A, \Gamma$ is obtained from a proof Π_1 of $\vdash [\Delta], A, \Gamma$ by a $\perp$-rule, then $\Pi^{\bullet}$ is defined in terms of $\Pi_1^{\bullet}$:*
  - *in $\Pi_1^{\bullet}$ replace any wire $p_B(t, u) \mapsto p_A(t', u)$ with a wire $p_B(t, u) \mapsto p_{\perp}(t', u)$;*
  - *add to the result of this replacement the wire $p_{\perp}(x, y) \mapsto p_A(x, y)$.*

Remark 14

- One should prove the analogue of theorem 4 which can be done without difficulty. In case of extreme laziness, observe that a (slightly more complicated) geometry of interaction of multiplicative neutrals can easily be given by means of the second-order translation of $\perp$ as $\exists\alpha(\alpha \otimes \alpha^{\perp})$, for which the previous section yields a geometry of interaction. Our modified $\perp$-rule can be translated in second-order linear logic by

$$\dfrac{\dfrac{\vdash \Gamma, A \quad \dfrac{}{\vdash A^{\perp}, A}ID}{\vdash \Gamma, A \otimes A^{\perp}, A}\otimes}{\vdash \Gamma, \exists\alpha(\alpha \otimes \alpha^{\perp}), A}\exists$$

- In terms of proof-nets, the "natural choice" for $\bot$ corresponds to proof-nets where the weakened formula ($\bot$ or $?A$) is physically disconnected. There is no reasonable correctness criterion for such nets :
  - In the case of a multiplicative formula A using only multiplicative units as atoms, such a choice would lead to identify all proofs of A ;
  - Hence the correctness problem for such proof-nets contains the decision problem for the constant-only multiplicative fragment, which is known to be **NP**-complete, see [LW92].
  - But the general shape of the known criterions is **coNP** [8], hence the existence of a criterion of the same shape is very unlikely...

  Our solution corresponds to a version of proof-nets in which weakened formulas are attached to a formula of the net.

- The geometry of interaction of $\bot$ (which is the geometry of the weakening rule) shows for the first time a non-contrived non-commutativity : the formula to the left of $\bot$ in the conclusion of the rule actually matters. One could imagine another version in which $\bot$ is physically linked to its rightmost neighbor. This non-commutative reading is controversial anyway : we could also decide to say that the rule involves two formulas, $\bot$ and A.

- The neutrals display another originality w.r.t. geometry of interaction : in the absence of $\bot$ and 1, Π^* is always a sum of two adjoint wirings $W + W^*$. The interpretation of the axiom for 1 is a single wire, and the $\bot$-rule introduces non self-adjoint wirings...

The additive neutrals will be easily interpreted, provided the usual axiom for $\top$ is replaced with the rule

$$\frac{\vdash \Gamma, \flat}{\vdash \Gamma, \top}\top$$

When Π is obtained from Σ by means of a $\top$-rule, then

$$\Pi^\bullet = \Sigma^\bullet + p_\top(x, y)$$

This definitely clarifies the relation between $\top$ and $\flat$; the addition of the message $p_\top(x, y)$ ensures $\Pi^\bullet \neq 0$. The cut-elimination procedure is extended in the following way : a cut between the flat axiom $\vdash \Phi, 0, \flat$ and the conclusion $\vdash \Psi, \top$ of a $\top$-rule is replaced with the proof of $\vdash \Psi, \Phi, \flat$ obtained from the premise $\vdash \Psi, \flat$ of the $\top$-rule by "inflation" (see remark 12). Soundness is almost immediate.

[8]Typically : π is a multiplicative proof-net iff for all switchings $\mathcal{S}$ the resulting graph is connected and acyclic ; this **coNP** turns out to be polytime, more precisely quadratic.

4 The case of exponentials

4.1 An alternative version of exponentials

We describe here a modification of the exponential rules ; the modification is close to some extant experimental systems, see e.g. [A93, MM93]. The system has a weak form of promotion, which is corrected by an additional rule for ?.

$$\frac{\vdash \Gamma, A}{\vdash ?\Gamma, !A}\,! \qquad \frac{\vdash A, \Gamma}{\vdash ?A, \Gamma}\,d? \qquad \frac{\vdash B, \Gamma}{\vdash ?A, B, \Gamma}\,w? \qquad \frac{\vdash ?A, ?A, \Gamma}{\vdash ?A, \Gamma}\,c? \qquad \frac{\vdash ??A, \Gamma}{\vdash ?A, \Gamma}\,??$$

These rules are respectively called *(weak) promotion, dereliction, contraction, digging.* They only ensure a variant of the subformula property ($??A$ subformula of $?A$) which may look strange at first sight, but which is not more artificial than the familiar definition which allows $A[t/x]$ to be a subformula of $\forall x A$. This modification of the notion of subformula yields infinitely many subformulas for a propositional linear formula, in accordance with the undecidability of propositional linear logic, [LMSS90]. The rule for weakening explicitly mentions a formula B of the context, in accordance with what we did for $\perp$. The rule of promotion is understood in the obvious way : if Γ is $A_1, \ldots, A_n$, then $?\Gamma$ is $?A_1, \ldots, ?A_n$. We are therefore interpreting a slight modification of linear logic, corresponding to extant experimental systems. If we adopt here the modifications of [A93, MM93], this is because geometry of interaction is particularly simple in this setting. But also, since geometry of interaction is in many senses the most natural semantics, it might be seen as backing the syntactical variants proposed in these works.

Exercise 2 *Prove the cut-elimination theorem for this variant of the exponential rules. Of course one has to define an* ad hoc *cut-elimination procedure.*

4.2 The pattern

The general pattern is as follows : we want to allow reuse ; the absolutely weakest form of reuse is expressed by the principle $\vdash ?A^{\perp}, A \otimes A$. The interpretation $\Pi^{\bullet}$ of the (natural) proof of this principle must have the following property : if U interprets a proof of A and $!U$ is the result of the promotion of U, then a cut between $!U$ and $\Pi^{\bullet}$ yields after normalisation a variant of V, where V is obtained by means of definition 10.4 from U and U in the respective roles of $\Pi_1^{\bullet}$ and $\Pi_2^{\bullet}$. In the formation of V the essential step is the formation of $\otimes_1(U) + \otimes_2(U)$, which is the sum of two variants. The execution formula is able to extract two variants of U from the sole $!U$. Building variants basically involves non-laminated wirings, and therefore the execution formula is unable to achieve the task. In particular the solution $!U = U$ is inadequate.

Of course, the problem is easily solved if we could drop lamination..., but this would destroy some essential features of our construction. But we can also mimic non-laminated wires by laminated ones : we are eventually lead to shift the dialectal components ux of $\otimes_1(U)$ from the private part to the public part. The context-free promotion $!U$ is interpreted as follows : in U replace all atoms $p_A(t,u)$ with atoms $p_{!A}(txu,z)$; This construction is the result of two steps :

- We first form $\otimes_1(U)$, by replacing in U all atoms $p_B(t,u)$ with atoms $p_B(t,xu)$.
- Then we replace in $\otimes_1(U)$ all atoms $p_B(t,xu)$ with atoms $p_{!A}(txu,z)$. The final z is present only because our predicates need a second argument : in reality there is no dialect in this case.

4.3 Interpretation of the exponential rules

In what follows $t_1 \ldots \widehat{t_i} \ldots t_n$ is short for $t_1 \ldots t_{i-1}t_{i+1} \ldots t_n$.

Definition 17 (def. 10 cont^d)

Case 17.1 *Assume that the proof* Π *of* $\vdash [\Delta], ?\Gamma, !A$ *is obtained from a proof* Π_1 *of* $\vdash [\Delta], \Gamma, A$ *by a promotion rule ; we define* $\Pi^\bullet$ *in terms of* $\Pi_1^\bullet$. *Let us assume that* Γ *is* $C_1, \ldots, C_n$ *(in this order). Replace in* $\Pi_1^\bullet$ *:*

- *all atoms* $p_A(t,u)$ *with atoms* $p_{!A}(txuy_1 \ldots y_n, z)$ *;*
- *all atoms* $p_B(t,u)$ *with atoms* $p_B(txuy_1 \ldots y_n, z)$ *when* B *is in* Δ *;*
- *all atoms* $p_{C_i}(t,u)$ *with atoms* $p_{?C_i}(t(uy_1 \ldots \widehat{y_i} \ldots y_n x)y_i, z)$.

($x, y_1, \ldots, y_n, z$ *are fresh variables). as usual* x, x', x'', z *are fresh variables. The result is by definition* $\Pi^\bullet$.

Case 17.2 *Assume that the proof* Π *of* $\vdash [\Delta], ?A, \Gamma$ *is obtained from a proof* Π_1 *of* $\vdash [\Delta], A, \Gamma$ *by a dereliction rule ; we define* $\Pi^\bullet$ *by replacing in* $\Pi_1^\bullet$ *:*

- *All wires* $p_C(t,u)$ *with atoms* $p_{?A}(tgz, uz)$ *;*
- *All atoms* $p_B(t,u)$ *(when* $B \neq A$*) with atoms* $p_B(t,uz)$.

Case 17.3 *Assume that the proof* Π *of* $\vdash [\Delta], ?A, B, \Gamma$ *is obtained from a proof* Π_1 *of* $\vdash [\Delta], B, \Gamma$ *by a weakening rule ; we define* $\Pi^\bullet$ *in terms of* $\Pi_1^\bullet$.

- *in* $\Pi_1^\bullet$ *replace any wire* $p_C(t,u) \mapsto p_B(t',u)$ *with a wire* $p_C(t, u(xx')y) \mapsto p_{?A}(xt'y, u(xx')y)$ *;*

- *in* $\Pi_1^\bullet$ *replace any wire* $p_C(t,u) \mapsto p_D(t',u)$, *with* $D \neq B$ *with a wire* $p_C(t, u(xx')y) \mapsto p_D(t', u(xx')y)$;

- *add to the result of this replacement the wire* $p_{?A}(x'zy', z'(xx')y) \mapsto p_B(z, z'(xx')y)$.

Case 17.4 *Assume that the proof* Π *of* $\vdash [\Delta], ?A, \Gamma$ *is obtained from a proof* Π_1 *of* $\vdash [\Delta], ?A_1, ?A_2, \Gamma$ *by a contraction rule (we note* $?A_1, ?A_2$ *two distinct occurrences of* A*) ; we define* $\Pi^\bullet$ *in terms of* $\Pi_1^\bullet$.

- *First put the atoms* $p_{?A_i}(t,u)$ *"in the form* $p(tt't'',u)$*" ; this means that we form*

$$W = \sum_{i=1}^{2} p_{?A_i}(xx'x'', y) + \sum_{C \in \Gamma} p_C(x,y)$$

and replace $\Pi_1^\bullet$ *with* $U = W\Pi_1^\bullet W$.

- *Then we replace in* U :

  - *All atoms* $p_{?A_1}(tt't'',u)$ *with atoms* $p_{?A}(t(gt')t'',u)$;
  - *All atoms* $p_{?A_2}(tt't'',u)$ *with atoms* $p_{?A}(t(dt')t'',u)$.

the result is by definition $\Pi^\bullet$.

Case 17.5 *Assume that the proof* Π *of* $\vdash [\Delta], ?A, \Gamma$ *is obtained from a proof* Π_1 *of* $\vdash [\Delta], ??A, \Gamma$ *by a digging rule ; we define* $\Pi^\bullet$ *in terms of* $\Pi_1^\bullet$.

- *First put the atoms* $p_{??A}(t,u)$ *"in the form* $p((tt't'')uu',v)$*" ; this means that we form*

$$W = p_{??A}((xx'x'')yy', z) + \sum_{C \in \Gamma} p_C(x,y)$$

and replace $\Pi_1^\bullet$ *with* $U = W\Pi_1^\bullet W$.

- *Then we replace in* U *all atoms* $p_{??A}((tt't'')uu',v)$ *with atoms* $p_{?A}(t(t'uu')t'',v)$;

Remark 15

The rule of promotion is the first violation of the principle that all our constructions are compatible with $\sim$. This is perhaps the ultimate meaning of the "!-box".

4.4 Basic examples

We shall treat certain basic cuts

$$\dfrac{\dfrac{\cdots}{\vdash \Gamma, !A}R \quad \dfrac{\cdots}{\vdash ?A^{\perp}, \Delta}S}{\vdash \Gamma, \Delta}CUT$$

on an exponential $!A$, in certain cases :

- Γ is empty and R is the promotion rule applied to a cut-free proof Π of $\vdash A$.
- S is a $!, d?, w?, c?$ or $??$-rule, applied to a cut-free proof Σ of $\vdash \Delta'$ and with main formula $?A^{\perp}$ in case $S \neq !$.

Example 2
Cut-elimination replaces the cut

$$\dfrac{\dfrac{\vdash A}{\vdash !A}! \quad \dfrac{\vdash A^{\perp}, B}{\vdash ?A^{\perp}, !B}!}{\vdash !B}CUT$$

with

$$\dfrac{\dfrac{\vdash A \quad \vdash A^{\perp}, B}{\vdash B}CUT}{\vdash !B}!$$

The interpretation of the original proof is $(U + V, \sigma_{!A,?A^{\perp};!B})$, where :

- U is obtained by replacing in $\Pi^{\bullet}$ all atoms $p_A(t, u)$ with atoms $p_{!A}(txu, zz')$;
- V is obtained by replacing in $\Sigma^{\bullet}$:
  - all atoms $p_{A^{\perp}}(t, u)$ with atoms $p_{?A^{\perp}}(t(ux)y, zz')$
  - all atoms $p_B(t, u)$ with atoms $p_{!B}(txuy, zz')$.

The interpretation of the modified proof is $(W + Y, \sigma_{A,A^{\perp};!B})$, where :

- W is obtained by replacing in $\Pi^{\bullet}$ all atoms $p_A(t, u)$ with atoms $p_A(txyu, z)$;
- Y is obtained by replacing in $\Sigma^{\bullet}$ all atoms $p_{A^{\perp}}(t, u)$ with atoms $p_{A^{\perp}}(txuy', z)$ and all atoms $p_B(t, u)$ with atoms $p_{!B}(txuy', z)$.

Observe that :

- $RES(U+V, \sigma_{!A,?A^\perp;!B}) = RES(U_1+V, \sigma_{!A,?A^\perp;!B})$, where U_1 is obtained from U by replacing all atoms $p_{!A}(txu, zz')$ with atoms $p_{!A}(t(x'x'')u, zz')$;
- $RES(U_1+V, \sigma_{!A,?A^\perp;!B}) = RES(Z, \sigma_{A,A^\perp;!B})$, where Z is obtained from U_1+V by replacing :
  - all atoms $p_{!A}(t(x'x'')u, zz')$ with atoms $p_A(tx''x'u, zz')$
  - all atoms $p_{?A^\perp}(t(ux)y, zz')$ with atoms $p_{?A^\perp}(tuxy, zz')$.
- $Z = \otimes_1(W+Y)$.

therefore $RES(U+V, \sigma_{!A,?A^\perp;!B}) = \otimes_1(RES(W+Y, \sigma_{A,A^\perp;!B}))$. This shows the soundness of this particular cut-elimination step.

Exercise 3
Extend example 2 to the more general case of n cuts between :

- *Cut-free proofs of the sequents* $\vdash !A_1, \ldots, \vdash !A_n$
- *A cut free proof of the sequent* $\vdash !A_1, \ldots, !A_n$ *ending with a promotion rule.*

Example 3
Cut-elimination replaces the cut

$$\dfrac{\dfrac{\vdash A}{\vdash !A}\,! \qquad \dfrac{\vdash A^\perp, \Phi}{\vdash ?A^\perp, \Phi}\,d?}{\vdash \Phi}\,CUT$$

with

$$\dfrac{\vdash A \qquad \vdash A^\perp, \Phi}{\vdash \Phi}\,CUT$$

The interpretation of the original proof is $(U+V, \sigma_{!A,?A^\perp;\Phi})$, where U is as in example 2 ; and V is obtained by replacing in $\Sigma^\bullet$

- All atoms $p_A^\perp(t,u)$ with atoms $p_{?A^\perp}(tgz, (uz)z')$;
- All atoms $p_B(t,u)$ (when $B \neq A^\perp$) with atoms $p_B(t, (uz)z')$.

whereas the modified proof is interpreted by $(\otimes_1(\Pi^\bullet) + \otimes_2(\Sigma^\bullet), \sigma_{A,A^\perp;\Phi})$. Now observe that :

- $RES(U+V, \sigma_{!A,?A^\perp;\Phi}) = RES(U+V, \sigma_{!A,?A^\perp;\Phi})$

- $RES(U+V, \sigma_{!A,?A^\perp;\Phi}) = RES(U_1+V, \sigma_{!A,?A^\perp;\Phi})$, where U_1 is obtained from U by replacing all atoms $p_{!A}(txu, zz')$ with atoms $p_{!A}(tgu, (z_1 u)z')$;

- $RES(U_1+V, \sigma_{!A,?A^\perp;\Phi}) = RES(W, \sigma_{A,A^\perp;\Phi})$, where W is obtained from U_1+V by replacing all atoms $p_{!A}(tgu, (z_1 u)z')$ with atoms $p_A(t, (z_1 u)z')$ and all atoms $p_{?A^\perp}(tgz, (uz)z')$ with atoms $p_{A^\perp}(t, (uz)z')$;

- $W = \otimes_2(\otimes_1(\Pi^\bullet) + \otimes_2(\Sigma^\bullet))$.

therefore $RES(U+V, \sigma_{!A,?A^\perp;\Phi}) = \otimes_2(RES(\otimes_1(\Pi^\bullet) + \otimes_2(\Sigma^\bullet), \sigma_{A,A^\perp;\Phi}))$. This shows the soundness of this particular cut-elimination step.

Example 4
Cut-elimination replaces the cut

$$\dfrac{\dfrac{\vdash A}{\vdash !A}! \qquad \dfrac{\vdash B, \Phi}{\vdash ?A^\perp, B, \Phi}w?}{\vdash B, \Phi}CUT$$

with the proof Σ. The interpretation of the original proof is $(U+V, \sigma_{!A,?A^\perp;\Phi})$, where U is as in example 2 ; and V is is defined in terms of $\Sigma^\bullet$:

- In $\Sigma^\bullet$ replace any wire $p_C(t,u) \mapsto p_B(t',u)$ by a wire $p_C(t, (u(xx')y)z'') \mapsto p_{?A^\perp}(xt'y, (u(xx')y)z'')$;

- in $\Sigma^\bullet$ replace any wire $p_C(t,u) \mapsto p_D(t',u)$, with $D \neq B$ by a wire $p_C(t, (u(xx')y)z'') \mapsto p_D(t', (u(xx')y)z'')$;

- add to the result of this replacement the wire $p_{?A^\perp}(x'zy', (z'(xx')y)z'') \mapsto p_B(z, (z'(xx')y)z'')$.

Now observe that nilpotency $\sigma_{!A,?A^\perp;\Phi}(U+V)$ is immediate, and that $RES(U+V, \sigma_{!A,?A^\perp;\Phi}$ is the sum of the following wires :

- all wires $p_C(t, (u(xx')y)z'') \mapsto p_D(t', (u(xx')y)z'')$, where $p_C(t,u) \mapsto p_D(t',u)$ is a wire in $\Sigma^\bullet$ and $D \neq B$

- all wires $p_C(t, (u(ww')v)z'') \mapsto p_B(t', (u(ww')v)z'')$, where $p_C(t,u) \mapsto p_B(t',u)$ is a wire in $\Sigma^\bullet$, and $p_{!A}(wxv, zz') \mapsto p_{!A}(w'xv, zz')$ is a wire in U.

This wiring is easily shown to be a variant of $\Sigma^\bullet$; however, the fact that $\Sigma^\bullet \neq 0$ plays an essential role : without at least one wire $p_{!A}(wxv, zz') \mapsto p_{!A}(w'xv, zz')$ in U,
$RES(U + V, \sigma_{!A,?A^\perp;\Phi}$ cannot keep any track of the wires $p_C(t,u) \mapsto p_B(t',u)$ of $\Sigma^\bullet$. This shows the soundness of this particular cut-elimination step.

Example 5
Cut-elimination replaces the cut

$$\dfrac{\dfrac{\vdash A}{\vdash !A}! \qquad \dfrac{\vdash ?A_1^\perp, ?A_2^\perp, \Phi}{\vdash ?A^\perp, \Phi}c?}{\vdash \Phi}CUT$$

with

$$\dfrac{\dfrac{\vdash A}{\vdash !A_2}! \qquad \dfrac{\dfrac{\vdash A}{\vdash !A_1}! \quad \vdash ?A_1^\perp, ?A_2^\perp, \Phi}{\vdash ?A_2^\perp, \Phi}CUT}{\vdash \Phi}CUT$$

The interpretation of the original proof is $(U + V, \sigma_{!A,?A^\perp;\Phi})$, where U is as in example 2 ; and V is obtained from $\Sigma^\bullet$ as follows :

- First define $V' = \otimes_2(\Sigma^\bullet)$;
- First put the atoms $p_{?A_i}(t,u)$ of V' "in the form $p(tt't'', u)$" ; this means that we form as in definition 17.4

  $$W = p_{?A_1}(xx'x'', y) + p_{?A_2}(xx'x'', y) + \sum_{C\in\Phi} p_C(x,y)$$

  and replace V' with $V'' = WV'W$;
- Then we replace in V'' :
  - All atoms $p_{?A_1}(tt't'', u)$ with atoms $p_{?A}(t(gt')t'', u)$;
  - All atoms $p_{?A_2}(tt't'', u)$ with atoms $p_{?A}(t(dt')t'', u)$.

  the result is called V.

The interpretation of the modified proof is $(X_1 + X_2 + Y, \sigma_{!A_1,?A_1^\perp,!A_2,?A_2^\perp;\Phi})$ where :

- $X_1 = U_1$

- $X_2 = \otimes_2(U_2)$
- $Y = \otimes_2(\otimes_2(\Sigma^\bullet))$

where U_i is obtained from U by replacing the predicate $p_{!A}$ by $p_{!A_i}$. Now observe that U (which is of the form $\otimes_1(U')$ is also of the form $\otimes_2(\otimes_2(U''))$. $X_1 + X_2 + Y$ is therefore of the form $\otimes_2(\otimes_2(Z))$, which is a variant of $\otimes_2(Z)$ by proposition 2. Therefore by proposition 1 we can take as interpretation of our modified proof the pair $(X'_1 + X'_2 + Y', \sigma_{!A_1,?A_1^\perp,!A_2,?A_2^\perp;\Phi})$, where :

- $X_1 = \otimes_2(X'_1)$
- $X'_2 = U_2$
- $Y = \otimes_2(\Sigma^\bullet)$

But it is immediate to see that

$$RES(X'_1 + X'_2 + Y', \sigma_{!A_1,?A_1^\perp,!A_2,?A_2^\perp;\Phi}) = RES(X''_1 + X'_2 + Y', \sigma_{!A_1,?A_1^\perp,!A_2,?A_2^\perp;\Phi})$$

where X''_1 is obtained from X'_1 by replacing all atoms $p_{A_1}(t,z)$ with atoms $p_{A_1}(t,zz')$; but observe that $X''_1 = \otimes_2(X'_1)$, hence $X''_1 = X_1 = U_1$. Now observe that all the wires in X_1 and X'_2 are of the form $p_{A_i}(tt't'', u)$, hence

$$RES(X_1 + X'_2 + Y', \sigma_{!A_1,?A_1^\perp,!A_2,?A_2^\perp;\Phi}) = RES(X_1 + X'_2 + Y'', \sigma_{!A_1,?A_1^\perp,!A_2,?A_2^\perp;\Phi}),$$

where Y'' is obtained by putting the atoms $p_{?A_i}(t,u)$ of Y' "in the form $p(tt't'', u)$" ; but then $Y'' = V''$. We are left with
$RES(U_1 + U_2 + V'', \sigma_{!A_1,?A_1^\perp,!A_2,?A_2^\perp;\Phi})$. Obviously this expression is unchanged if we merge the $p_{?A_i}$ into $p_{?A}$ and the $p_{!A_i}$ into $p_{!A}$. This merge changes V'' into V, $U_1 + U_2 +$ into ZUZ and $\sigma_{!A_1,?A_1^\perp,!A_2,?A_2^\perp;\Phi}$ into $Z\sigma_{!A,?A^\perp;\Phi}Z$, where Z is the projection

$$Z = p_{?A}(x(gx')x'', y) + p_{?A}(x(dx')x'', y) + \sum_{C \in \Gamma} p_C(x,y)$$

Observe that $V = ZVZ$, hence

$$\begin{aligned} RES(ZUZ + V, Z\sigma_{!A,?A^\perp;\Phi}Z) &= Z \cdot RES(U + V, \sigma_{!A,?A^\perp;\Phi}) \cdot Z = \\ &= RES(U + V, \sigma_{!A,?A^\perp;\Phi}) \end{aligned}$$

and thus we have eventually proved soundness.

Example 6
Cut-elimination replaces the cut

$$\dfrac{\dfrac{\vdash A}{\vdash !A}\,! \qquad \dfrac{\vdash ??A^\perp, \Phi}{\vdash ?A^\perp, \Phi}\,??}{\vdash \Phi}\,CUT$$

with

$$\dfrac{\dfrac{\dfrac{\vdash A}{\vdash !A}!}{\vdash !!A}! \qquad \vdash ??A^{\perp}, \Phi}{\vdash \Phi} CUT$$

The interpretation of the original proof is $(U + V, \sigma_{!A,?A^{\perp};\Phi})$, where U is as in example 2 ; and V is obtained from $\Sigma^{\bullet}$ as follows :

- First put the atoms $p_{??A}(t, u)$ "in the form $p((tt't'')uu', v)$" ; this means that we form

$$W = p_{??A}((xx'x'')yy', z) + \sum_{C \in \Phi} p_C(x, y)$$

as in definition 17.5 and replace $\Pi_1^{\bullet}$ with $V' = W\Sigma^{\bullet}W$.

- Then we replace in V' all atoms $p_{??A}((tt't'')uu', v)$ with atoms $p_{?A}(t(t'uu')t'', vy)$;

the result is by definition V. The modified proof is interpreted as $(X + Y, \sigma_{!!A,??A^{\perp};\Phi})$, with :

- X is obtained from $\Pi^{\bullet}$ by replacing all atoms $p_A(t, u)$ with atoms $p_{!!A}((txu)yz, z')$;
- $Y = \otimes_2(\Sigma^{\bullet})$

We first observe that $RES(X + Y, \sigma_{!!A,??A^{\perp};\Phi}) = RES(X' + \otimes_2(V'), \sigma_{!A,?A^{\perp};\Phi})$, where X' is obtained from $\Pi^{\bullet}$ by replacing all atoms $p_A(t, u)$ with atoms $p_{!A}(t(xyz)u, z')$. This is because $X' = ZXZ^*$ and $\otimes_2(V') = ZYZ^*$ with

$$Z = p_{!!A}((xx'x'')yy', z) \mapsto p_{!A}(x(x'yy')x'', z) + \\ p_{??A^{\perp}}((xx'x'')yy', z) \mapsto p_{??A^{\perp}}(x(x'yy')x'', z) + \sum_{C \in \Phi} p_C(x, y)$$

Now observe that

$$RES(X' + \otimes_2(V'), \sigma_{!A,?A^{\perp};\Phi}) = RES(X' + WVW, \sigma_{!A,?A^{\perp};\Phi})$$

but

$$RES(X' + WVW, \sigma_{!A,?A^{\perp};\Phi}) = RES(WX'W + V, \sigma_{!A,?A^{\perp};\Phi}) = \\ = RES(U + V, \sigma_{!A,?A^{\perp};\Phi})$$

This proves soundness in this last case.

4.5 Nilpotency for exponentials

It remains to extend the nilpotency theorem to the full case ; this offers no difficulty of principle (basically our previous treatment of exponentials in [G88] suitably modified in the spirit of section 3). For instance the nilpotency theorem basically needs an extension of definition 12 :

Definition 18 (def. 12 cont^d)

Case 18.1 *If X is a weak type, then we define the weak type $!X$ as $Z^{\perp\perp}$, where Z consists of all $!U$ obtained from some $U \in X$ by means of definition 17.1 (in the simplified case $\Gamma = \Delta = \emptyset$).*

Case 18.2 *If X is a weak type, then we define the weak type $?X$ as $(!X^{\perp})^{\perp}$.*

This definition is the key to an unproblematic extension of the nilpotency theorem to the full case (left without hypocrisy to the reader).

Theorem 5 (Nilpotency)
$\sigma\Pi^\bullet$ is nilpotent.

4.6 Soundness for exponentials, a sketch

Soundness is more delicate ; in fact we can only prove soundness under a strong restriction on Γ.

Theorem 6 (Limited soundness)
If a cut-free proof Π of $\vdash \Gamma$ is obtained from a proof Σ of $\vdash [\Delta]\Gamma$ by means of the transformations sketched in definition 15 (suitably extended to accommodate exponentials, see below), and if Γ contains no exponential and no second order existential quantifier, then $\Pi^\bullet$ is a variant of $RES(\Sigma^\bullet, \sigma_{\Delta;\Gamma})$.

PROOF: the proof is an imitation of theorem 1 ii) of [G88]. We give some hints :

- The restriction on Γ makes it possible to consider a limited form of cut-elimination, where an exponential cut is eliminated only when the premise containing $!A$, let us say $\Phi, !A$, comes from a !-rule, with Φ empty. One can show that cut-elimination works with this restricted algorithm.

- The examples 2 (including its n-ary generalization of exercise 3), 3, 4, 5, 6 are the basic steps of the verification of the soundness of this limited cut-elimination algorithm.

- Among the specific technicalities of this extension, we need to develop the $\flat$-calculus in presence of exponentials. For instance the promotion rule in presence of $\flat$ is understood as follows : since $?\flat$ is an illegal expression we can form $?\Gamma$ only when Γ does not contain $\flat$. One also needs to define cut-elimination between a flat axiom and an exponential rule. This is straightforward :
  - A cut between the flat axiom $\vdash \Gamma, !A, \flat$ and a $d?$-rule is replaced with a cut between the flat axiom $\vdash \Gamma, A, \flat$ and the premise of the $d?$-rule ;
  - A cut between the flat axiom $\vdash \Gamma, !A, \flat$ and a $w?$-rule, with premise $\vdash B, \Gamma$ is replaced with a cut between the flat axiom $\vdash \Gamma, B^{\perp}, \flat$ and the premise of the $w?$-rule ;
  - A cut between the flat axiom $\vdash \Gamma, !A, \flat$ and a $c?$-rule is replaced with two cuts between the flat axioms $\vdash \Gamma, !A, \flat$ and $\vdash !A, \flat$ and the premise of the $c?$-rule ;
  - A cut between the flat axiom $\vdash \Gamma, !A, \flat$ and a $??$-rule is replaced with a cut between the flat axiom $\vdash \Gamma, !!A, \flat$ and the premise of the $??$-rule ;
  - A cut between a promotion $!A$ and the flat axiom $\vdash \Gamma, ?A^{\perp}, \flat$ is replaced with the flat axiom $\vdash \Gamma, \flat$. □

Remark 16

- One should also prove the soundness w.r.t. the full cut-elimination procedure. The only way to do so would be to work with *proof-nets* in some variant of [G94] (with boxes for !), then to prove a Church-Rosser property, and the fact that the interpretation of a cut-free proof of Γ depends (up to variance) only on the associated net. This seems to be unproblematic, but we quailed in the face of the burden...

- In [G88] we were able to prove a little more, namely ! was allowed in Γ. The problem here is that reduction above a promotion rule would need a more general notion of variant. If we come back to definition 6, we see that the essential point is the possible choices for W and W'. We can perfectly define, *for each type* A a set of wirings (inducing an equivalence $\sim_A$, coarser than $\sim$). For instance for $!A$ we could consider all wirings $\sum p_{!A}(xyt, z) \mapsto p_{!A}(xyu, z)$ to enlarge the notion of variant in that case. No doubt that with this extension, we can allow ! in Γ.

- What about a complete soundness ? Surely, one needs a liberalized notion of variant, as above, typically to take care of rather arbitrary choices made in the case of the ?-rules, but is this enough ? Presumably not, and we honestly don't know whether the ♭-calculus (perhaps improved with additional principles) can cope with the situation. In fact we have conflicting intuitions :

- Dr Jekill thinks that the syntax can be adapted to cope with the geometrical interpretation ;
- Mr Hyde thinks that there is something basically infinite in exponentials and that for this reason, there is an irreducible global configuration, the !-box.

- Anyway, we do not want to achieve soundness by replacing the rather natural notion of being variants (with the possibility of modifying the possible W, W') with a notion of observational equivalence : the definition of variance should remain rather elementary, if possible.

5 Appendix : Hilbert spaces and related topics

5.1 Hilbert spaces

Definition 19
A prehilbertian *space is a complex vector space $\mathcal{H}$ equipped with a positive hermitian form, i.e. a function $(x \mid y)$ from $\mathcal{H} \times \mathcal{H}$ to C such that :*

- $(\alpha x + \beta y \mid z) = \alpha(x \mid z) + \beta(y \mid z)$, *for* $x, y, z \in \mathcal{H}$, $\alpha, \beta \in C$
- $(x \mid y) = \overline{(y \mid x)}$ *for* $x, y \in \mathcal{H}$ *; in particular* $(z \mid \alpha x + \beta y) = \bar{\alpha}(z \mid x) + \bar{\beta}(z \mid y)$ *and* $(x \mid x)$ *is always real.*
- $(x \mid x) \geq 0$ *for* $x \in \mathcal{H}$

Among the immediate properties of such spaces, let us mention the famous Cauchy-Schwarz inequality : $|(x \mid y)|^2 \leq (x \mid x)(y \mid y)$, which implies that $\|x\| = (x \mid x)^{1/2}$ defines a semi-norm on $\mathcal{H}$. Another classic is the "median identity" $\|x + y\|^2 + \|x - y\|^2 = 2\|x\|^2 + 2\|y\|^2$.

Definition 20
$\mathcal{H}$ is said to be a Hilbert space when $\| \cdot \|$ is a norm, i.e. when $(x \mid x) > 0$ for all $x \neq 0$ and $\mathcal{H}$ is complete (i.e. is a so-called Banach space) w.r.t. $\| \cdot \|$.

Every prehilbertian space $\mathcal{H}$ can be transformed into a Hilbertian space $\mathcal{H}''$; the process involves two steps :

- Separation : quotient $\mathcal{H}$ by the subspace consisting of vectors with a null semi-norm. This induces a hermitian form on the quotient $\mathcal{H}'$, which is a norm, i.e. is such that $\|x\| = 0$ implies $x = 0$.

- Completion : add limits for all Cauchy sequences, and extend the hermitian form to this extended space $\mathcal{H}''$.

Example 7
Let I be a set ; we define $\ell^2(I)$ to consist of all square-summable sequences of complex numbers indexed by I, i.e. of all families $(\lambda_i)_{i\in I}$ (sometimes noted $\Sigma\lambda_i.i$) such that $\Sigma_{i\in I} \quad |\lambda_i|^2 < +\infty$. We define a hilbertian form on $\mathcal{H}$ by $((\lambda_i) \mid (\mu_i)) = \Sigma_{i\in I} \quad \lambda_i\overline{\mu_i}$ (the series is shown to be absolutely convergent by a direct proof of the Cauchy-Schwarz inequality). $\ell^2(I)$ is easily shown to be a Hilbert space ; in fact general (and quite easy) results on Hilbert space shows that any Hilbert space is isomorphic to some $\ell^2(I)$. Since the isomorphism class of $\ell^2(I)$ only depends on the cardinality of I, we see that there are three cases :

- If I is finite, then the vector space is finite dimensional ; such Hilbert spaces are too small in practice

- If I is infinite but not denumerable, then the space is too big for most applications.

- If I is denumerable, then we get the main Hilbert space, "the" Hilbert space. It must be noticed that although in practice most Hilbert spaces will fall in this equivalence class, the isomorphism might be non-trivial, i.e. there is basically one space, but it may appear through very unlikely disguises.

Definition 21
If $\mathcal{H}$ and $\mathcal{H}'$ are Hilbert spaces, then we can form a new space $\mathcal{H}'' = \mathcal{H} \oplus \mathcal{H}'$ by considering the set of all formal sums $x \oplus x', x \in \mathcal{H}, x' \in \mathcal{H}'$, and define a hermitian form by $(x \oplus x' \mid y \oplus y') = (x \mid x') + (y \mid y')$. $\mathcal{H}''$ is easily seen to be a Hilbert space, the direct sum *of $\mathcal{H}$ and $\mathcal{H}'$. $\mathcal{H}$ and $\mathcal{H}'$ can be identified with subspaces of $\mathcal{H}''$.*

In general, we shall write $\mathcal{H}'' = \mathcal{H} \oplus \mathcal{H}'$ to speak of an isomorphic situation : $\mathcal{H}$ and $\mathcal{H}'$ are closed subspaces of $\mathcal{H}$ (hence Hilbert spaces), $(x \mid y) = 0$ for $x \in \mathcal{H}, x' \in \mathcal{H}'$, and every vector $x'' \in \mathcal{H}''$ can (uniquely) be written $x'' = x + x'$ for some $x \in \mathcal{H}$ and $x' \in \mathcal{H}'$. Observe that, given $\mathcal{H}$, $\mathcal{H}'$ is uniquely determined : $\mathcal{H}' = \{x'; \forall x \in \mathcal{H} \quad (x \mid x') = 0\}$.

Definition 22
If $\mathcal{H}$ and $\mathcal{H}'$ are Hilbert spaces, then we can form a new space $\mathcal{H}'' = \mathcal{H} \otimes \mathcal{H}'$ by considering the vector space of all finite linear combinations (with coefficients in C) of formal expressions $x \otimes x'$ (with $x \in \mathcal{H}, x' \in \mathcal{H}'$). If one defines $(\Sigma_i \alpha_i x_i \otimes x'_i \mid \Sigma_j \beta_j y_j \otimes y'_j) = \Sigma_{ij} \alpha_i \beta_j (x_i \mid y_j)(x'_i \mid y'_j)$, then it is easily shown that this is actually a hermitian form. $\mathcal{H}''$ is obtained from this prehilbertian space by separation and completion.

Observe that separation amounts to quotient the vector space by the space of vector with a null semi-norm, which is exactly the space generated by the following vectors :

- $x \otimes (x' + y') - x \otimes x' - x \otimes y'$
- $(x + y) \otimes x' - x \otimes x' - y \otimes x'$
- $(\alpha x) \otimes x' - x \otimes (\alpha x')$
- $(\alpha x) \otimes x' - \alpha(x \otimes x')$

5.2 Bounded operators

Definition 23
If $\mathcal{H}$ and $\mathcal{H}'$ are Hilbert spaces, then a map u from $\mathcal{H}$ to $\mathcal{H}'$ is said to be a bounded operator *when the following hold :*

- *u is linear, i.e. $u(\alpha x + \beta y) = \alpha u(x) + \beta u(y)$ for $\alpha, \beta \in C$ and $x, y \in \mathcal{H}$.*
- *The quantity $\|u\| = \sup_{\|x\| \leq 1} \|u(x)\|$ is finite ; $\|u\|$ is the* norm *of u.*

If u and v are bounded operators from $\mathcal{H}$ to $\mathcal{H}'$, if $\alpha \in C$, then one can define bounded operators $u + v$ and αu from $\mathcal{H}$ to $\mathcal{H}'$, by means of $(u + v)(x) = u(x) + v(x)$, $(\alpha u)(x) = \alpha u(x)$. If u and v are bounded operators from $\mathcal{H}'$ to $\mathcal{H}''$ and $\mathcal{H}$ to $\mathcal{H}'$ respectively, then one can define a bounded operator uv from $\mathcal{H}$ to $\mathcal{H}''$, by means of $(uv)(x) = u(v(x))$. Observe that $\|u + v\| \leq \|u\| + \|v\|$, $\|\alpha u\| \leq |\alpha| \|u\|$, $\|uv\| \leq \|u\| \|v\|$. The operators 0 and 1 (the null operator and the identity) have respective norms 0 and 1.

Proposition 6
Assume that $u_i, i = 1, 2$ are bounded operators from $\mathcal{H}_i$ to $\mathcal{H}'_i$; then

- *There is a unique bounded operator $u_1 \oplus u_2$ from $\mathcal{H}_1 \oplus \mathcal{H}_2$ to $\mathcal{H}'_1 \oplus \mathcal{H}'_2$ such that $(u_1 \oplus u_2)(x_1 \oplus x_2) = u_1(x_1) \oplus u_2(x_2)$*
- *There is a unique bounded operator $u_1 \otimes u_2$ from $\mathcal{H}_1 \otimes \mathcal{H}_2$ to $\mathcal{H}'_1 \otimes \mathcal{H}'_2$ such that $(u_1 \otimes u_2)(x_1 \otimes x_2) = u_1(x_1) \otimes u_2(x_2)$*

Definition 24
A bounded operator u from $\mathcal{H}$ to $\mathcal{H}'$ is said to be an isometry *of $\mathcal{H}$ into $\mathcal{H}'$ when it preserves the norm, i.e. $(u(x) \mid u(x)) = (x \mid x)$ for all $x \in \mathcal{H}$; this condition is easily seen to imply the more general condition $(u(x) \mid u(y)) = (x \mid y)$. Among typical isometries, let us mention rotations (in the Hilbert space C^n) and the maps which identify $\mathcal{H}$ and $\mathcal{H}'$ with subspaces of $\mathcal{H} \oplus \mathcal{H}'$. An isometry has norm 1 (except when the source space is reduced to the null vector). A surjective isometry from $\mathcal{H}$ onto $\mathcal{H}'$ turns out to be an isomorphism of structures.*

Example 8
Assume that f is a partial injective map from a subset of I into J ; then one can define a bounded operator u_f from $\ell^2(I)$ into $\ell^2(J)$ by $u_f((x_i)) = (y_j)$, where the sequence (y_j) is defined by $y_{f(i)} = x_i$, $y_j = 0$ when $j \notin rg(f)$. (With friendler notations :
$u_f(\Sigma\lambda_i.i) = \Sigma\lambda_i.f(i))$. The norm of u_f is equal to 1 (except when the domain of f is empty) ; furthermore in case f is total, then u_f is an isometry of $\ell^2(I)$ into $\ell^2(J)$, and in case f is also surjective, then the isometry is onto. Observe that, if f, g are partial injections from respectively a subset of J into K and a subset of I into J, then $u_{fg} = u_f u_g$. Similarly if the graph of the partial function f is the union of the graphs of the partial functions g and h, then $u_f = u_g + u_h$.

5.3 The adjoint of an operator

The main elementary property of the Hilbert space is the following :

Proposition 7
The dual $\tilde{\mathcal{H}}$ of $\mathcal{H}$ is (semi)-isomorphic to $\mathcal{H}$.

PROOF: We first explain the meaning of the result : if $a \in \mathcal{H}$, then the map φ_a from $\mathcal{H}$ to C defined by $\varphi_a(x) = (x \mid a)$ is linear and continuous (use Cauchy-Schwarz), i.e. is a member of the topological dual of $\mathcal{H}$. Conversely, any element of $\tilde{\mathcal{H}}$ is indeed of the form φ_a. Therefore the map which sends a to φ_a is a bijection of $\mathcal{H}$ onto $\tilde{\mathcal{H}}$. But this map is not quite linear, since $\varphi_{\alpha a+\beta b} = \bar{\alpha}\varphi_a + \bar{\beta}\varphi_b$, i.e. it is linear up to complex conjugation, this is why it is styled "semi-linear". The proposition is established as follows : if φ is a nonzero continuous form on $\mathcal{H}$, then one can show (using the median identity and the completeness of the space) that the set $\{x; \varphi(x) = 1\}$ has exactly one element a of minimum norm. Then one easily shows that $\varphi = \varphi_b$, with $b = a/(a \mid a)$. □

If u is a bounded operator from $\mathcal{H}$ to $\mathcal{H}'$, then u induces a linear map $\tilde{u}$ from the dual $\tilde{\mathcal{H}}'$ to the dual $\tilde{\mathcal{H}}$, by means of $\tilde{u}(\varphi) = \varphi \circ u$. By proposition 7, $\tilde{u}$ induces in turn a map u^* from $\mathcal{H}'$ to $\mathcal{H}$ defined by : $\varphi_a \circ u = \varphi_{u^*(a)}$, in other terms :

Definition 25
If u is a bounded operator from $\mathcal{H}$ to $\mathcal{H}'$, then we define the adjoint u^* *of u, a map from $\mathcal{H}'$ to $\mathcal{H}$ by : $(u(a) \mid b) = (a \mid u^*(b))$ for all $a, b \in \mathcal{H}$.*

Example 9
If f is a partial injective function from the subset X of I onto the subset Y of J, let g be its inverse ; then with the notations of example 8, we get $u_f{}^* = u_g$.

Proposition 8
The adjoint of a bounded operator is still a bounded operator ; furthermore the following hold :

- $(\alpha u + \beta v)^* = \bar{\alpha} u^* + \bar{\beta} v^*$
- $1^* = 1$
- $(uv)^* = v^* u^*$
- $u^{**} = u$
- $\|u^*\| = \|u\|$
- $\|uu^*\| = \|u\|^2$

PROOF: Most of the properties are immediate ; the last one follows from the Cauchy-Schwarz inequality. □

Proposition 9
If u is an isometry from $\mathcal{H}$ into $\mathcal{H}'$, then $u^* u = 1$; if u is surjective, then $uu^* = 1$.

PROOF: $(x \mid y) = (u(x) \mid u(y)) = (x \mid u^* u(y))$; then $(x \mid u^* u(y) - y) = 0$ for all $x \in \mathcal{H}$, which implies (take $x = u^* u(y) - y$) $u^* u(y) = y$; the second half of the proposition is immediate. Observe that we may have $u^* u = 1$, but $uu^* \neq 1$: consider u_f when f is a non-surjective injection of I into J. □

5.4 C^*-algebras

Definition 26
A C^*-algebra A consists in the following data :

- *A complex Banach algebra, with unit ; in particular A enjoys the properties $u \neq 0$ implies $\|u\| \neq 0$, $\|\alpha u\| = |\alpha| \|u\|$, $\|1\| = 1$, $\|u+v\| \leq \|u\| + \|v\|$, $\|uv\| \leq \|u\| \|u\|$, and A is complete w.r.t. $\| \cdot \|$.*

- *A unary operation* $(\cdot)^*$, *called the* adjunction, *or the involution, and which must satisfy the properties of proposition 8.*

Example 10
The algebra $\mathcal{B}(\mathcal{H})$ of bounded operators from $\mathcal{H}$ to itself is the most typical C^*-algebra. But there are other examples, typically the commutative algebra $\mathcal{C}(X)$ of continuous complex-valued functions on a compact space X (with pointwise addition and multiplication, the adjunction being complex conjugation etc.). A famous theorem states that any commutative C^*-algebra is isomorphic to some $\mathcal{C}(X)$. The general case is not that simple : however every C^*-algebra is isomorphic to a subalgebra of some $\mathcal{B}(\mathcal{H})$, i.e. it is always possible to represent the elements of a C^*-algebra as actual bounded operators ; when u is represented by an element of $\mathcal{B}(\mathcal{H})$ then we say that u *operates*, or acts on $\mathcal{H}$.

5.5 Zoology of operators

C^*-algebras generalize the field of complex numbers ; one can keep this in mind to understand the following zoology of operators :

- An operator u is *hermitian* when $u = u^*$ or equivalently $(u(x) \mid x)$ is real for all $x \in \mathcal{H}$. Hermitian operators clearly generalize real numbers. Among hermitian operators all projections and symmetries (see below), and all operators $u + u^*$ and uu^*. By the way, $v = uu^*$ enjoys a stronger property, namely that $(v(x) \mid x) \geq 0$ for $x \in \mathcal{H}$: such an operator is said to be a *positive* hermitian operator, and positive hermitians generalize positive reals.

- An operator u is said to be a *projection* when it is hermitian and $u^2 = u$; in such a case, $1 - u$ is also a projection. If u operates on $\mathcal{H}$, then the range E of u and the range F of $1 - u$ are such that $\mathcal{H} = E \oplus F$, and u corresponds to the orthogonal projection of $\mathcal{H}$ onto the subspace E : if $x = e + f$, $e \in E, f \in F$, then $u(x) = e$. Projections, which generalize the reals $0, 1$ therefore correspond to closed subspaces of the Hilbert space. Non-zero projections have norm 1.

- An operator u is said to be *unitary* when $uu^* = u^*u = 1$; unitary operators clearly generalize the unit circle. On Hilbert space, being unitary is equivalent to saying that $(u(x) \mid u(x)) = (x \mid x)$, and that u is surjective : i.e. unitary operators are represented by isometries of $\mathcal{H}$. Their norm is always 1.

- An operator u is said to be a *symmetry* when it is both hermitian and unitary ; this generalizes the reals $0, 1$. If u is a projection, then $2u - 1$

is a symmetry, and conversely if u is a symmetry, then $(1+u)/2$ is a projection. Symmetries are indeed yet another way to speak of closed subspaces of $\mathcal{H}$: instead of defining a projection from an orthogonal direct sum decomposition, one can introduce the symmetry $u(e+f) = e - f$.

- Hermitian and unitary operators share one property, namely that they are normal, i.e. that $uu^* = u^*u$. For normal operators, lot of results from finite dimensional algebra can be generalized, typically some forms of diagonalisation. However, one can easily meet non normal operators, especially in geometry of interaction, the typical example being partial isometries.

- u is said to be a *partial isometry* when uu^* is a projection. This condition indeed implies that u^*u is a projector (PROOF: consider $v = uu^*u - u$; then $vv^* = (uu^*)^3 - 2(uu^*)^2 + uu^*$; if uu^* is a projection, then then $vv^* = 0$, hence $\|v\|^2 = \|vv^*\| = 0$, hence $v = 0$. From $uu^*u = u$ one easily gets $(u^*u)^2 = u^*u$. $\square$) By symmetry, the conditions uu^* projection and u^*u projection are equivalent. On a Hilbert space, u acts as follows : if E and F are the subspaces corresponding to u^*u and uu^*, then u induces an isometry between E and F. Nonzero partial isometries have norm 1.

- A partial isometry which is also hermitian is called a *partial symmetry* ; equivalently $u^* = u$ and $u^3 = u$. Symmetries and projections are partial symmetries.

Example 11
Let us see how the operators u_f of example 8 react to our zoology : in general, if f is a partial injection of I into I, then u_f is a partial isometry of $\ell^2(I)$. u_f is normal when the domain and the range of f coincide. u_f is unitary when f is a bijection of I. u_f is hermitian (i.e. is a partial symmetry) when f is an involution of its domain. Finally u_f is a projection when f is the identity on its domain.

5.6 The algebra $\Lambda^*(L)$

We explain here how $\lambda^*(L)$ can be completed into a C^*-algebra. Here we directly refer to the definitions of subsection 2.2. This basically amounts, by topological generalities, to equip $\lambda^*(L)$ with C^*-seminorm, and to proceed as usual, separation/completion. A C^*-seminorm is exactly the same thing as a C^*-norm, except that it may take the value 0 on nonzero objects, and also that topological completeness is not required. By the way, observe that, if R

is a clause of $\lambda^*(L)$, then $\|R\| = 1$ or 0, (since RR^* is an idempotent, we get $\|RR^*\| = \|RR^*\|^2$, hence $\|RR^*\| = 1$ or 0, and $\|R\| = \sqrt{\|RR^*\|} = 1$ or 0) and therefore $\|\sum \alpha_i \cdot (P_i \mapsto Q_i)\|$ is bounded by $\sum |\alpha_i|$ for any C^*-seminorm. Then the pointwise supremum of any nonempty family of C^*-seminorms on $\lambda^*(L)$ is finite, and it is immediate that this supremum is also a C^*-seminorm ; hence we can define a unique such seminorm on $\lambda^*(L)$ as soon as there is at least one C^*-seminorm. For this its obviously suffices to show that $\lambda^*(L)$ operates on a Hilbert space.

Proposition 10
$\lambda^*(L)$ acts on the Hilbert space $\ell^2(G)$, where G is the set of ground propositions of L in such a way that the sum of $\lambda^*(L)$ is interpreted by sum of operators, the scalar multiplication of $\lambda^*(L)$ by scalar multiplication of operators, the product $\lambda^*(L)$ by composition of operators, and the involution * of $\lambda^*(L)$ by adjunction of operators.

PROOF: let G be the set of ground propositions in L ; to any clause $P \mapsto Q$ in L we can associate an injection $|P \mapsto Q|$ from a subset of G into G :

- If $g \in G$ unifies with Q, then the mgu θ yields closed values for all the variables in Q, which are the variables of P, hence $P\theta$ is also a ground formula : we set $|P \mapsto Q|(g) = P\theta$.

- If $g \in G$ does not unifies with Q, then $|P \mapsto Q|(g)$ is undefined.

It is immediate that $|P \mapsto Q|$ is a partial injection (this is because the variables of Q are all present in P), and that

$$|P \mapsto Q| \circ |P' \mapsto Q'| = |(P \mapsto Q) \cdot (P' \mapsto Q')|$$

an equation between partial functions that persists when resolution fails, if we interpret 0 as the fully undefined function. As in example 8 the partial injections $|P \mapsto Q|$ induce operators (partial isometries) of the Hilbert space $\ell^2(G)$:

$$|P \mapsto Q|(\sum \alpha_i.g_i) = \sum \alpha_i \cdot |P \mapsto Q|(g_i).$$

This immediately extends to all elements of $\lambda^*(L)$ which are therefore ascribed operators on $\ell^2(G)$, and the properties that we have stated immediately follow from definition 8 and example 9. □

Definition 27 (The completion)
$\Lambda^*(L)$ is defined to be the C^*-algebra obtained by completing $\lambda^*(L)$ w.r.t. its greatest C^*-seminorm.

Remark 17
It is not difficult to show that a nonzero element of $\lambda^*(L)$ induces a nonzero operator on $\ell^2(G)$, hence the greatest C^*-seminorm on $\lambda^*(L)$ is Hausdorff.

5.7 The execution formula

The execution formula can also be seen as the solution of a linear equation on a Hilbert space $\mathcal{H}$ on which $\Lambda^*(L)$ operates (see subsection 5.4) :

- U produces, given an input $h \in \mathcal{H}$ an output $U(h) \in \mathcal{H}$;
- σ feedbacks certain outputs h' of U to inputs $\sigma(h') \in \mathcal{H}$;
- σ^2 is a projection corresponding to the subspace $\mathcal{H}'$ on which the feedback is effective ; $1-\sigma^2$ is a projection corresponding to the subspace $\mathcal{H}''$ on which we want to observe the external behavior of the loop. Remark that $\mathcal{H} = \mathcal{H}' \oplus \mathcal{H}''$.
- The situation can be summarized by the following figure :

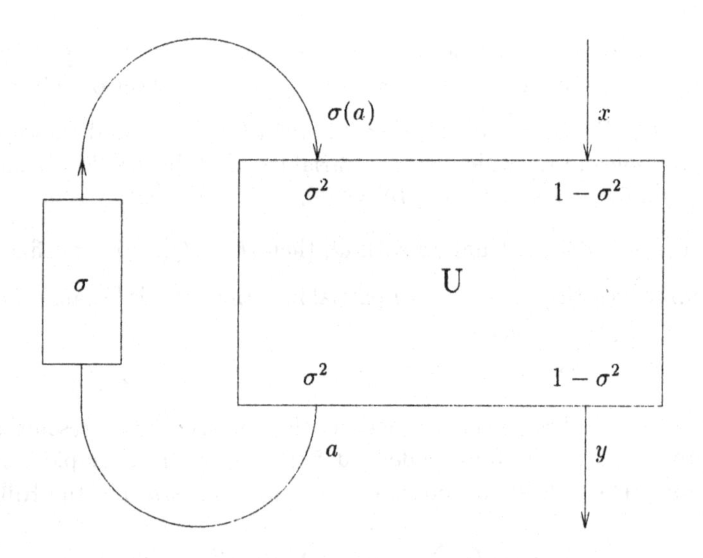

- In other terms, given $x \in \mathcal{H}''$ we are looking for $b \in \mathcal{H}$ such that $(1-\sigma^2)(b) = x$ (i.e. $b = c \oplus x$) and $(\sigma U)(b) = b - x (i.e. c = \sigma U(b))$; we eventually keep the output value $y = (1-\sigma^2)(U(h))$. In other terms, given $x \in \mathcal{H}''$ we look for $a \in \mathcal{H}'$ and $y \in \mathcal{H}''$ such that

$$U(\sigma(a) \oplus x) = a \oplus y$$

As a linear equation this can be written :

- $(\sigma U)(b) = b - x$, equivalently $x = (1 - \sigma U)(b)$;

- b is well defined as soon as $1-\sigma U$ is invertible, which is the case when σU is nilpotent ; in that case $b = (1-\sigma U)^{-1}(x)$, which can also be written $b = (1-\sigma U)^{-1}(1-\sigma^2)(x)$;
- Therefore $y = (1-\sigma^2)U(1-\sigma U)^{-1}(1-\sigma^2)(x)$, i.e. $y = RES(U,\sigma)(x)$.

Remark 18
A naïve way to solve the equation is to write U and σ as a 2×2-matrices (u_{ij}) and (σ_{ij}) (the only nonzero coefficient of σ_{ij} is σ_{11} and $\sigma_{11}^2 = 1$) ; the equation writes as a system :

$$y = u_{21}\sigma_{11}(a) + u_{22}(x) \qquad a = u_{11}\sigma_{11}(a) + u_{12}(x)$$

Successive replacements of a by its value given by the second equation yield :

$$y = u_{21}\sigma_{11}u_{11}\sigma_{11}(a) + u_{21}\sigma_{11}u_{12}(x) + u_{22}(x) =$$
$$= u_{21}\sigma_{11}u_{11}\sigma_{11}u_{11}\sigma_{11}(a) + u_{21}\sigma_{11}u_{11}\sigma_{11}u_{12}(x) + u_{21}\sigma_{11}u_{12}(x) + u_{22}(x)$$

which suggests the infinitary expansion

$$y = u_{11}(x) + \sum_{n\geq 0} u_{21}(\sigma_{11}u_{11})^n\sigma_{11}u_{12}(x)$$

This expansion is by the way legitimate when the sum is finite, which is the case when $\sigma_{11}u_{11}$ is nilpotent (which is the same as σU nilpotent). The expression can be rewritten as

$$y = u_{11}(x) + u_{21}(1-\sigma_{11}u_{11})^{-1}\sigma_{11}u_{12}(x)$$

which is exactly $RES(U,\sigma)$ (more precisely : its unique non-zero coefficient $RES(U,\sigma)_{22}$, which is equal to the coefficient $EX(U,\sigma)_{22}$).

5.8 Matrices and direct sums

Among examples of cstar-algebras, we have the classical example of the algebra $\mathcal{M}_n(\mathcal{C})$ of $n\times n$-matrices with complex coefficients, and more generally $\mathcal{M}_n(\mathcal{A})$ of $n\times n$-matrices with coefficients in a C^*-algebra $\mathcal{A}$. In that case, the adjoint of a matrix (a_{ij}) is the matrix $(a_{ji}{}^*)$. If the algebra $\mathcal{A}$ acts on a Hilbert space $\mathcal{H}$, then $\mathcal{M}_n(\mathcal{A})$ acts on the n-ary direct sum $\mathcal{H}\oplus\ldots\oplus\mathcal{H}$ by :

$$(a_{ij})(\bigoplus_{i=1}^{n} x_i) = \bigoplus_{i=1}^{n}\sum_{j=1}^{n} a_{ij}x_j$$

and this representation is compatible with summation, product, adjunction, etc.

The treatment of geometry of interaction in [G88] involves such matrices, more precisely, the index set consists of the formulas of the concluding sequent (and also of the cut-formulas). This means that in the case of a cut-free proof of $\vdash A_1, ..., A_n$, we see the proof as a linear input/output dependency between n inputs and n outputs (one for each A_i). Our presentation avoids any mention of matrices ; in fact this is because of the :

Proposition 11
$\lambda^*(T^m \cdot n)$ is (isomorphic to) the algebra of $n \times n$-matrices with entries in $\lambda^*(T^m)$.

PROOF: let us assume for instance $m = 2$; the basic idea is that the matrix M_{ij} whose entries are all zero but the one of index ij, which is 1 (i.e. the clause $p(x, y) \mapsto p(x, y)$), is represented by the clause $p_i(x, y) \mapsto p_j(x, y)$ of $\lambda^*(T^2 \cdot n)$. □
In other terms the predicate letters are used as the indexing of a square matrix. It is funny to observe that 0-ary predicates p, q, r enjoy

$$(p \mapsto q)(q' \mapsto r) = \delta_{qq'}.(p \mapsto r)$$

with $\delta_{qq'} = 1$ if $q = q'$, 0 otherwise, which makes precise the relation between unification and matrix composition which is behind the previous proposition.
When we deal with the rules for the binary connectives, we must in all cases merge two formulas A and B into a single formula C (which is $A \otimes B$, $A ⅋ B$, $A \& B$ or $A \oplus B$). This basically amounts to replace a matrix whose indices include A and B by another matrix in which these two indices are replaced by a single one, C. The basic property is that this replacement is a $*$-isomorphism, i.e. (we omit the precise definition) it is a map from a C^*-algebra into another one preserving the structure of C^*-algebra[9].
Hence in the case of a binary rule, when we replace the indices Γ, A, B with the indices Γ, C, we are seeking a $*$-isomorphism from $\mathcal{M}_{n+2}(\mathcal{A})$ into $\mathcal{M}_{n+1}(\mathcal{A})$. To understand how to construct such a map, imagine that u operates on a $n+2$-ary direct sum $\mathcal{H} \oplus \ldots \oplus \mathcal{H}$; then we would like to see u as acting on a $n+1$-ary sum. For this it is enough to merge isometrically the last two summands of the $n+2$-ary sum into one summand, by means of an isometry φ of $\mathcal{H} \oplus \mathcal{H}$ into $\mathcal{H}$.

Proposition 12
Let $\varphi(x \oplus y) = P(x) + Q(y)$ be a map from $\mathcal{H} \oplus \mathcal{H}$ into $\mathcal{H}$; then φ is an isometry iff the following hold :

[9] Except perhaps the identity 1 whose image is not requested to be 1 ; the other preservations force it to be a projection.

- $P^*P = Q^*Q = 1$
- $P^*Q = Q^*P = 0$

Furthermore, φ is surjective, i.e. is an isomorphism, when $PP^* + QQ^* = 1$.

PROOF: Easy from $\varphi^*\varphi = 1$ (and $\varphi\varphi^* = 1$ in case of surjectivity). □

Example 12
Consider the 3×3-matrix

$$u = \begin{bmatrix} a_{11} & a_{12} & a_{13} \\ a_{21} & a_{22} & a_{23} \\ a_{31} & a_{32} & a_{33} \end{bmatrix}$$

and let us merge the indices 2, 3 into the index 2 by means of the isometry φ defined from P, Q ; the result is easily shown to be :

$$v = \begin{bmatrix} a_{11} & a_{12}P^* + a_{13}Q^* \\ Pa_{21} + Qa_{31} & Pa_{22}P^* + Pa_{23}Q^* + Qa_{32}P^* + Qa_{33}Q^* \end{bmatrix}$$

And it is easy to check that this operation on matrices preserves sum, composition, adjunction ; the identity matrix is sent to

$$\begin{bmatrix} 1 & 0 \\ 0 & PP^* + QQ^* \end{bmatrix}$$

which is in general only a projection.

Now in order to bridge this with the main text, it is enough to remark that in the algebra $\lambda^*(T^2)$, the clauses $P = p(x,y) \mapsto p(xg,y)$ and $Q = p(x,y) \mapsto p(xd,y)$ satisfy the conditions of proposition 12. By the way observe that $PP^* + QQ^* \neq 1$: this is because there are terms that unify neither with xg nor with xd.

5.9 Tensorisation and arity

Of course there is another merge that falls into the previous analysis, namely the merge of contexts, in case of a &-rule. The only difference is that the pair (P', Q') chosen works on the second component : $P' = p(x,y) \mapsto p(x,yg)$ and $Q' = p(x,y) \mapsto p(x,yd)$. The fact that P', Q' coexist can only be explained in terms of tensorization : in general it is possible to define the tensor product $\mathcal{A} \otimes \mathcal{B}$ of C^*-algebras $\mathcal{A}$ and $\mathcal{B}$. This new algebra is obtained by completion of the algebra of formal linear combinations of tensors $a \otimes b$, $a \in \mathcal{A}$, $b \in \mathcal{B}$

with respect to a certain C^*-norm[10]. Without entering into the technicities of tensor products, the main idea is that, if a operates on $\mathcal{H}$ and b operates one $\mathcal{H}'$, then $a \otimes b$ operates on $\mathcal{H} \otimes \mathcal{H}'$ by : $(a \otimes b)(x \otimes y) = a(x) \otimes b(y)$.
Our way to cope with tensorisation of algebras $\Lambda^*(L)$ is to play with the arities :

Proposition 13
$\lambda^*(T^m)$ contains an isomorphic copy of the m-ary tensor power of $\lambda^*(T)$.

PROOF: typically if $m = 2$, p is unary and q is binary, then we can define an isomorphism ϕ from $\lambda^*(T) \otimes \lambda^*(T)$ into $\lambda^*(T^2)$, by

$$\phi((p(t) \mapsto p(u)) \otimes (p(t') \mapsto p(u'))) = q(t,t') \mapsto q(u,u')$$

This isomorphism is of course not surjective. □
What is behind the proposition is that the term tt' behaves like the tensor product $t \otimes t'$, which can be said pedantically as :

Proposition 14
Let G be the set of ground terms of a term language, including the binary function $\odot$; then the map $\varphi(g \otimes g') = gg'$ extends to an isometry of $\ell^2(G) \otimes \ell^2(G)$ into $\ell^2(G)$.

Proposition 13 enables us to use $*$-isomorphisms to replace the tensorization of algebras $\Lambda^*(L)$ by other algebras $\lambda^*(L)$. In the main text this opportunity is mainly used to induce commutations, since one of the basic facts about the tensor product $\mathcal{A} \otimes \mathcal{B}$ is that $u \otimes 1$ commutes with $1 \otimes v$. For instance, to come back to our discussion of the pairs (P,Q) and (P',Q'), we can introduce $P'' = p(x) \mapsto p(xg)$ and $Q'' = p(x) \mapsto p(xd)$ in $\lambda^*(T)$, and it is immediate that the operators $P'' \otimes 1, Q'' \otimes 1, 1 \otimes P'', 1 \otimes Q''$ are sent (by the $*$-isomorphism of proposition 13) on P, Q, P', Q' respectively, and this explains their good relative behavior. In the same way, the constructions of $\otimes_1(u)$ and $\otimes_2(u)$ of definition 7 involve tensorizations. And, last but not least, the treatment of exponentials strongly depends on the internalization of tensorization.

Remark 19
What is the meaning of lamination (definition 8 ? If we make W operate on $(\mathcal{H} \oplus \ldots \oplus \mathcal{H}) \otimes \mathcal{H}$ and if m is a message (definition 5), then $W(1 \otimes m) = (1 \otimes m)W$, i.e. that W belongs to the *commutant* of the set of messages of the form $1 \otimes m$.

References

[A93] **Asperti, A.** : Linear Logic, Comonads and Optimal reductions, *to appear in Fundamenta Informaticae, Special Issue devoted to categories in computer Science.*

[10] In general several C^*-norms are possible ; the choice includes a greatest and a smallest one

[B92] **Blass, A.** : A game semantics for linear logic, *Annals Pure Appl. Logic 56, pp. 183-220, 1992.*

[DR93] **Danos, V. & Regnier, L.** : Proof-nets and the Hilbert space : a summary of Girard's geometry of interaction, *this volume.*

[G86] **Girard, J.-Y.** : Linear Logic, *Theoretical Computer Science 50.1, pp. 1-102, 1987.*

[G86A] **Girard, J.-Y.** : Multiplicatives, *Rendiconti del Seminario Matematico dell'Università e Policlinico di Torino, special issue on Logic and Computer Science, pp. 11-33, 1987.*

[G87] **Girard, J.-Y.** : Towards a Geometry of Interaction, *Categories in Computer Science and Logic, Contemporary Mathematics 92, AMS 1989, pp. 69-108.*

[G88] **Girard, J.-Y.** : Geometry of Interaction I : interpretation of system F, *Proceedings Logic Colloquium '88, eds. Ferro & al., pp. 221-260, North Holland 1989.*

[G88A] **Girard, J.-Y.** : Geometry of Interaction II : deadlock-free algorithms, *Proceedings of COLOG-88, eds Martin-Löf & Mints, pp.76-93, SLNCS 417.*

[G94] **Girard, J.-Y.** : Proof-nets for additives, *manuscript, 1994.*

[GAL92] **Gonthier, G. & Abadi, M. & Levy, J.-J.** : The Geometry of Optimal Lambda Reduction, *Proc. 19-th Annual ACM Symposium on Principles of Programming Languages, Albuquerque, New Mexico, ACM Press, New York, 1992.*

[L93] **Lafont, Y.** : From proof-nets to interaction nets, *this volume.*

[LMSS90] **Lincoln, P. & Mitchell, J. & Scedrov, A. & Shankar N.** : Decision Problems in Propositional Linear Logic, *Annals of Pure and Applied Logic 56, 1992, pp. 239-311.*

[LW92] **Lincoln, P. & Winkler, T.** : Constant-only Multiplicative Linear Logic is NP-complete, *to appear in Theoretical Computer Science.*

[MR91] **Malacaria, P. & Regnier, L.** : Some results on the interpretation of λ-calculus in operator algebras, *Proc. of LICS '91, IEEE Computer Society Press, 1991, pp. 63-72.*

[MM93] **Martini, S. & Masini A.** : On the fine structure of the exponential rules, *this volume.*

Printed in the United States
By Bookmasters